The Chromosome

The Chromosome includes selected reviews from the Tenth John Innes Symposium held from September 7 to 10, 1992, in an edited and modified form. Previous publications from the series of John Innes Symposia are as follows:

The Generation of Subcellular Structures. Amsterdam, Elsevier, 1973.
Proceedings of the First John Innes Symposium, 1972.

Modification of the Information Content of Plant Cells. Amsterdam, North-Holland, 1974.
Proceedings of the Second John Innes Symposium, 1973.

Structure-function relationships of proteins. Amsterdam, North-Holland, 1976.
Proceedings of the Third John Innes Symposium, 1975.

The Plant Genome. Norwich, John Innes Charity, 1980.
Proceedings of the Fourth John Innes Symposium and Second Haploid Conference, 1979.

Genetic Rearrangement. London, Croom Helm, 1983.
Proceedings of the Fifth John Innes Symposium, 1982.

The Cell Surface in Plant Growth and Development. *Journal of Cell Science*, Supplement 2, Cambridge, Company of Biologists, 1985.
Proceedings of the Sixth John Innes Symposium, 1984.

Virus Replication and Genome Interactions. *Journal of Cell Science*, Supplement 7, Cambridge, Company of Biologists, 1987.
Proceedings of the Seventh John Innes Symposium, 1986.

Protein Targeting. *Journal of Cell Science*, Supplement 11, Cambridge, Company of Biologists, 1989.
Proceedings of the Eighth John Innes Symposium, 1988.

Molecular and Cellular Basis of Pattern Formation. *Development*, Supplement 1, Cambridge, Company of Biologists, 1991.
Proceedings of the Ninth John Innes Symposium, 1990.

The Chromosome

J.S. Heslop-Harrison and R.B. Flavell
John Innes Centre, Colney Lane, Norwich NR4 7UH, UK

© BIOS Scientific Publishers Limited, 1993

First published in the United Kingdom 1993 by
BIOS Scientific Publishers Limited,
St Thomas House, Becket Street, Oxford OX1 1SJ, UK.

A CIP catalogue record for this book is available from the British Library.

ISBN 1 872748 32 5

Typeset by MFK Typesetting Ltd, Hitchin, UK.
Printed by Henry Ling Ltd at the Dorset Press, Dorchester, UK.

Preface

The title of this volume is well chosen. *The Chromosome* encapsulates those aspects of the genome which are concerned with its own biology. Aspects which have perhaps been moved from centre stage in recent years as attention has focussed on the sequences of the genes. Proposals to sequence genomes are driven by the technological challenge and by the access that gene sequences provide for all aspects of the biology of an organism. Although the grand sequencing projects are not neglected in this volume, most of the papers are concerned with the biology of the genome and with the processes which affect the genome directly; replication, recombination, segregation and gene expression which still present many unsolved and fundamental problems. As we shall see, there has been rapid progress across this broad front. The papers follow a natural progression from simple through to complex genomes, from molecular to cytogenetic studies.

The great power that simple systems have brought to analysing genetic processes over the decades is illustrated by the elegant work of Stahl. While conceding that phage do not mate in the same way as fruit flies, he shows how complex phenomena, such as recombination, can be explored in model systems in a way which is not yet open to analysis in complex organisms, and from this starting point provides a model for the phenomenon of interference. Recombination provides new combinations of genes and is involved in DNA repair. Sherratt and colleagues propose a further function – an involvement of site-specific recombination in chromosome partition in *Escherichia coli*, a process which these authors and Schaechter and von Freiesleben liken to mitosis in eukaryotes.

Although much is known about enzymes involved in DNA replication, some aspects of the control of replication are not well understood. In *E. coli* and in many viruses, replication starts at a specific site; there are indications that in eukaryotes such as *Saccharomyces cerevisiae* with 'simple' genomes there is a need for specific sequences in the DNA to initiate DNA replication, but the need for such sequences in higher eukaryotes is not yet proven. Besides the DNA, it appears that many other factors are involved in replication; the reader will be greatly helped by the review of DePamphilis in fighting through the complexities. One aspect of chromosome replication has been greatly clarified in recent years. To replicate the ends without loss clearly required a mechanism different from that which uses oligoribonucleotide priming on the lagging strand. The DNA sequences found in special structures at the ends of eukaryotic chromosomes – the telomeres – are described by Richards *et al*. and the mechanism by which they are replicated, by Greider *et al.* who also propose that the regulation of telomere synthesis may be involved in cellular senescence, since telomerase becomes reactivated in cells which escape senescence to become immortal.

The degree of chromatin condensation in eukaryotic chromosomes is associated with several functions. Highly condensed chromatin is late to replicate and not transcribed. No doubt some of the complex and subtle genetic phenomena, such as position effect variegation and *trans*-inactivation discussed by Henikoff *et al.* and parental imprinting described by Ferguson-Smith and Surani, are associated with properties of the chromatin effective over a long range, but attempts to find the molecular basis for these genetic phenomena have not yet produced clear cut results. Involvement of medium-range effects on gene expression is well illustrated by the work of Dillon *et al.* who show that the presence of a locus control region (LCR) is necessary for the expression of genes in the β-globin cluster. The LCR may be 50 kb from the gene that it activates; presumably the LCR acts by looping round so that it may interact with factors bound at the gene's promoter. Even in this well-studied case, there are unsolved problems. For example, it is not yet known what determines the order of expression of the β-like genes.

What factors determine the state of chromatin condensation? Leno *et al.* show that decondensation of sperm chromosomes is mediated by an acidic protein, nucleoplasmin, previously identified as a molecular chaperone which accompanies the histones H2A and H2B during nucleosome assembly. In this 'remodelling' of the sperm nucleus, the basic X and Y proteins in the sperm are replaced by histones in the pronucleus. What is it that then determines the different degrees of condensation in different types of chromatin? There is no doubt that in some chromosomes there is an involvement of the DNA sequence but cytological studies of mouse meiotic chromosomes described by Moens and Pearlman show that the underlying DNA sequence is not sufficient to determine the state of condensation of a satellite DNA; in these chromosomes the minor satellite is centromeric and condensed in *Mus domesticus* and pericentric and dispersed in the closely related *Mus spretus*. In *M. domesticus* × *M. spretus* hybrids, the parental patterns of condensation are retained in the paired homologues. Methylation is one way in which the DNA of vertebrates can be modified to influence its interactions with proteins. Antequera and Bird describe the large body of evidence associating methylation with chromatin condensation and suppression of gene activity. However, they emphasize that methylation may not be the primary determinant of the inactive state but introduced to perpetuate this state once established by other means.

Technical developments are revolutionizing genome studies. Rapid sequencing methods have been used to determine the base sequence of an entire chromosome of *S. cerevisiae*. Oliver *et al.* describe how this was achieved in a collective effort, marshalling human resources in the way the ancient Chinese did to build the Great Wall. This collection of sequences and the collection from *E. coli* described by Blattner *et al*. are providing a tally of the genetic functions needed to create a unicellular organism. Already there are a number of surprises. The more complex genomes of multicellular organisms need different approaches and as Flavell *et al*. show, the daunting task of complete molecular analysis starts with a large-scale map. But working with higher organisms brings its rewards, and one great advantage is the opportunity to see chromosomes under the microscope. Shaw *et al*. and Heslop-Harrison *et al*. show how modern microscopy combined with molecular hybridization still illuminates the structural and functional organization of the nucleus.

E.M. Southern (*Oxford*)

Acknowledgements

The editors are grateful to the following for reviewing manuscripts: Dave Laurie, David Hopwood, Mervin Bibb, Noel Ellis, Peter Shaw, Roy P Dunford, Richard Jorgensen, and, particularly, Caroline Dean, Keith Chater and Trude Schwarzacher.

We are also grateful to Gill Harrison and Lindsay Powles for assistance with literature checking, editing and formatting of papers and the index.

Contents

Contents

Contributors

Allsopp, Rich C. McMaster University, Hamilton, Ontario L8N 3Z5, Canada

Antequara, Francisco. Institute of Cell and Molecular Biology, University of Edinburgh, King's Buildings, Edinburgh EH9 3JR, UK

Antoniou, Michael. Laboratory of Gene Structure and Expression, National Institute for Medical Research, The Ridgeway, Mill Hill, London NW7 1AA, UK

Autexier, Chantal. Cold Spring Harbor Laboratory, PO Box 100, Cold Spring Harbor, New York 11724, USA

Avilion, Ariel A. Cold Spring Harbor Laboratory, PO Box 100, Cold Spring Harbor, New York 11724, USA

Bacchetti, Silvia. McMaster University, Hamilton, Ontario L8N 3Z5, Canada

Bird, Adrian. Institute of Cell and Molecular Biology, University of Edinburgh, King's Buildings, Edinburgh EH9 3JR, UK

Blakely, Garry. Institute of Genetics, Glasgow University, Glasgow G11 5JS, UK.

Blattner, Fredrick R. Laboratory of Genetics, University of Wisconsin, 445 Henry Mall, Madison, Wisconsin 53706, USA

Burke, Mary. Institute of Genetics, Glasgow University, Glasgow G11 5JS, UK

Burland, Valerie D. Laboratory of Genetics, University of Wisconsin, 445 Henry Mall, Madison, Wisconsin 53706, USA

Chuang, S. Laboratory of Genetics, University of Wisconsin, 445 Henry Mall, Madison, Wisconsin 53706, USA

Collins, Kathleen. Cold Spring Harbor Laboratory, PO Box 100, Cold Spring Harbor, New York 11724, USA

Colloms, Sean. Institute of Genetics, Glasgow University, Glasgow G11 5JS, UK

Counter, Chris M. McMaster University, Hamilton, Ontario L8N 3Z5, Canada

Daniels, Donna L. Laboratory of Genetics, University of Wisconsin, 445 Henry Mall, Madison, Wisconsin 53706, USA

Dean, Caroline. John Innes Centre, Colney Lane, Norwich NR4 7UH, UK

DeBoer, Ernie. Laboratory of Gene Structure and Expression, National Institute for Medical Research, The Ridgeway, Mill Hill, London NW7 1AA, UK

DePamphilis, Melvin L. Roche Institute of Molecular Biology, Roche Research Center, Nutley, New Jersey 07110, USA

Dillon, Niall. Laboratory of Gene Structure and Expression, National Institute for Medical Research, The Ridgeway, Mill Hill, London NW7 1AA, UK

Drabek, Dubravka. Laboratory of Gene Structure and Expression, National Institute for Medical Research, The Ridgeway, Mill Hill, London NW7 1AA, UK

Dreesen, Thomas D. Department of Zoology and Physiology, Louisiana State University, Baton Rouge, Louisiana 70803, USA

Ellis, Jamie. Laboratory of Gene Structure and Expression, National Institute for Medical Research, The Ridgeway, Mill Hill, London NW7 1AA, UK

Ferguson-Smith, Anne C. Wellcome/CRC Institute of Cancer and Developmental Biology, University of Cambridge, Tennis Court Road, Cambridge CB2 1QR, UK

Flavell, Richard B. John Innes Centre, Colney Lane, Norwich NR4 7UH, UK

Fraser, Peter. Laboratory of Gene Structure and Expression, National Institute for Medical Research, The Ridgeway, Mill Hill, London NW7 1AA, UK

von Freiesleben, Ulrik. Department of Molecular Biology and Microbiology, Tufts University School of Medicine, Boston, Massachusetts 02111, USA

Gent, Manda E. Manchester Biotechnology Centre, University of Manchester Institute of Science and Technology, PO Box 88, Manchester M60 1QD, UK

Greider, Carol W. Cold Spring Harbor Laboratory, PO Box 100, Cold Spring Harbor, New York 11724, USA

Grosveld, Frank. Laboratory of Gene Structure and Expression, National Institute for Medical Research, The Ridgeway, Mill Hill, London NW7 1AA, UK

Hanscombe, Olivia. Laboratory of Gene Structure and Expression, National Institute for Medical Research, The Ridgeway, Mill Hill, London NW7 1AA, UK

Harley, Calvin B. McMaster University, Hamilton, Ontario L8N 3Z5, Canada

Harrington, Lea A. Cold Spring Harbor Laboratory, PO Box 100, Cold Spring Harbor, New York 11724, USA

Henikoff, Steven. Howard Hughes Medical Institute, Basic Sciences Division, Fred Hutchinson Cancer Research Center, Seattle, Washington 98104, USA

Heslop-Harrison, J.S. (Pat). Karyobiology Group, John Innes Centre, Colney Lane, Norwich NR4 7UH, UK

Higgins, Christopher F. Imperial Cancer Research Fund, Institute of Molecular Medicine, University of Oxford, John Radcliffe Hospital, Oxford OX3 9DU, UK

Highett, Martin. Department of Cell Biology, John Innes Institute, Colney Lane, Norwich NR4 7UH, UK

Imam, Ali. Laboratory of Gene Structure and Expression, National Institute for Medical Research, The Ridgeway, Mill Hill, London NW7 1AA, UK

Indge, Keith J. Department of Biochemistry and Applied Molecular Biology, University of Manchester Institute of Science and Technology, PO Box 88, Manchester M60 1QD, UK

James, Carolyn M. Manchester Biotechnology Centre, University of Manchester Institute of Science and Technology, PO Box 88, Manchester M60 1QD, UK

Koken, Marie-Ange. Laboratory of Gene Structure and Expression, National Institute for Medical Research, The Ridgeway, Mill Hill, London NW7 1AA, UK

Laskey, Ronald A. Wellcome/CRC Institute of Cancer and Developmental Biology, University of Cambridge, Tennis Court Road, Cambridge CB2 1QR, UK

Leitch, Andrew R. School of Biological Sciences, Queen Mary and Westfield College, University of London, London E1 4N5, UK

Leno, Gregory H. Department of Biochemistry, University of Mississippi Medical Center, 2500 North State Street, Jackson, Mississippi 39216-4505, USA

Leslie, Nick. Institute of Genetics, Glasgow University, Glasgow G11 5JS, UK

Lindenbaum, Michael. Laboratory of Gene Structure and Expression, National Institute for Medical Research, The Ridgeway, Mill Hill, London NW7 1AA, UK

Loughney, Kate. ICOS Corporation, Bothell, Washington 98021, USA

Mantell, Lin L. Cold Spring Harbor Laboratory, PO Box 100, Cold Spring Harbor, New York 11724, USA

May, Gerhard. Institute of Genetics, Glasgow University, Glasgow G11 5JS, UK

McCulloch, Richard. Institute of Genetics, Glasgow University, Glasgow G11 5JS, UK

Meijer, Dies. Laboratory of Gene Structure and Expression, National Institute for Medical Research, The Ridgeway, Mill Hill, London NW7 1AA, UK

Moens, Peter B. Department of Biology, York University, Downsview, Ontario M3J 1P3, Canada

Moore, Graham. John Innes Centre, Colney Lane, Norwich NR4 7UH, UK

Oliver, Stephen G. Manchester Biotechnology Centre, University of Manchester Institute of Science and Technology, PO Box 88, Manchester M60 1QD, UK

Pearlman, Ronald E. Department of Biology, York University, Downsview, Ontario M3J 1P3, Canada

Philipsen, Sjaak. Laboratory of Gene Structure and Expression, National Institute for Medical Research, The Ridgeway, Mill Hill, London NW7 1AA, UK

Philpott, Anna. Massachusetts General Hospital Cancer Center, Laboratory of Molecular Genetics, Building 149, 13th Street, Charlestown, Massachusetts 02129, USA

Plunkett, Guy III. Laboratory of Genetics, University of Wisconsin, 445 Henry Mall, Madison, Wisconsin 53706, USA

Prowse, Karen R. Cold Spring Harbor Laboratory, PO Box 100, Cold Spring Harbor, New York 11724, USA

Pruzina, Sara. Laboratory of Gene Structure and Expression, National Institute for Medical Research, The Ridgeway, Mill Hill, London NW7 1AA, UK

Rawlins, David. Department of Cell Biology, John Innes Institute, Colney Lane, Norwich NR4 7UH, UK

Richards, Eric J. Department of Biology, Washington University, St Louis, Missouri 63130, USA

Roberts, Jennifer. Institute of Genetics, Glasgow University, Glasgow G11 5JS, UK

Schaechter, Moselio. Department of Molecular Biology and Microbiology, Tufts University School of Medicine, Boston, Massachusetts 02111, USA

Schwarzacher, Trude. Karyobiology Group, John Innes Centre, Colney Lane, Norwich NR4 7UH, UK

Shaw, Peter. Department of Cell Biology, John Innes Institute, Colney Lane, Norwich NR4 7UH, UK

Sherratt, David J. Institute of Genetics, Glasgow University, Glasgow G11 5JS, UK

Smith, Stephanie K. Cold Spring Harbor Laboratory, PO Box 100, Cold Spring Harbor, New York 11724, USA

Southern, Edward M. Department of Biochemistry, University of Oxford, South Parks Road, Oxford OX1 3QU, UK

Stahl, Franklin W. Institute of Molecular Biology and Department of Biology, University of Oregon, Eugene, Oregon 97403-1229, USA

Strouboulis, John. Laboratory of Gene Structure and Expression, National Institute for Medical Research, The Ridgeway, Mill Hill, London NW7 1AA, UK

Surani, M. Azim. Wellcome/CRC Institute of Cancer and Developmental Biology, University of Cambridge, Tennis Court Road, Cambridge CB2 1QR, UK

Talbot, Dale. Laboratory of Gene Structure and Expression, National Institute for Medical Research, The Ridgeway, Mill Hill, London NW7 1AA, UK

Vaziri, Homayoun. McMaster University, Hamilton, Ontario L8N 3Z5, Canada

Whyatt, David. Laboratory of Gene Structure and Expression, National Institute for Medical Research, The Ridgeway, Mill Hill, London NW7 1AA, UK

Chapter 1

Genetic recombination: thinking about it in phage and fungi

Franklin W Stahl

Abstract

Mechanisms of genetic recombination are similar between phage and fungi. In both creatures, recombination is initiated by double-chain breaks. Repair of these breaks is effected by interaction between the broken chromosome and an intact homologue. When the break and repair occur at a marked locus, gene conversion is a visible consequence. A conspicuous fraction of these conversions are accompanied by crossing over. In eukaryotes, at least, crossovers tend not to occur close to each other – they 'interfere' with each other. Conversions show no interference with each other. Interference between crossovers is inversely related to distance, and the relevant metric of distance appears to be 'genetic' rather than 'physical'. A simple stochastic model efficiently accounts for the quantitative relationships between interference and linkage map distance.

1. Introduction

In 1952, I decided to study bacteriophages despite being warned by my research advisor, Don Charles, that such a step could ruin my career. At that time, most biologists considered phages to be a special case. They might amuse but they could certainly never instruct in matters of "real genetics". Were I to persist in my foolishness, Charles said, the best I could hope for in my career would be "some job in some bacteriology department in some medical school, somewhere". The implication was clear – I was doomed to intellectual exile. In the event, I persisted in my foolishness, and can now contemplate the consequences.

By the late 1950s, it was already clear that Charles, a mouse geneticist and statistician, was wrong about the relevance of phage studies to genetics. Phage have contributed importantly to all aspects of genetics. In the subdiscipline of genetic recombination, developments of the past ten years have revealed an extraordinary commonality of mechanism in phage (λ) and fungi (*Saccharomyces*). With but a few caveats, we can say that recombination proceeds as follows in both creatures.

2. The mechanism of generalized, homologous recombination

The primary (and perhaps unique) initiating event for recombination is a double chain break delivered to one of the participants. In λ, these events are exclusively at the termini of the chromosomes when DNA replication is blocked, but they occur elsewhere when replication is permitted or when a double chain break is artificially provided (Fig. 1). In yeast, meiosis-specific double chain breaks arise where the chromatin in vegetative and meiotic cells is DNase I hypersensitive. Some of these sites are demonstrated transcription promoters (M Lichten, personal communication).

The breaks are 'processed' by exonucleases that digest double-stranded (ds) DNA in the 5' to 3' direction. Such digestion produces '3' overhangs', which become coated with RecA or RecA-like proteins. The 3' ends, too, can be somewhat digested.

The coated DNA explores ('invades') other DNA throughout the cell until a stretch of 10^2 base-pairing partners is found. In eukaryotes, the two DNA ends created by the initiating double chain break contrive, by unknown mechanism, to find homology on a common chromatid. In λ, the two ends often find homology on different duplexes.

The invading 3' ends prime DNA synthesis that uses the invaded (intact) duplex as template. This synthesis serves to replace DNA lost by processing of the initiating double chain break. When the events fall upon a genetically marked site,

---

Figure 1. The distribution of exchanges along λ chromosomes when DNA replication is blocked. λ strains marked by near-terminal mutations were crossed lytically in the absence of DNA replication. One parent was composed of heavy isotopes of carbon and nitrogen (thick chromosome in diagrams), while the other was composed of ordinary isotopes (thin chromosome in diagrams). In crosses C and D, the *R* mutant parent was cut *in vivo* by a restriction endonuclease (*Xho*I) active at a site about two-thirds of the physical distance between the left and right ends of λ. The lysates resulting from the crosses were centrifuged to near equilibrium in a caesium formate density gradient. Density fractions were collected through a needle hole in the bottom of the tube. These fractions were assayed for total λ phage (open triangles) and for $B^+ R^+$ recombinants (circles). Among the recombinants, plaques were distinguished as *c* (clear, filled circles), c^+ (turbid, open circles) or c/c^+ (mottled, half-filled circles).

In panels A and C, the *B c* parent is heavy-labelled, while in panels B and D the $c^+ R$ parent is heavy-labelled. When the $c^+ R$ parent is cut by *Xho*I (panels C and D), a new density peak of $B^+ c R^+$ phages arises at a density position predicted by the location of the *Xho*I site on the λ chromosome. Figure is reproduced from Stahl *et al.* (1990) with permission from the Genetics Society of America.

the loss of DNA from one chromosome and its replacement using a homologue as a template alters the allele ratio. In fungi, such an alteration may be perceived as a tetrad with a 6:2 or a 5:3 allele ratio (conversion and half-conversion tetrads, respectively.)

Events transpire such that the initiating break is repaired and the DNA on either side has or has not undergone reciprocal exchange (crossing over). (Later

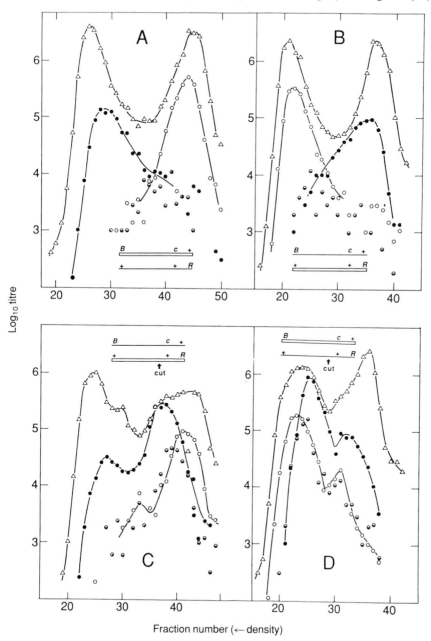

we will return to this concept of alternative outcomes of double chain break repair. It looks as though they are responsible for that chestnut, chiasma interference.)

The products of resolution of the intermediates are similar between phage and fungi. Regions of hybrid DNA mark the site of the recombination event. Hetero-duplexes resulting from genetic differences between the two participants in these hybrid regions are often resolved by mismatch-correction enzymes. In fungi, such correction, mediated by a battery of enzymes, can result in a half-conversion being converted into a full conversion. In λ, mismatch correction occurs by enzymes acting in essentially the same ways as do their homologues in yeast.

This view of homologous recombination in λ and yeast, though not yet secured in every respect, is not likely to be wrong in any important way. Many of its features are supported by congruent genetic and biochemical analyses.

3. Formal phage genetics

In 1952, the possibility that phage recombination might merely amuse was implied by the manner in which the results of simple phage crosses differed from those of eukaryotes (see Hershey 1959, and Stahl 1979 for fuller treatments, with refe-rences, of formal phage genetics):

(1)   The numbers of recombinants vary among individual bacterial cells infected lytically with two genotypes of a T-even phage, but the numbers of comple-mentary recombinants released by these individual cells are essentially independent of each other. This lack of correlation argued that recombina-tion in phage was non-reciprocal, unlike crossing over in meiosis.

(2)   When bacterial cells were infected simultaneously with three different geno-types of phage, the progeny contained particles with markers from each of the three parents.

(3)   When infected cells were lysed prematurely by artificial means, the (few) progeny phage particles they contained manifested a lower frequency of recombinants than did particles released from the cells at the normal time.

(4)   Whereas crossing over in meiosis often manifests interference, phage crosses are characterized by 'negative interference'; the detected presence of a crossover in one marked interval increases the likelihood for each other marked interval on the same phage particle that it will be recombinant.

Thus, in several respects, genetic recombination in phage was seen to be different from meiotic recombination. In an apparent effort to force phage into a meiotic framework, Visconti and Delbrück (1953) proposed that phages 'mate' pairwise, just as chromosomes synapse in meiosis. With this proposal, they declared a fundamental similarity between phages and those organisms that were well established in genetic studies. The differences, they suggested, were of secondary importance. These differences reflected Visconti and Delbrück's view that the infected cells in a phage cross are like small *Drosophila* population cages inoculated by males of one pure genotype and females of another. Successively matured particles are like samples of fly gametes drawn from the cage after varying numbers of generations. Add to that the lack of sexual differentiation in phage (so that particles 'mate' at random with respect to genotype), and the differences between phages and flies were rationalized. In particular, the negative interference of phage crosses appeared not to describe individual 'matings'. Instead, it was a consequence of heterogeneity in mating experience among the members of the population.

Subsequent analyses revealed that Delbrück's dipterocentrism was misplaced. Interactions in T-even phage are not disciplined by discrete, successive pairwise 'matings'. Instead, short segments of the chromosome indulge in recombinational activities independently of each other. Furthermore, the T-even linkage maps proved to be circular, unlike any ordinary fly linkage map. Some of the 'negative interference' in T-even crosses is attributable to this circularity – in a circular map exchanges cannot be Poisson-distributed, since each mature particle must (eventually) have had an even number of them.

Stahl *et al.* (1964) proposed that multiple matings in T-even phages occur between circularly permuted rods and that the number of exchanges per mating is exactly one. The reasoning in that analysis is not entirely straightforward, but, taken at face value, the analysis implies total (positive) interference in each mating. Thus, Hershey (1959) was prescient when he said about phage crosses, "In a not quite trivial sense it may be permissible to say that positive interference is not found because it is obscured by negative interference".

Lytic cycle crosses of phage λ manifest the same sorts of non-meiotic phenomena that characterize T-even phages (except for linkage circularity). In particular, λ has negative interference that is demonstrably due to heterogeneity in mating experience. The possibility of positive interference in hypothetical individual mating acts remains open.

4. Chiasma interference in eukaryotes

Positive interference certainly *does* characterize meiotic recombination in most eukaryotes. The primary observation of interference, made in *Drosophila* by Sturtevant (1915) and Muller (1916), is that random gametes selected for being recombinant in a marked interval are less likely to be recombinant in a second, nearby marked interval than is a gamete not so selected. Since recombination occurs after premeiotic DNA synthesis (in the 'four-strand stage'), this interference could represent the operation of either (or both) of two non-random processes. (1) Exchanges could be distributed non-randomly with respect to each other along the length of the paired homologues (chiasma interference). (2) The selection of homologous chromatids to be exchanged could be influenced by the choice made at nearby or neighbouring exchanges (chromatid interference). Since chromatid interference is typically weak or lacking, we shall confine our thoughts to chiasma interference.

What kind of long-range force could inhibit exchange in the neighbourhood of an exchange? Several possibilities have been suggested. An early one (see Muller, 1916) supposed that chromosomes were rather inflexible so that they could not 'cross' each other twice in a short distance. More recently, King and Mortimer (1990) proposed that an exchange initiates a polymerization that spreads in both directions from the exchange. The polymer aborts incipient exchanges that it encounters. Each of these models implies that interference is a function of physical distance between chiasmata, since the processes mediating the interference are dependent on physical processes of bending or polymerization.

5. Interference as a function of linkage distance

The assumption that interference is directly related to physical distance is so sensible as to have escaped challenge until recently (Foss *et al.*, 1993). However, there is a conspicuous weakness with the assumption – interference extends over widely different physical distances in different organisms. An alternative to interference as a function of physical distance is interference as a function of *genetic* distance. The strength of this assumption is that interference extends over approximately the same genetic distance in the organisms for which we have examined data. The coefficient of coincidence, which is an inverse measure of interference, is about zero for very close intervals and approximates unity when the linkage map distance is about 40 cM (Fig. 2). The coefficient of coincidence referred to here is (observed frequency of double recombinants)/(frequency of doubles expected in the absence of interference) for two short intervals separated from each other by a map distance X. (The measure is 'inclusive' in the sense that it is not concerned with exchanges that may be occurring in other intervals, including the interval that separates the two marked intervals.)

How might interference be directly dependent on genetic rather than physical distance? How might the meiotic cell measure linkage map distance? Foss *et al.* (1993) propose that the cell does not measure, it counts. The objects counted by the system are recombination events (Cs) that do not result in crossing over (Co) (Mortimer and Fogel, 1974).

Mortimer and Fogel (1974) demonstrated for yeast that Cs (identified as gene conversions) are randomly distributed with respect to each other, i.e. they do not interfere with each other. Cxs, conversions that are accompanied by crossing over, do, of course, interfere with nearby crossing over. Mortimer and Fogel proposed that interference is a consequence of rules governing the outcome of randomly initiated recombination events, Cs. They suggested that Cxs and Cos alternate. The notion of alternation was in accord with observations that about

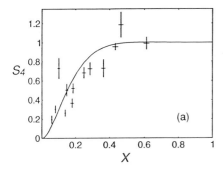

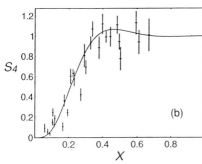

Figure 2. Coefficient of coincidence (S_4) versus linkage map distance (X). Data points are measured coefficients of coincidence for intervals separated by differing linkage map distances in Morgans. (a) *Neurospora crassa*. Data from Perkins (1962) and Strickland (1961). (b) *Drosophila melanogaster*. Data from Bridges and Curry (1935). The curves are appropriately selected from the set of theoretical curves in Fig. 4 as described in the text.

half of the conversions (Cs) in yeast are accompanied by crossing over of flanking markers (Cxs).

Subsequent analyses (Fogel *et al.*, 1983) estimated the Cx/C ratio in yeast as 0.35 instead of 0.5. The Mortimer and Fogel (1974) concept can be salvaged by proposing that two Co events intervene between Cxs (Stahl, 1979). Foss *et al.* (1993) have deduced the equation relating the coefficient of coincidence to map distance when any stated fixed number of Cos intervene between Cxs (Fig. 3).

With the help of David Perkins, we (Foss *et al.*, 1993) have applied the model to data from *Neurospora*, for which the ratio Cx/C is estimated (by averaging available literature reports) to be 0.30. That estimate predicts that for *Neurospora*, also, two Cos must intervene between Cxs. That view is tested by comparing the $Cx(Co)^2$ model with data on interference, likewise culled from the *Neurospora* literature. The fit, which is remarkably good, is shown in Fig. 2.

In *Drosophila*, the Cx/C ratio has been estimated (Hilliker and Chovnick, 1981; Hilliker *et al.*, 1991) at about 0.20. If our counting model for interference

$$S_4 = (m+1)e^{-y} \sum_{i=0}^{\infty} \frac{y^{m+(m+1)i}}{[m+(m+1)i]!}; \qquad y = 2(m+1)X$$

Figure 3. Coefficient of coincidence (S_4) as a function of map distance (X, in Morgans) for the model in which Cs are Poisson-distributed and m Cos must intervene between neighbouring Cxs. y is the mean number of Cs in the segment between the two (short) test intervals. The equation, which is from Foss *et al.* (1993), is graphed in Fig. 4.

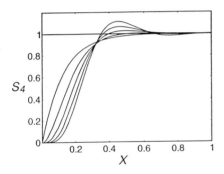

Figure 4. Coefficient of coincidence (S_4) versus map distance (X, in Morgans) for $m = 1$–5 (from left to right). When m is zero, there is no interference (horizontal line at $S_4 = 1$). Figure is reproduced from Foss *et al.* (1993) with permission from the Genetics Society of America.

applies, four Cos must intervene between Cxs in the fruit fly. We (Foss *et al.*, 1993) have tested the model against the X-chromosome data of Bridges and Curry (1935). The fit of theory to these data, too, is highly satisfying (Fig. 2) and we conclude that the model must be taken seriously. It might be said 'very seriously' in view of the fact that the dependence of interference on map distance is correctly predicted by an equation with only one parameter (m), and that parameter, rather than being adjusted to fit the interference data, is estimated by a distinct operation.

What is the point of interference? In eukaryotes, disjunction of homologues at the first meiotic division is apt to fail if those homologues have enjoyed no exchanges. One way to minimize the no-exchange class is to have a large mean number of randomly (Poisson) distributed events. A large number of chiasmata, however, may lead to difficulties in disjunction as a result of multiple entanglements between bivalents (Merriam and Frost, 1964). Our counting model provides a solution. A large mean number of Poisson-distributed C events will 'guarantee' that the zero-class for C events is small. If interference is established after one C event has been committed to be Cx, the zero class for Cxs will be equally small. The mean number of Cxs per bivalent is then determined by the ratio Cx/C.

Foss *et al.* (1993) discuss the possibility that in the autosomes of *Drosophila*, which enjoy an effective distributive pairing system, the four Cs nearest the centromere are always Cos. The fifth event is a Cx, and there can be a second Cx only after the next four Cs have been committed to be Cos. Such a rule would account for the low rate of crossing over near the centromere and the lack of interference across the centromere. If small recombination nodules are taken to be Cs or Cos, while large nodules signify Cxs, this rule would also account for the high ratio of small to large nodules near the centromere (Carpenter, 1979).

By what mechanism could a chromosome count Cos? Contributions will be gratefully accepted. The only model I have to offer is a 'sweeping' model, and I apply it to *Drosophila*. I suppose that after the C events are randomly laid down, they are visible as early recombination nodules (small, elipsoidal nodules). A broom (made of spindle fibres?) sweeps from the centromere outward, moving the small nodules into piles (of five, in *Drosophila*). Such a pile is a late (large, spherical) nodule. Each early nodule is associated with a C; the late nodules (aggregates of five early nodules) signal the sites of Cxs.

One prediction of this fantasy is that the ratio (size of late nodules)/(size of early nodules) should be larger for *Drosophila* than it is for *Neurospora*. Measurements by Carpenter (1979) and by Bojko (1989), respectively, support this prediction.

References

Bojko M. (1989) Two kinds of "recombination nodules" in *Neurospora crassa. Genome*, **32:** 697-709.
Bridges CB, Curry V. (1935) *Yearbook Carnegie Institute* 34 (Morgan TH, Bridges CB, Schultz J, eds). p 287.
Carpenter ATC. (1979) Synaptonemal complex and recombination nodules in wild-type *Drosophila melanogaster* females. *Genetics*, **92:** 511-541.
Fogel S, Mortimer RK, Lusnak K. (1983) Meiotic gene conversion in yeast: molecular and experimental perspectives. In: *Yeast Genetics* (Spencer JFT, Spencer DM, Smith ARW, eds). New York: Springer Verlag.
Foss E, Lande R, Stahl FW, Steinberg CM. (1993) Chiasma interference as a function of genetic distance. *Genetics*, **133:** 679–689.
Hershey AD. (1959) The production of recombinants in phage crosses. *Cold Spring Harbor Symp. Quant. Biol.* **23:** 19-46.
Hilliker AJ, Chovnick A. (1981) Further observations on intragenic recombination in *Drosophila melanogaster. Genet. Res.* **38:** 281-296.
Hilliker AJ, Clark SH, Chovnick A. (1991) The effect of DNA sequence polymorphisms on intragenic recombination in the rosy locus of *Drosophila melanogaster. Genetics*, **129:** 779-781.
King JS, Mortimer RK. (1990) A polymerization model of chiasma interference and corresponding computer simulation. *Genetics*, **126:** 1127-1138.
Merriam JR, Frost JN. (1964) Exchange and nondisjunction of the X-chromosomes in female *Drosophila melanogaster. Genetics*, **49:** 109-122.
Mortimer RK, Fogel S. (1974) Genetical interference and gene conversion. In: *Mechanisms in Recombination* (Grell RF, ed). New York: Plenum Publishing. pp 263-275.
Muller HJ. (1916) The mechanism of crossing over. *Am. Nat.* **50:** 193-221 and ff.
Perkins DD. (1962) Crossing-over and interference in a multiply marked chromosome of *Neurospora. Genetics*, **47:** 1253-1274.
Stahl FW. (1979) *Genetic Recombination, Thinking about it in Phage and Fungi.* San Francisco: WH Freeman.
Stahl FW, Edgar RS, Steinberg J. (1964) The linkage map of bacteriophage T4. *Genetics*, **50:** 539-552.
Stahl FW, Fox MS, Faulds D, Stahl MM. (1990) Break-join recombination in phage λ. *Genetics*, **125:** 463-474.
Strickland WN. (1961) Tetrad analysis of short chromosome regions of *Neurospora crassa. Genetics*, **46:** 1125-1141.
Sturtevant AH. (1915) The behavior of the chromosomes as studied through linkage. *Z. Abstam. Vererbung.* **13:** 234-287.
Visconti N, Delbrück M. (1953) The mechanism of genetic recombination in phage. *Genetics*, **38:** 5-33.

Chapter 2

The bacterial chromosome: DNA topology, chromatin structure and gene expression

Christopher F Higgins

1. Introduction

The enteric bacterium, *Escherichia coli*, has a single circular chromosome carrying about 4000 genes. As *E. coli* and closely related species, such as *Salmonella typhimurium* are the best characterized bacterial species, they are considered almost exclusively in this chapter. It should, however, be remembered that the bacterial kingdom is highly diverse. Genome sizes can vary considerably between species: the smallest known genome is about 600 kb (*Chlamydia*) and the largest 13 000 kb (*Calothrix*). Some bacterial species have two chromosomes. Others have large plasmids which encode essential functions, often associated with pathogenicity. Although most plasmids are circular, some are linear such as the plasmids of the Lyme disease pathogen *Borrelia* (Hinnebusch *et al.*, 1990). In short, it should not necessarily be assumed that *E. coli* is representative of all bacterial species: it is simply the best studied and provides a conceptual framework against which other species can be compared and contrasted.

2. Chromosome organization

The *E. coli* genome is a single circular molecule of about 4700 kb. A complete physical map of the chromosome has been derived (Kohara *et al.*, 1987; see Blattner *et al.*, Chapter 4, this volume). Physical maps of other bacterial chromosomes (eg: *Neisseria gonorrhoeae*, Bihlmaier *et al.*, 1991; *S. typhimurium*, Liu and Sanderson, 1992) have also been obtained. About 30% of the *E. coli* chromosome has been sequenced by a random approach, with individual groups sequencing the specific genes in which they are interested (Medigue *et al.*, 1991). More recently, directed sequencing has been aimed at determining the sequence of large segments of the chromosome, with the eventual aim of completing the entire genome (Daniels *et al.*, 1992). From the available sequence, a number of general conclusions can be drawn (for further detail see Blattner *et al.*, Chapter 4, this volume). There is little 'unoccupied' DNA: the bulk of the chromosome consists of coding sequences with relatively short intergenic regions. Even in regions where functional genes have not been identified there are open reading frames which give every impression of encoding a functional product. The close packing of the genome is not unexpected, as the time which it takes for the chromosome to replicate effectively limits the rate at which cell division can occur. As replication is from a single origin, anything which increases the length of the chromosome will, potentially, decrease the growth rate and, presumably, put the cell at a selective disadvantage.

As far as can be assessed, there is nothing particularly unusual about the order of genes around the chromosome: with some exceptions (see below), the position of a given gene on the chromosome appears to be relatively unimportant. Thus, apart from operons in which genes of similar function are cotranscribed, genes encoding components of a particular biochemical pathway are often scattered around the chromosome rather than clustered. Coding sequences are found on both DNA strands, although there may be some merit in certain genes being transcribed in the same direction as DNA replication. The finding that the common laboratory *E. coli* strain W3110 has about 20% of the genome inverted with respect to wild-type without being at an obvious disadvantage implies that the precise position of many genes is unimportant. Similarly, although the order of genes on the *S. typhimurium* chromosome is similar to that of *E. coli*, a large inversion of 11% of the genome appears to have little or no biological significance (Riley and Sanderson, 1990). The gene order in more distantly related species is often very different from that of *E. coli*. Finally, many genes can be moved around the chromosome with little obvious effect on expression or fitness. The most important exception is genes located near the origin of replication. As chromosome replication often reinitiates before the first replication cycle is complete, genes located near the origin are overrepresented in the cell compared with genes located near the terminus of replication (Schmid and Roth, 1987). However, while the positioning of many genes seems to be less than critical, certain chromosomal inversions are not possible (Segall *et al.*, 1988), implying some organizational constraints. Whether these are to do with the position of a gene affecting its expression, or to structural constraints within the nucleoid is unknown.

It is also important to remember that the genome is not static: apart from point mutations, gross re-arrangements such as insertions, deletions or duplications of DNA are common, even between isolates of the same species. One obvious example is the IS elements, the number, type and position of which vary between *E. coli* isolates. The transposition of ISs can move genes within or even between species. Thus, although the chromosomes of *E. coli* and *S. typhimurium* can be aligned over most of their length, there are over twenty regions where genes have inserted into the chromosome of one species (or deleted from the other). The classic example is the *lac* operon of *E. coli*. Amusingly, although this is one of the most intensively studied of all *E. coli* operons it is not a 'true' *E. coli* gene and has only recently been aquired by this species. Comparison of the *E. coli* genome with that of *S. typhimurium* and other enterics shows that, together with about 40 kb of DNA flanked by potential IS elements, the region of the chromosome including the *lac* operon has inserted into the *E. coli* genome relatively recently in evolution (Buvinger *et al.*, 1984). In addition to insertions and deletions, duplications are frequent. It has been estimated that within a population of supposedly genetically identical bacteria, a few percent may carry duplications at a given locus (Anderson and Roth, 1981). Generally, these spontaneous duplications are lost by recombination; they do, however, provide scope for evolutionary variation if cells are placed under appropriate selective pressure. They also provide one of the clearest indications that the genome is not static, even within a clonal population of cells.

3. Repetitive sequences

Until recently it was assumed that bacteria had little or no repetitive DNA, equivalent to Alu sequences or other 'junk' DNA which occupies a high proportion of the genomes of many eukaryotic species. While it is true that such elements do not occupy large segments of the bacterial genome, repetitive sequences are probably present in the genomes of most bacteria. The best example is the REP (repetitive extragenic palindromic) sequence of *E. coli* (Higgins *et al.*, 1982; Stern *et al.*, 1984; also called PUs, Gilson *et al.*, 1984). The REP sequence is a highly conserved element of about 35 nucleotides which includes an inverted repeat sequence. Over 500 copies of REP are present on the *E. coli* chromosome, occurring either singly or in tandem arrays of several copies. REP sequences are always found within transcribed sequences, either in intergenic regions or in the 3' untranslated regions of a transcript; their absence from coding sequences presumably reflects the constraints they would place on protein sequence. There is no apparent correlation between the function of a gene and whether or not it is associated with a REP sequence. Furthermore, REP sequences appear to be more-or-less randomly distributed around the chromosome although none has, as yet, been identified on a plasmid or 'phage DNA.

What is the function of these elements? Many suggestions have been advanced (reviewed by Higgins *et al.*, 1988a). It has been demonstrated that the presence of a REP sequence in a transcript can influence mRNA stability, and hence, gene expression (Newbury *et al.*, 1987a,b). However, this is solely by virtue of the fact that REP sequences include an inverted repeat and can form a stable stem-loop structure in RNA when transcribed: any stem-loop structure has similar effects on mRNA stability and this cannot, therefore, explain the high degree of primary

sequence conservation between REP sequences. In certain chromosomal locations REP sequences can affect translational coupling between genes (Stern *et al.*,
1988) or terminate transcription (Gilson *et al.*, 1986), but most REP sequences do
not perform these functions. Proteins have been reported to bind to REP
sequences in DNA, such as DNA gyrase and DNA polymerase I (Gilson *et al.*,
1990; Yang and Ames, 1988). None of these functions, however, can yet provide
an adequate explanation for several properties of the REP sequences. First, their
variable locations: if *S. typhimurium* and *E. coli* are compared their distribution is
often different as if they have inserted into, or deleted from, specific chromosomal
sites (although no transposition has been detected). If they served a specific role at
these sites they might be expected to be conserved between the two species.
Second, REP sequences do not appear to be conserved (at least at the same
frequency) in all enterobacteria, let alone more distantly related species, arguing
against an important role as a protein binding site, in gene expression or chromosome architecture. Thus, no function has yet been identified which can adequately
explain the distribution and sequence conservation between REP sequences. It
has, therefore, been suggested that REPs might be the bacterial equivalent of
'selfish' DNA maintained by a gene conversion event rather than by a selective
pressure on the host (Higgins *et al.*, 1988a). This view is strengthened by the
identification of a second repeated element in *E. coli*, with similar characteristics
to REP although it is a larger element (125 bp) and its sequence is entirely
different: the ERIC (Hulton *et al.*, 1991; also called IRU, Sharples and Lloyd,
1990). Additionally, it is becoming apparent that many (and perhaps most)
bacterial species have repetitive elements, the specific sequence of the element
being in large part species specific (see references in Hulton *et al.*, 1991).

Another recently identified element in bacterial genomes is the retroposon
(Inouye and Inouye, 1991; Lim *et al.*, 1990). These encode a hybrid, branched
RNA-DNA complex (msDNA) of unknown function as well as a reverse transcriptase necessary for msDNA synthesis. The identification of bacterial reverse
transcriptases has many implications for genome flexibility and evolution. The
distribution in *E. coli* strains suggests they are mobile. Interestingly, although
found in *E. coli* B, these elements are absent from *E. coli* K12: hence the original
erroneous conclusion that bacteria do not have reverse transcriptases!

4. Chromatin: secondary and tertiary organization of the chromosome

The linear arrangement of genes on the chromosome is only one aspect of
chromosome organization. If it is considered that an *E. coli* cell is 1 μm long, while
the extended chromosome is 1 mm long, there is clearly a formidable packaging
problem. This problem is emphasized by the calculation that the length of a single
gene is about 0.25 μm, a size not dissimilar to the diameter of the cell. Given that

chromosome replication proceeds almost continuously through the cell cycle, that about 20% of chromosomal genes are transcribed at any one time, that the genes are more or less evenly spaced throughout the genome (there are no large segments of silent DNA), and that genes can be turned on and off almost instantaneously, chromosome packaging would seem to be an almost insoluble problem.

A second important aspect of chromosome organization is that DNA is under net torsional stress. It has been known for many years that DNA in the bacterial cell is negatively supercoiled (Sinden *et al.*, 1980). More recent *in vivo* studies have shown that about half the free energy of negative supercoiling is constrained by bound proteins: the remainder is available to influence many processes such as transposition, replication, recombination and transcription (Lilley, 1986).

The *E. coli* chromosome can be seen, by electron microscopy, to be condensed into a nucleoid, and this nucleoid can be isolated by gentle cell lysis and differential centrifugation. The isolated nucleoid is fragile, as it is not delimited by a membrane, and is composed of DNA, RNA and is enriched for a number of specific proteins (the 'histone-like' proteins; Drlica and Rouviere-Yaniv, 1987). Electron microscopy of the spread chromosome shows it to consist of many loops and there is strong biochemical evidence that the chromosome consists of about 50 independently supercoiled, domains. Thus, about 50 'nicks' are required to completely relax the *E. coli* chromosome (Worcel and Burgi, 1972; Sinden and Pettijohn, 1981). What constrains and delimits these domains is unknown. Indeed, it is possible that the ends of the domains are not fixed but that the boundaries of these domains are flexible. As these domains are independently supercoiled, an attractive possibility is that they are differentially supercoiled and that this might play a role in gene expression (see below). However, recent studies based on the transposition of supercoiling-sensitive promoters to many different chromosomal locations, have indicated that the supercoiling of all the domains is equivalent (G Pavitt and CF Higgins, unpublished data).

Several histone-like proteins are associated with the *E. coli* nucleoid (reviewed by Drlica and Rouviere-Yaniv, 1987). The two most abundant of these are HU and H-NS (also called H1). There is estimated to be 20 000–40 000 molecules of each of these proteins in the cell, sufficient to bind once every 400 bp of chromosomal DNA. Two other presumptive histone-like proteins, HLP-1 and protein H have recently been shown to be an outer membrane protein and a ribosomal subunit, respectively (Bruckner and Cox, 1989; Hrivas *et al.*, 1990). There is every reason to suppose that HU and H-NS are the two most important proteins involved in organizing the bacterial nucleoid. HU is a heterodimer of two small basic subunits and binds DNA in a non-specific fashion. Purified HU can wrap DNA into nucleosome-like particles and alter its helical pitch (Broyles and Pettijohn, 1986; Rouviere-Yaniv *et al.*, 1979).

H-NS was first purified as a major component of the bacterial genome, independently by several groups, and variously named B1 (Varshavsky *et al.*, 1977), H1 (Busby *et al.*, 1979) and H-NS (Gualerzi *et al.*, 1986). Mutations in the gene encoding H-NS (variously designated *osmZ*, *bglY*, *pilG*, *virR*, *drdX* in the past; now designated *hns*) have been isolated many times, based on a wide variety of regulatory phenotypes (see Higgins *et al.*, 1990a, b, for review). *hns* mutations affect expression of a variety of genes including genes under osmotic control (Graeme-Cook *et al.*, 1989; Higgins *et al.*, 1988b; Hulton *et al.*, 1990), temperature control (Dorman *et al.*, 1990; Goransson *et al.*, 1990; Maurelli and Sansonetti, 1988), cryptic genes such as the β-glucoside operon (Defez and DeFelice, 1981) as well as affecting re-arrangements such as the *fimA* site-specific recombination events (Higgins *et al.*, 1988a, b; Kawula and Orndorff, 1991), and the frequencies of chromosomal deletion (Lejeune and Danchin, 1990). The *hns* gene is located at 27 min on the *E. coli* chromosome, between *tdk* and *galU*, transcribed as a monocistronic operon (Goransson *et al.*, 1990; Hulton *et al.*, 1990; May *et al.*, 1990). The H-NS protein is about 16.5 kDa, probably functions as a dimer (Falconi *et al.*, 1988), and exists as three isoforms (Spassky *et al.*, 1984; A Seirafi and CF Higgins, unpublished results) although the nature of the modification and its biological significance (if any) remains unknown. The DNA binding properties of H-NS are unusual. It binds to any DNA fragment irrespective of sequence. However, it has a somewhat higher affinity for curved DNA (Owen-Hughes *et al.*, 1992; Yamada *et al.*, 1990). Furthermore, the interaction with curved DNA plays a role in determining the specificity of its effects on gene expression *in vivo* (Owen-Hughes *et al.*, 1992). Upon binding DNA, H-NS appears to polymerize, along the molecule in a cooperative fashion (Rimsky and Spassky, 1990; T Owen-Hughes, DS Santos and CF Higgins, unpublished data). This can also be observed by electron microscopy. Thus, in the presence of excess H-NS all DNA molecules are covered by the protein: in the presence of excess DNA some molecules are covered from end-to-end while others remain naked (DS Santos and CF Higgins, unpublished data). How this cooperative binding contributes to the mechanism of action of H-NS is still unclear. Mutants defective in H-NS have altered levels of DNA supercoiling (Higgins *et al.*, 1988b; Hinton *et al.*, 1992; Owen-Hughes *et al.*, 1992), consistent with a structural role in the nucleoid. The phenotypes of *hns* mutants are probably a secondary consequence of altering this structure. Thus, several promoters affected by H-NS are sensitive to DNA supercoiling. At other promoters H-NS appears to repress or 'silence' gene expression (Gorannson *et al.*, 1990). A model consistent with the available data is that H-NS provides a structural framework, reducing the flexibility of chromosomal DNA and restricting promoter function. Transcription in the presence of H-NS then requires additional factors to present an appropriate DNA conformation for RNA polymerase to function, either a change in DNA topology (see later) or specific accessory (regulatory) proteins.

5. DNA supercoiling and gene expression

Chromosomal and plasmid DNA in bacterial cells is negatively supercoiled. DNA supercoiling is maintained primarily by the opposing activities of two enzymes, DNA gyrase and DNA topoisomerase I. The free energy of negative supercoiling can potentially influence many cellular processes including DNA replication, transposition and transcription (reviewed by Drlica, 1992). It has frequently been assumed that the absolute level of DNA supercoiling is such an important parameter that homeostatic mechanisms will operate to maintain a constant level of supercoiling in the cell. However, studies with mutants deficient in DNA gyrase, topoisomerase I, and with inhibitors of DNA gyrase activity, have shown that perturbation of DNA supercoiling is not incompatible with cellular function and can be tolerated. More recently, a considerable body of evidence indicates that the cellular level of supercoiling is a highly dynamic parameter and is altered by transcription (Wu *et al.*, 1988) and in response to a variety of environmental signals (Higgins *et al.*, 1990b). Indeed, given the economy of function in bacterial cells it would be surprising if the scope for heterogeneity in the structure and function of DNA provided by DNA supercoiling were not used to advantage in regulating gene expression.

There is now a considerable body of evidence indicating that DNA supercoiling varies in response to environmental signals. Growth of cells at high osmolarity increases the mean linking number of plasmid DNA (Higgins *et al.*, 1988b; Hseih *et al.*, 1991a) and a similar increase is seen for cells grown anaerobically (Dorman *et al.*, 1988; Hseih *et al.*, 1991b). Supercoiling may also vary with growth phase (Dorman *et al.*, 1988), temperature (Goldstein and Drlica, 1984) and nutrient availability (Balke and Gralla, 1987). Similar conclusions, at least for anaerobic growth, have been drawn from studies of chromosomal supercoiling (Hseih *et al.*, 1991b). The above studies were carried out on isolated DNA and provide a somewhat indirect measure of DNA supercoiling as they require the measurement of linking number of DNA purified free of any protein. It is conceivable that the constraining influence of proteins bound to the DNA *in vivo* compensates for the differences in linking number observed for isolated DNA such that torsional stress *in vivo* is maintained at a constant level. However, this potential objection cannot be sustained: direct measurement of superhelicity *in vivo* has shown real increases in net superhelicity in response to both changes in osmolarity and oxygen availability (Deyn *et al.*, 1990; McClellan *et al.*, 1990).

Given that DNA supercoiling does change in response to environmental signals, might this play a role in the regulation of gene expression? It is well established that artificial perturbation of DNA supercoiling (with gyrase inhibitors or in *gyr* or *topA* mutants) alters the expression of many genes. Furthermore, the range of supercoiling changes induced in such experiments is not dissimilar to that observed in response to environmental signals. Thus, it would be difficult to envisage a scenario in which environmentally induced changes in DNA supercoiling do not alter gene expression. There is now a considerable body of evidence, albeit indirect, that environmentally induced changes in supercoiling do play a role in the regulation of gene expression (Higgins *et al.*, 1988b; NiBhriain *et al.*, 1989). A good example is the osmoregulated *proU* locus which encodes a transport system for the uptake of the osmoprotectant glycine betaine. The *proU* promoter is sensitive to the perturbation of DNA supercoiling and selection of mutations which alter *proU* expression has yielded only *gyr*, *topA*, and *osmZ* (*hns*) lesions (Higgins *et al.*, 1988b): no evidence for a classical regulatory protein has been obtained, despite extensive searching. More direct evidence of a role for DNA supercoiling in the regulation of this promoter has recently been presented (Owen-Hughes *et al.*, 1992). Thus, osmoregulation *in vivo* requires a curved sequence located downstream of the *proU* promoter: H-NS interacts at this curved sequence *in vivo* to induce changes in DNA topology and alter expression of *proU*.

Of course, environmentally induced changes in DNA supercoiling are unlikely to be the sole regulatory mechanism for most promoters. More probably, they provide an underlying regulatory circuit, upon which more specific regulatory mechanisms are superimposed. For example, the genes encoding the major outer membrane porins, *ompF* and *ompC*, respond both to environmentally induced changes in supercoiling and to the specific regulatory proteins OmpR and EnvZ (Graeme-Cook *et al.*, 1989). Similarly, promoters under stringent (ppGpp) control may also respond to changes in supercoiling *in vivo* (Ohlsen and Gralla, 1992).

6. How are environmentally induced changes in DNA supercoiling mediated?

As yet there is little information on the mechanisms by which environmental signals influence DNA supercoiling. There is also no reason to suppose that a single mechanism operates. One possibility is a change in the activities of DNA gyrase or topoisomerase I. As environmentally induced topological changes can occur rapidly and in the absence of protein synthesis, this requires an alteration in the activities of these enzymes, rather than their synthesis. It has been suggested that changes in ATP–ADP ratios in response to aerobic–anaerobic shifts influence gyrase activity and, hence, supercoiling (Hseih *et al.*, 1991b). Changes in the interaction of histone-like proteins with DNA could also alter supercoiling. It is known that HU wraps DNA *in vitro* and affects its topology (Rouviere-Yaniv *et al.*, 1979), H-NS-deficient strains show altered supercoiling *in vivo* (Hinton *et al.*, 1992) and the interaction of H-NS with curved sequences influences DNA topology *in vivo* (Owen-Hughes *et al.*, 1992). How an alteration in the binding of these proteins to DNA might be achieved is unclear. It is possible, for example, that increased K^+ concentrations in the cell in response to osmotic upshock (Sutherland *et al.*, 1986), might influence the binding of the histone-like proteins to DNA. Alternatively, specific sensors might modify these proteins covalently, and, hence, their DNA binding properties. Clearly, this is a fundamental question to which we presently have few answers.

It should also be remembered that a change in DNA supercoiling (linking number) is a parameter that can be measured. Changes in DNA topology are, however, a reflection of changes in chromatin structure which cannot easily be assessed in any other way at present. Thus, we have a 'chicken and egg' scenario. Which comes first? Do changes in DNA topology, mediated by gyrase or topo-isomerase, alter chromatin structure and, hence gene expression? Or, do changes in the binding of histone-like proteins influence gyrase activity and, hence, DNA supercoiling? New methodologies to enable the three-dimensional organization of chromatin to be studied are required. Nevertheless, the important message is that promoters (or indeed any other piece of DNA) should not be viewed as a simple, linear string of nucleotides, with specific regulatory proteins binding to defined sequences. In the cell the DNA exists as a complex nucleoprotein structure. Any factor which perturbs this complex may influence transcription and *vice versa*. The study of promoter function needs to be considered as a complex, three-dimensional problem.

7. What is the role of the 'supercoiling response'?

At first sight, changes in DNA supercoiling and/or chromatin structure appear to be a rather crude mechanism for regulating gene expression. We are used to thinking of gene expression as a tightly controlled process, the cell aiming to adjust the precise level of expression of each gene in response to its environmental needs. However, it may be equally important for the cell to be imprecise in regulating certain genes, creating variability within the population. One circumstance where such variability might be important is during the interaction of a bacterial pathogen with its host cell. Within a population of bacterial cells, if every cell were identical then host resistance would develop rapidly. Furthermore, as host cells vary, most would be inaccessible to the bacterium. If, however, expression of those genes required for interaction with the host cell varied in a semi-random fashion, a few cells within the population would have the correct configuration of expression to interact successfully with the host. Thus, it would seem to be important for genetically identical cells to exhibit 'physiological' or 'random' variability. This need for random variability in the expression of a number of genes may underly the supercoiling response. This hypothesis is strengthened by the observation that many of the genes whose expression is influenced by changes in DNA supercoiling (and are *hns*-dependent) are peripheral genes associated with virulence functions rather than 'housekeeping' genes. Additionally, those signals which influence the supercoiling response, osmolarity, oxygen availability, temperature, etc. are precisely those which are encountered during the bacteria–host interaction. Finally, the virulence of *hns* (*virR*) mutants of *Salmonella* and *Shigella* is highly attenuated. What about other bacterial species? Intriguingly, several proteins identified as regulating virulence gene expression are histone-like. Thus, alginate synthesis in *Pseudomonas* which is crucial to the virulence of this species, is influenced by a protein highly related to eukaryotic histone H1 (Kato *et al.*, 1990). Similarly, a histone H1 homologue in *Chlamydia* appears to play a role in regulating virulence (Perara *et al.*, 1992; Tao *et al.*, 1991). The YmoA protein of *Yersinia* which influences virulence is also histone-like (Cornelis *et al.*, 1991), and is related to the *hha* gene product which regulates haemolysin synthesis in *E. coli* (Nieto *et al.*, 1991). Clearly, there is still much to be learnt about chromatin organization in bacteria. Equally clearly, the concept that, because there are no regular nucleosomes, bacterial chromatin is simple and uninteresting is untenable and an understanding of the organization and dynamics of bacterial chromatin is fundamental to our understanding of gene expression and cellular function.

Acknowledgements

I am grateful to all the members of my laboratory who have contributed to the experiments and concepts discussed here: Jay Hinton, Chris Hulton, Tom Owen-Hughes, Graham Pavitt, Diogenes Santos and Alex Seirafi. My laboratory is funded by the Imperial Cancer Research Fund.

References

Anderson RP, Roth JR. (1981) Spontaneous tandem genetic duplications in *Salmonella typhimurium* arise by unequal recombination between rRNA (*rrn*) cistrons. *Proc. Natl Acad. Sci. USA*, **78:** 3113-3117.

Balke VL, Gralla JD. (1987) Changes in linking number of supercoiled DNA accompany growth transitions in *Escherichia coli*. *J. Bacteriol.* **169:** 4499-4506.

Bihlmaier A, Romling U, Meyer TF, Tummler B, Gibbs CP. (1991) Physical and genetic map of the *Neisseria gonorrhoeae* strain MS11-N198 chromosome. *Mol. Microbiol.* **5:** 2529-2539.

Broyles SS, Pettijohn DE. (1986) Interaction of the *Escherichia coli* HU protein with DNA. *J. Mol. Biol.* **187:** 47-60.

Bruckner RC, Cox MM. (1989) The histone-line H protein of *Escherichia coli*. *Nucl. Acids Res.* **17:** 3145-3161.

Busby S, Kolb A, Buc H. (1979) Isolation of plasmid protein complexes from *Escherichia coli*. *Eur. J. Biochem.* **99:** 105-111.

Buvinger WE, Lampel KA, Bojanowski RJ, Riley M. (1984) Location and analysis of nucleotide sequences at one end of a putative lac transposon in the *Escherichia coli* chromosome. *J. Bacteriol.* **159:** 618-623.

Cornelis GR, Sluiters C, Delor I, Geib D, Kaniga K, Rouvroit CL, Sory M-P, Vanooteghem J-C, Michiels T. (1991) *ymoA*, a *Yersinia enterolitica* chromosomal gene modulating the expression of virulence functions. *Mol. Microbiol.* **5:** 1023-1034.

Daniels DL, Plunkett G III, Burland V, Blattner FR. (1992) Analysis of the *Escherichia coli* genome: DNA sequence of the region from 84.5 to 86.5 minutes. *Science*, **257:** 771-778.

Deyn A, Malkhosyan S, Duzhy D, Panchenko Y, Mirkin S (1991) Formation of $(dA-dT)_n$ cruciforms in *Escherichia coli* under different environmental conditions. *J. Bacteriol.* **173:** 2658-2664

Defez R, DeFelice M. (1981) Cryptic operon for β-glucoside metabolism in *Escherichia coli* K12. *Genetics*, **97:** 11-25.

Dorman CJ, Barr GC, NiBhriain N, Higgins CF. (1988) DNA supercoiling and the anaerobic and growth phase regulation of *tonB* gene expression. *J. Bacteriol.* **170:** 2816-2826.

Dorman CJ, NiBhriain N, Higgins CF. (1990) DNA supercoiling and environmental regulation of virulence gene expression in *Shigella flexneri. Nature*, **344:** 789-792.

Drlica K. (1992) Control of bacterial supercoiling. *Mol. Microbiol.* **6:** 425-433.

Drlica K, Rouviere-Yaniv J. (1987) Histonelike proteins of bacteria. *Microbiol. Rev.* **51:** 301-319.

Falconi M, Gualtieri MT, LaTeana A, Losso MA, Pon CL. (1988) Proteins from the prokaryotic nucleoid: primary and quarternary structure of the 15 kDa *Escherichia coli* DNA binding protein H-NS. *Mol. Microbiol.* **2:** 323-329.

Gilson E, Clement J-M, Brutlag D, Hofnung M. (1984) A family of dispersed repetitive extragenic palindromic DNA sequences in *E. coli. EMBO J.* **3:** 1417-1422.

Gilson E, Rousset JP, Clement J-M, Hofnung M. (1986) A subfamily of *E. coli* palindromic units implicated in transcription termination? *Ann. Inst. Pasteur. Microbiol.* **137B:** 259-270.

Gilson E, Perrin D, Hofnung M. (1990) DNA polymerase I and a protein complex bind specifically to *E. coli* palindromic unit highly repetitive DNA. *Nucl. Acids Res.* **18:** 3941-3952.

Goldstein E, Drlica K. (1984) Regulation of bacterial DNA supercoiling: plasmid linking numbers vary with growth temperature. *Proc. Natl Acad. Sci. USA*, **81:** 4046-4050.

Goransson M, Sonden B, Nilsson P, Dagberg B, Forsman K, Emanuelsson K, Uhlin B-E. (1990) Transcriptional silencing and thermoregulation of gene expression in *Escherichia coli. Nature*, **344:** 682-685.

Graeme-Cook KA, May G, Bremer E, Higgins CF. (1989) Osmotic regulation of porin expression; a role for DNA supercoiling. *Mol. Microbiol.* **3:** 1287-1294.

Gualerzi CO, Losso MA, Lammi M, Friedrich K, Pawlik RT, Canonaco MA, Pingoud A, Pon CL. (1986) Proteins from the prokaryotic nucleoid. In: *Bacterial Chromatin* (Gualerzi CO, Pon CL, eds). Heidelberg: Springer Verlag. pp 101-134.

Higgins CF, Ames GF-L, Barnes WM, Clement J-M, Hofnung M. (1982) A novel inter-cistronic regulatory element of prokaryotic operons. *Nature*, **298:** 760-762.

Higgins CF, McLaren RS, Newbury SF. (1988a) Repetitive extragenic palindromic sequences, mRNA stability and gene expression: evolution by gene conversion? A review. *Gene*, **72:** 3-14.

Higgins CF, Dorman CJ, Stirling DA, Waddell L, Booth IR, May G, Bremer E. (1988b) A physiological role for DNA supercoiling in the osmotic regulation of gene expression in *S. typhimurium* and *E. coli. Cell*, **52:** 569-584.

Higgins CF, Hinton JCD, Hulton CSJ, Owen-Hughes T, Pavitt GD, Seirafi A. (1990a) Protein H1: a role for chromatin structure in the regulation of bacterial gene expression and virulence? *Mol. Microbiol.* **4:** 2007-2012.

Higgins CF, Dorman CJ, NiBhriain N. (1990b) Environmental influences on DNA super-coiling: a novel mechanism for the regulation of gene expression. In: *The Bacterial Chromosome* (Drlica K, Riley M, eds). Washington DC: American Society for Microbiology Press.

Hinnebusch J, Bergstrom S, Barbour AG. (1990) Cloning and sequence analysis of linear plasmid telomeres of the bacterium *Borrelia burgdorferi. Mol. Microbiol.* **4:** 811-820.

Hinton JCD, Santos DS, Seirafi A, Hulton CSJ, Pavitt GD, Higgins CF. (1992) Expression and mutational analysis of the nucleoid-associated protein H-NS of *Salmonella typhimurium. Mol. Microbiol.* **6:** 2327-2337.

Hrivas L, Coleman J, Koski P, Vaara M. (1990) Bacterial 'histone-like protein I' (HLP-I) is an outer membrane constituent. *FEBS Lett.* **262:** 123-126.

Hseih L-S, Burger RM, Drlica K. (1991a) Bacterial DNA supercoiling and [ATP]/[ADP] changes associated with a transition to anaerobic growth. *J. Mol. Biol.* **219:** 443-450.

Hseih L-S, Rouviere-Yaniv J, Drlica K. (1991b) Bacterial DNA supercoiling and [ATP]/ [ADP] ratio: changes associated with salt shock. *J. Bacteriol.* **173:** 3914-3917.

Hulton CSJ, Seirafi A, Hinton JCD, Sidebotham JM, Waddell L, Pavitt DG, Owen-Hughes T, Spassky A, Buc H, Higgins CF. (1990) Histone-like protein H1 (H-NS), DNA supercoiling and gene expression in bacteria. *Cell,* **63:** 631-642.

Hulton CSJ, Higgins CF, Sharp PM. (1991) ERIC sequences: a novel family of repetitive elements in the genomes of *E. coli, S. typhimurium* and other enterobacteria. *Mol. Microbiol.* **5:** 825-834.

Inouye M, Inouye S. (1991) Retroelements in bacteria. *Trends Biochem. Sci.* **16:** 18-21.

Kato J, Misra TK, Chakrabarty AM. (1990) AlgR3, a protein resembling eukaryotic histone H1, regulates alginate synthesis in *Pseudomonas aeruginosa. Proc. Natl Acad. Sci. USA,* **87:** 2887-2891.

Kawula TH, Orndorff PE. (1991) Rapid site-specific DNA inversion in *Escherichia coli* mutants lacking the histone-like protein H-NS. *J. Bacteriol.* **173:** 4116-4123

Kohara Y, Akiyama K, Isono K. (1987) The physical map of the whole *E. coli* chromosome: application of a new strategy for rapid analysis and sorting of a large genomic library. *Cell,* **50:** 495-508.

Lejeune P, Danchin A. (1990) Mutations in the *bglY* gene increase the frequency of spontaneous deletions in *E. coli. Proc. Natl Acad. Sci. USA,* **87:** 360-363.

Lilley DMJ. (1986) Bacterial chromatin: a new twist to an old story. *Nature,* **320:** 14-15.

Lim D, Gomes TAT, Maas WK. (1990) Distribution of msDNAs among serotypes of enteropathogenic *E. coli. Mol. Microbiol.* **4:** 1711-1714.

Liu S-L, Sanderson KE. (1992) A physical map of the *Salmonella typhimurium* LT2 genome made by using *Xba*I analysis. *J. Bacteriol.* **174:** 1662-1672.

Maurelli AT, Sansonetti PJ. (1988) identification of a chromosomal gene controlling temperature regulated expression of *Shigella* virulence. *Proc. Natl Acad. Sci. USA,* **85:** 2820-2824.

May G, Dersch P, Haardt M, Middendorf A, Bremer E. (1990) The *osmZ* (*bglY*) gene encodes the DNA-binding protein H-NS (H1a), a component of the *Escherichia coli* K12 nucleoid. *Mol. Gen. Genet.* **224:** 81-90.

McClellan JA, Boublikova P, Palecek E, Lilley DMJ. (1990) Superhelical torsion in cellular DNA responds directly to environmental and genetic factors. *Proc. Natl Acad. Sci. USA,* **87:** 8373-8377.

Medigue C, Viari A, Henaut A, Danchin A. (1991) *Escherichia coli* molecular genetic map (1500 kbp): update II. *Mol. Microbiol.* **5:** 2629-2640.

Newbury SF, Smith NH, Robinson EC, Hiles ID, Higgins CF. (1987a) Stabilisation of translationally active mRNA by prokaryotic REP sequences. *Cell,* **48:** 297-310.

Newbury SF, Smith NH, Higgins CF. (1987b) Differential mRNA stability controls relative gene expression within a polycistronic operon. *Cell,* **51:** 1131-1143.

NiBhriain N, Dorman CJ, Higgins CF. (1989) An overlap between osmotic and anaerobic stress responses: a potential role for DNA supercoiling in the coordinate regulation of gene expression. *Mol. Microbiol.* **3:** 933-942.

Nieto JM, Carmona M, Bolland S, Jubete Y, de la Cruz F, Juarez A. (1991) The *hha* gene modulates haemolysin expression in *Escherichia coli. Mol. Microbiol.* **5:** 1285-1293.

Ohlsen KL, Gralla JD. (1992) Interrelated effects of DNA supercoiling, ppGpp, and low salt on melting within the *Escherichia coli* ribosomal RNA *rrn*BP$_1$ promoter. *Mol. Microbiol.* **6:** 2243-2251.

Owen-Hughes TA, Pavitt GD, Santos DS, Sidebotham JM, Hulton CSJ, Hinton JCD, Higgins CF. (1992) The chromatin-associated protein H-NS interacts with curved DNA to influence DNA topology and gene expression. *Cell,* **71:** 255–265.

Perara E, Ganem D, Engel JN. (1992) A developmentally regulated chlamydial gene with apparent homology to eukaryotic histone H1. *Proc. Natl Acad. Sci. USA,* **89:** 2125-2129.

Riley M, Sanderson KE. (1990) Comparative genetics of *E. coli* and *S. typhimurium*. In: *The Bacterial Chromosome* (Drlica K, Riley M, eds). Washington DC: American Society for Microbiology Press. pp 85-95.

Rimsky S, Spassky A. (1990) Sequence determinants for H1 binding on *E. coli lac* and *gal* promoters. *Biochemistry*, **29:** 3765-3771.

Rouviere-Yaniv J, Yaniv M, Germond J-E. (1979) *E. coli* DNA binding protein HU forms nucleosome-like structure with circular double stranded DNA. *Cell*, **17:** 265-274.

Schmid MB, Roth JR. (1987) Gene location affects expression level in *Salmonella typhimurium*. *J. Bacteriol.* **169:** 2872-2875.

Segall A, Mahan MJ, Roth JR. (1988) Rearrangement of the bacterial chromosome: forbidden inversions. *Science*, **241:** 1341-1348.

Sharples GJ, Lloyd RG. (1990) A novel repeated DNA sequence located in the intergenic regions of bacterial chromosomes. *Nucl. Acids Res.* **18:** 6503-6508.

Sinden RR, Carlson J, Pettijohn DE. (1980) Torsional tension in the double helix measured with trimethylpsoralen in living *E. coli* cells. *Cell*, **21:** 773-783.

Sinden RR, Pettijohn DE. (1981) Chromosomes in living *Escherichia coli* cells are segregated into domains of supercoiling. *Proc. Natl Acad. Sci. USA*, **78:** 224-228.

Spassky A, Rimsky S, Garreau H, Buc H. (1984) H1a, an *E. coli* DNA-binding protein which accumulates in stationary phase, strongly compacts DNA *in vitro*. *Nucl. Acids Res.* **12:** 5321-5340.

Stern MJ, Ames GF-L, Smith NH, Robinson EC, Higgins CF. (1984) Repetitive extragenic palindromic sequences; a major component of the bacterial genome. *Cell*, **37:** 1015-1026.

Stern MJ, Prossnitz E, Ames GF-L. (1988) Role of the intercistronic region in post-transcriptional control of gene expression in the histidine transport operon of *Salmonella typhimurium*: involvement of REP sequences. *Mol. Microbiol.* **2:** 141-152.

Sutherland L, Cairney J, Elmore MJ, Booth IR, Higgins CF. (1986) Osmotic regulation of transcription: induction of the *proU* glycine betaine transport gene is dependent on accumulation of intracellular potassium. *J. Bacteriol.* **168:** 805-814.

Tao S, Kaul R, Wenman WM. (1991) Identification and nucleotide sequence of a developmentally regulated gene encoding a eukaryotic histone H1-like protein from *Chlamydia trachomatis*. *J. Bacteriol.* **173:** 2828-2832.

Varshavsky AJ, Nedospasov A, Bakayev VV, Bakayeva TG, Giorgiev G. (1977) Histone like proteins in the purified *Escherichia coli* deoxyribonucleoprotein. *Nucl. Acids Res.* **4:** 2725-2745.

Worcel A, Burgi E. (1972) On the structure of the folded chromosome of *Escherichia coli*. *J. Mol. Biol.* **71:** 127-147.

Wu H-Y, Shyy S, Wang JC, Liu LF. (1988) Transcription generates positively and negatively supercoiled domains in the template. *Cell*, **53:** 433-440.

Yamada H, Muramatsu S, Mizuno T. (1990) An *Escherichia coli* protein that preferentially binds to sharply curved DNA. *J. Biochem.* **106:** 420-425.

Yang Y, Ames GF-L. (1988) DNA gyrase binds to the family of prokaryotic repetitive extragenic palindromic sequences. *Proc. Natl Acad. Sci. USA*, **85:** 8850-8854.

Chapter 3

Site-specific recombination and the partition of bacterial chromosomes

David J Sherratt, Garry Blakely, Mary Burke, Sean Colloms, Nick Leslie, Richard McCulloch, Gerhard May and Jennifer Roberts

Abstract

Normal partition of the *Escherichia coli* chromosome to daughter cells at cell division requires a functional site-specific recombination event at a site, *dif*, within the replication terminus region. This recombination requires the products of two unlinked genes (*xerC* at 4024 kb on the *E. coli* map and *xerD* at 3050 kb). Mutation of *xerC*, *xerD*, or both genes; or deletion of the *dif* site leads to visible defects in chromosome partition and cell division. Both XerC and XerD proteins have sequence homology to the integrase-family of site-specific recombinases, and share 37% amino acid identity. The two proteins bind cooperatively to the recombination site and each is assumed to catalyse a distinct part of the recombination reaction. The same recombinase proteins bind to and recombine related sequences present in natural multicopy plasmids and ensure their stable inheritance at cell division. The role of this recombination in stable plasmid inheritance appears to be through its ability to convert plasmid multimers (arising through homologous recombination), to monomers; only monomers are stably inherited. We hypothesize that this site-specific recombination has a similar function in chromosomal partition; chromosome dimers that could arise by an odd number of homologous exchanges between newly replicated sister chromosomes need to be converted to monomers if sister chromosomes are to be partitioned normally to daughter cells during cell division. Consistent with this is the observation that the *xerC* mutant phenotype is largely suppressed in cells deficient in homologous recombination.

1. Introduction

In the bacterial cell cycle, the process of chromosome replication from initiation to termination needs to be in concert with other cell growth and division param-

eters. Though substantial information is known about cell growth, septum formation, and the proteins involved in cell division (e.g. see de Boer *et al.*, 1990; and
multiple articles in: *The Bacterial Cell Cycle: Structural and Molecular Aspects.
Res. Micro.* **142:** 113–354), the precise ways in which initiation and termination of
replication are coordinated temporally and positionally with respect to the cell
and the cell cycle remain unclear. At steady state, each cell division event requires
a prior initiation and termination event. In *E. coli* cells growing at generation
times of 60 minutes or less, each complete replication event (i.e. time from
initiation of a fork to the termination of that replication fork) takes 40 minutes
with a division event occurring 20 minutes later. In addition to completion of
chromosome replication, cell division requires that cells have grown to a particular minimum length. Newborn cells contain a single nucleoid, positioned
centrally. During a cell's growth, nucleoid separation normally occurs soon after
replication termination and requires *de novo* protein synthesis. The sister
nucleoids are positioned at a quarter and three-quarters of the cell length from a
given pole (Begg and Donachie, 1991; Donachie and Begg, 1989; Hiraga *et al.*,
1990). Though the mechanism that moves sister chromosomes apart prior to cell
division is unknown, conditional-lethal mutants in the partition process have been
isolated (Hiraga *et al.*, 1989). One class of mutant is defective in a gene that
encodes a protein MukB, with some similarity to eukaryotic kinesin, suggesting
that some mitosis-like event might be involved in chromosome partition (Niki *et
al.*, 1991).

Before newly replicated chromosomes can be partitioned, they need to be
decatenated, since the linking numbers of the parental DNA strands have to be
reduced to zero before separation. Whereas DNA gyrase is thought to be involved
in maintaining a constant superhelix density ahead of the replication fork, it now
appears that a specialized topoisomerase, topoisomerase IV, is involved in the
final decatenation at replication termination (Kato *et al.*, 1990; Schmid *et al.*,
1990). Our own studies, summarized in this report, suggest that in addition to
decatenation, a site-specific recombination event in the terminus region is
required to ensure that newly replicated chromosomes can be partitioned to
daughter cells as topologically separate entities (Blakely *et al.*, 1991).

In various studies of bacterial plasmid stability, it has been demonstrated that
plasmid multimers (arising by homologous inter-molecular recombination) are
not stably inherited. Moreover, it has been demonstrated that many plasmids use
site-specific recombination to convert plasmid multimers to monomers, hence
ensuring their stable inheritance (Austin *et al.*, 1981; Summers and Sherratt,
1984). Work from our own laboratory has now demonstrated that the stabilizing
site-specific recombination system used by multicopy plasmids related to ColE1 is
used by the bacterial chromosome to ensure its stable partition at cell division.
The chromosome has a recombination site, *dif*, in the region of replication
termination, which is acted on by two chromosomally encoded recombinase

proteins, XerC and XerD. Deletion of *dif* or mutation of either or both of the recombinase genes led to cell populations that, though viable, contained substantial proportions of cells with aberrant chromosomes contained within filamentous cells (Fig. 1, see p. 173; Blakely *et al.*, 1991; Kuempel *et al.*, 1991). This mutant phenotype is dependent on a functional homologous recombination system, consistent with the view that this site-specific recombination system acts to remove chromosomal dimers that are formed by homologous recombination. Other bacteria encode related recombinases (G Blakely, unpublished) and we believe it likely that site-specific recombination plays a role in the partition at cell division of most circular replicons.

2. Site-specific recombination: two families of site-specific recombinase mediate a wide range of programmed DNA rearrangements

Site-specific recombination is a precise conservative break–join reaction that is involved in a wide variety of programmed DNA rearrangements in prokaryotes and eukaryotes (Sadowski, 1986). For example: in the integration and excision of the genomes of bacterial viruses into and out of bacterial chromosomes (Landy, 1989); in converting the initial products of replicative inter-replicon transposition to final products (Arthur and Sherratt, 1979); in inversion gene switches that lead to alternate expression of microbial cell surface proteins and bacteriophage tail fibre proteins (Johnson, 1991); and in the copy number control and stable inheritance of circular replicons in yeast and in bacteria (Blakely *et al.*, 1991; Colloms *et al.*, 1990; Futcher, 1986). Less precise yet related rearrangements occur in the process of *conjugative transposition* and in the activities of *integrons*, bacterial genetic elements that are responsible for the movement of antibiotic resistance genes into and out of plasmids and transposable elements (Murphy, 1989; Schmidt *et al.*, 1989).

Site-specific recombination systems can be classified into one of two families, the lambda integrase family and the resolvase/DNA invertase family (Argos *et al.*, 1986; Sadowski, 1986; Stark *et al.*, 1992). The various biological roles of site-specific recombination can be mediated by members of both families, despite the fact that the catalytic mechanisms used by recombinases of each family are very different. Recombinases of the resolvase/DNA invertase family are small proteins (about 185 amino acids) that show much amino acid conservation (Hatfull and Grindley, 1983; Sherratt, 1989). The enzymes contain two separable functional domains; the C-terminal third of the molecule is involved in specific binding to the recombination site while the N-terminal two-thirds contains the catalytic site and determinants for monomer–monomer interactions.

Integrase family recombinases have been found in yeast and many bacterial species (Landy, 1989; Sadowski, 1986). The recombinases vary in size from about

280 amino acids to in excess of 350. They show substantial amino acid divergence, though two conserved domains (domains I and II) are present in the C-terminal region of the protein (Fig. 2). Even here only four amino acid residues are completely conserved in all family members (a R in domain I and a R, H and Y in domain II; Argos *et al.*, 1986; Chen *et al.*, 1992).

The *core* recombination site, within which strand exchange occurs, is approximately 30 bp in size for each class of recombinase, though additional accessory sequences may also be necessary. Each core recombination site is thought to bind two recombinase monomers; in recombination between a pair of sites the four strand exchanges are also almost certainly catalysed by four recombinase monomers. Participating recombination sites are brought together into a functional recombinational synapse by protein–DNA and protein–protein interactions; Watson–Crick base-pairing between the participating duplexes is not involved in synapsis and is not required for the initial strand exchanges. For the integrase-family recombinases, we have suggested, on biochemical and mechanistic grounds, that correctly synapsed sites are arranged in an antiparallel orientation (Stark *et al.*, 1989a). Details of the strand exchange mechanism for integrase family recombinases are described below.

3. Resolution selectivity

A pair of identical recombination sites can be arranged in three configurations with respect to each other. They may be in separate DNA molecules where they provide the substrates for inter-molecular recombination. When they are in the same molecule, they can be inverted or directly repeated with respect to each other and will be substrates for intramolecular inversion or intramolecular resolution, respectively (Fig. 3). Many site-specific recombination systems act efficiently on only one arrangement of recombination site. For example, transposon Tn3 resolvase shows *resolution selectivity*, catalysing recombination intramolecularly only on directly repeated recombination sites (Stark *et al.*, 1989a, b, 1992; Kanaar and Cozzarelli, 1992). In contrast, the related DNA invertases show *inversion selectivity*, since they only recombine inverted sites intramolecularly (Johnson, 1991). Both classes of enzymes fail to recombine intermolecularly. This selectivity is essential for the effective biological functioning of a given site-specific recombination system (see later). How can a recombinase detect how a pair of identical recombination sites are arranged with respect to each other? This selectivity for resolution or inversion is observed both *in vivo* and *in vitro*. In the latter case, all that is required for functional recombination is the DNA substrate containing a pair of appropriately arranged recombination sites (preferably super-coiled), recombinase, a simple solution of salts, and in the case of invertase, the accessory protein FIS. From studies of such reactions *in vitro*, it became clear that the productive recombinational synapse for a particular enzyme has a fixed and specific topology. For example, the resolvase/*res* synapse contains three

(a)

```
XerC   1  MTDLHTDVERYLRYLSVERQLSPITLLNYQRQLEAIINFASENGLQSWQQ  50
          : :    :*  :*   * : *:  *   **  * *:*  ::::    **   :
XerD   1  MKQDLARIEQFLDALWLEKNLAENTLNAYRRDLSMMVEWLHHRGL-TLAT  49


      51  CDVTMVRNFAVRSRRKGLGAASLALRLSALRSFFDWLVSQNELKANPAKG 100
          :   :     * * * * ***:* :*::* :    :*
      50  AQSDDLQALLAERLEGGYKATSSARLLSAVRRLFQYLYREKFREDDPSAH  99


     101  VSAPKAPRHLPKNIDVDDMNRLLDID-INDPLAVRDRAMLEVMYGAGLRL 149
          : ** * :***:: ::;***: *::** :**:*****:*: ***:
     100  LASPKLPQRLPKDLSEAQVERLLQAPLIDQPLELRDKAMLEVLYATGLRV 149
λ Int     conserved domain I                         TGQRV


     150  SELVGLDIKHLDLESGEVWVMGKGSKERRLPIGRNAVAWIEHWLDL---R 196
          ****** :   : *   * * *:*:*** *** :*:* :** *:* :*:   :
     150  SELVGLTMSDISLRQGVVRVIGKGNKERLVPLGEEAVYWLETYLEHGRPW 199
λ Int     GDLCEMKWSD


     197  DLFGSEDDALFLSKLGKRISARNVQKRFAEWGIKQGLNNH-VHPHKLRHS 245
          * *    * ** *   :  :      *:  ::: *::   : ** ***
     200  LLNGVSIDVLFPSQRAQQMTRQTFWHRIKHYAVLAGIDSEKLSPHVLRHA 249
λ Int     conserved domain II                       HELRSL


     246  FATHMLESSGDLRGVQELLGHANLSTTQIYTHLDFQHLASVYDAAHPRAKRGK 299
          ****:*: ::*** ** **** :*********:  ::*   : :  ****
     250  FATHLLNHGADLRVVQMLLGHSDLSTTQIYTHVATERLRQLHQQHHPRA    299
λ Int     SA-RLYEKQISDKFAQHLLGHKSDTMASQYRD
```

(b)

```
                 Domain I                    Domain II

XerC         AGLRLSELVGLDIKH...HKLRHSFATHMLESS-GDLRGVQELLGHAN-LSTT-QIYTH
XerD         TGLRVSELVGLTMSD...HVLRHAFATHLLNHG-ADLRVVQMLLGHSD-LSTT-QIYTH
Lambda Int   TGQRVGDLCEMKWSD...HELRSLSA-RLYEKQ-ISDKFAQHLLGHKS-DTMA-SQYRD
Flp          NCGRFSDIKNVDPKS...HIGRHLMTSFLSMKGLTELTNVVGNWSDKRASAVARTTYTH
Cre          TLLRIAEIARIRVKD...HSARVGAARDMARAG-VSIPEIMQAGGWTN-VNIV-MNYIR
P22 Int      TGLRRSNIINLEWQQ...HDLRHTWASWLVQAG-VPISVLQEMGGWES-IEMV-RRYAH
R46 ORF      TGMRISEGLQLRVKD...HTLRHSFATALLRSG-YDIRTVQDLLGHSD-VSTT-MIYTH
Tn554 TnpA   GGLRIGEVLSLRLED...HMLRHTHATQLIREG-WDVAFVQKRLGHAHVQTTL-NTYVH
Tn554 TnpB   CGMRISELCTLKKGC...HAFRHTVGTRMINNG-MPQHIVQKFLGHES-PEMT-SRYAH
FimB         HGFRASEICRLRISD...HMLRHSCGFALANMG-IDTRLIQDYLGHRN-IRHT-VWYTA
FimE         HGMRISELLDLHYQD..HNLRHACGYELAERG-ADTRLIQDYLGHRN-IRHT-VRYTA
```

Figure 2. (a) Comparison of the amino acid sequence of XerC and XerD. Overall there is 37% identity throughout the 298 amino acids of each protein, with highest identity in the conserved domains I and II. Identical amino acids are indicated by an asterisk and conservative changes by a colon. The N-terminal sequences have been confirmed by N-terminal protein sequencing.
(b) Conserved domains I and II of integrase family recombinases. The tetrad of conserved amino acids is in bold face. The conserved tyrosine initiates the phosphodiester attack, whilst the other residues are required for initial phosphodiester cleavage or strand transfer.

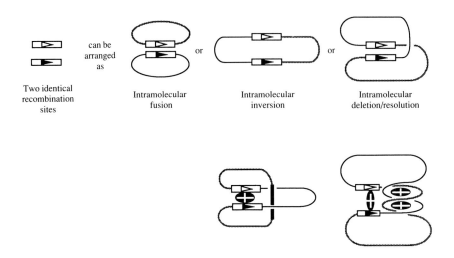

Figure 3. The molecular basis of resolution or inversion selectivity. The top panel of the figure cartoons the three ways in which a pair of recombination sites can be arranged (arbitrarily the sites are oriented in a parallel configuration). Molecules are depicted as circles; in reality, the model requires that molecules have no *free ends*. Topologically restrained domains within linear molecules also meet this criterion. For molecules containing inverted recombination sites, the sites can be synapsed in a parallel orientation without the duplex on one side of the sites (one domain; —), crossing the duplex on the other side of the sites (the other domain; ᴑᴑᴑᴑᴑᴑᴑᴑᴑᴑ). Since there are no *crossings* or *nodes* in plane projection, we describe this as a 0 synapse. Generally, for inverted sites aligned in a parallel configuration, 0 or any even number (2,4,6 . . .) of *nodes* can be present. A molecule with two interdomainal nodes is shown below. For molecules containing directly repeated recombination sites aligned in a parallel configuration, 1, or any *odd* number of interdomainal nodes must be placed between the synapsed sites in a plane projection. Molecules with 1 and 3 interdomainal nodes are shown. Note that nodes can have a (+) or a (−) sign. Those shown are (−) and would be favoured by (−) supercoiling. The structure of the recombination site and particularly the presence of accessory sequences in the resolvase and DNA invertase systems lead to productive recombinational synapses requiring (−2) interdomainal nodes (invertase), or (−3) interdomainal nodes (resolvase) as cartooned in the lower part of the figure. Recombinase molecules are diagrammed as ovals containing four monomers. In the inversion system the accessory protein FIS is depicted as a solid bar, whereas with resolvase, two tetramers of resolvase act as accessory protein. In both systems, four recombinase molecules are thought to bind the *core* recombination site, each catalysing a single strand exchange. Inverted sites cannot be arranged into a (−3) synapse (or indeed any *odd*-noded synapse) in parallel alignment without introducing energetically unfavourable nodes or tangle. Hence, they fail to support recombination. Similarly, directly repeated sites cannot be favourably arranged into a (−2) or *even*-noded synapse in parallel alignment and therefore fail to support inversion. Topology then provides a *filter* to select against unfavoured reactions. For a more detailed discussion see Stark *et al.* (1989a; 1992) and references therein.

negative interdomainal supercoils while the DNA invertase Gin/*gix* synapse contains two negative supercoils (see Fig. 3 for definitions). Figure 3 outlines how this requirement for a fixed synapse topology provides the molecular basis of selectivity. Since accessory proteins and accessory DNA sequences provide the requirements for specific recombination synapse topology, their presence is essential in systems that exhibit selectivity. Similarly, site-specific recombination systems that do not exhibit selectivity have no mandatory requirements for accessory sequences and proteins.

4. *Escherichia coli* chromosomal genes encode proteins required for site-specific recombination at *dif* in the chromosomal terminus region and at *cer* and its homologues in natural plasmids

The first evidence that small multicopy plasmids related to ColE1 use site-specific recombination to ensure their stable inheritance came from experiments of Summers and Sherratt (1984), in which plasmid stability was inversely correlated with plasmid multimerization. A recombination site, *cer*, was identified in ColE1 and shown to be necessary for site-specific resolution of multimers to monomers. *cer* is about 250 bp long with the 32 bp core recombination site, within which strand exchange occurs at one end. The other 220 bp function as necessary accessory sequences (Fig. 4). No evidence for plasmid-encoded recombination enzymes was found and so transposon mutagenesis of the *E. coli* chromosome was used to identify those chromosomal genes that encode proteins involved in site-specific

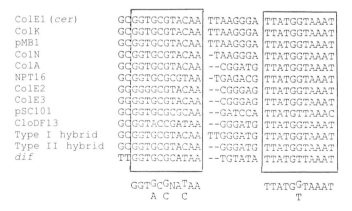

Figure 4. The *core* recombination sites of *cer*, other natural homologues of *cer* and the chromosomal site *dif*. The left-hand enclosed box indicates the XerC-binding region and the right-hand enclosed box indicates the XerD-binding region. The type I and type II hybrids were products of recombination between *cer* and its CloDF13 analogue. The type I hybrid shows resolution selectivity whereas the type II hybrid recombines all configurations of site.

recombination at *cer*. *Reporter* plasmids containing two copies of a directly repeated *cer* flanking a selective marker (usually Cm^R or *lacZ*), an origin of replication and another selective marker that would not be deleted during Xer-mediated deletion (Stirling *et al.*, 1988a), were used to select mutants (Xer⁻) that failed to delete the reporter marker. Such experiments, using transposon Tn5 as the mutagen, initially led to the identification and characterization of two genes designated *xerA* and *xerB*. Subsequently they were shown to encode the ArgR and PepA proteins respectively (Stirling *et al.*, 1988b; 1989); both proteins have an accessory though obligate role in site-specific recombination at *cer*. Subsequently, an unlinked gene, *xerC*, located at 3710 bp on the *E. coli* map, and encoding a lambda integrase family protein was identified and characterized (Colloms *et al.*, 1990). Cell extracts derived from cells over-expressing XerC, and partially purified XerC showed weak yet specific binding activity to DNA fragments containing *cer* and *dif*. We were not able to show recombination *in vitro* with such preparations. Though the initial Tn5 insertion that led to the identification of XerC, gave only a partial mutant phenotype because the transposon was inserted in an upstream gene of the same transcriptional unit, other insertions and deletions in the *xerC* gene itself have shown that a functional *xerC* gene is not necessary for *E. coli* viability. Nevertheless cultures of *xerC* mutants show extensive filamentation and aberrant nucleoids (Blakely *et al.*, 1991; Fig. 1, see p. 173), consistent with a role of XerC in the partition of *E. coli* chromosomes. Most recently we have demonstrated that a second related recombinase (which we have designated XerD) also has an essential role in site-specific recombination at *cer* and *dif* (unpublished). The *xerD* gene, was originally named *xprB*, maps at 3050 kb and is linked to *recJ*. It was originally identified as an ORF (open reading frame) that encodes a putative integrase-like protein (Lovett and Kolodner, 1991). The similarity of the putative XprB protein to XerC (37% identity; more similar than any other known integrase-like protein) led us to test whether XprB function is required for recombination at *cer* and *dif*. A bacterial strain containing a mini-Tn10 within the *xprB* gene (supplied by S Lovett) failed to support recombination at *cer* and *dif* and showed an aberrant cell division, and chromosome partition phenotype that was indistinguishable from that of *xerC* mutants. This led to the re-designation of *xprB* as *xerD*. *xerC xerD* double mutants have a phenotype similar to that of the single mutants; complementation of the double mutant phenotype requires both XerC and XerD in *trans*.

In vitro protein–DNA binding experiments have shown that XerC binds to the left half of the core recombination sites of *cer* and *dif*, whilst XerD binds to the right half (Fig. 4). Binding of XerC and XerD to *cer* or *dif* is cooperative. In contrast, artificial pseudo-sites containing either two XerC-binding sites or two XerD-binding sites flanking a central region fail to permit recombination and fail to bind either XerC or XerD cooperatively, demonstrating that the cooperative interactions are confined to interactions between XerC and XerD.

Though site-specific recombination at *cer* and *dif* requires both XerC and XerD, and both proteins show similar binding characteristics to *cer* and *dif* sites, there are substantial differences in recombination at these two sites. For example, XerC/D mediated recombination at *cer*, and its natural plasmid homologues, shows strong resolution selectivity, recombination occurring exclusively intramolecularly. This is crucial for the role of XerC/D-*cer* recombination in stable plasmid inheritance; since only plasmid monomers are stably inherited it is important that XerC/D recombination converts multimers to monomers rather than making multimers from monomers. Recombination at *cer* also requires about 220 bp of specific accessory sequences located adjacent to the core *cer* site (located to the left of *cer* as presented in Fig. 4) and at least two accessory proteins, ArgR and PepA (Stirling *et al.*, 1988b, 1989). ArgR binds to a specific position in the accessory sequences and the precise role of PepA is unknown, though genetic experiments have implicated both ArgR and PepA in the selectivity process (Summers, 1989). Site-specific recombination between *cer* and its homologue in plasmid CloDF13, though inefficient, generated two hybrid sites (the type I and type II hybrids; Fig. 4). The type I hybrid with eight nucleotides in its central region has recombination properties indistinguishable from *cer*. In contrast, the type II hybrid with just two base pairs fewer in the central region, recombines intermolecularly and intramolecularly (i.e. it has lost resolution selectivity); recombines in the absence of accessory sequences and in the absence of the accessory proteins ArgR and PepA. The chromosomal site *dif* shows similar properties, at least when present in multicopy plasmids (i.e. it shows no selectivity and no requirement for accessory proteins and DNA sequences). How then *dif* functions in ensuring that chromosomes are monomeric at cell division remains unclear, though various hypotheses are discussed later.

These observations suggest to us a general molecular mechanism for selectivity. We propose that recombination systems showing no selectivity, can function with a simple core recombination site and recombinase. Recombinase–recombination core site and recombinase–recombinase interactions are sufficient to form a productive recombinational synapse, irrespective of the configuration of the sites with respect to each other. Therefore such recombination will not show selectivity (recombination at *dif* and the type II hybrids fall into this category). In systems showing selectivity, for example, XerC/D acting at *cer*; resolvase acting at *res*; and the DNA invertase Gin at *gix*, we propose that recombinase–core site and recombinase–recombinase interactions are insufficiently strong to give a productive synapse. The addition of accessory sequences and accessory proteins that act on them, provides the extra 'glue' to make a stable synapse, and imposes a specific synapse topology. The requirement for a fixed synapse topology provides the *topological filter* that ensures selectivity during recombination (Fig. 3).

5. The mechanism of site-specific recombination by integrase family recombinases

Recombinases of the integrase and resolvase/DNA invertase families catalyse site-specific recombination by a series of transesterifications that use protein-DNA phosphoesters as reaction intermediates. There is no hydrolysis of phospho-diester bonds and therefore no high energy co-factors are required in the phos-phodiester joining reaction. For catalysis of strand exchange, integrase family recombinases use an active site tyrosine close to the C-terminus of the protein to act as an initial nucleophile in the first transesterification reaction, to form a 3' phosphoester with the target DNA (Figs 5 and 6). In the characterized systems this transesterification occurs as one of a pair of such reactions occurring at

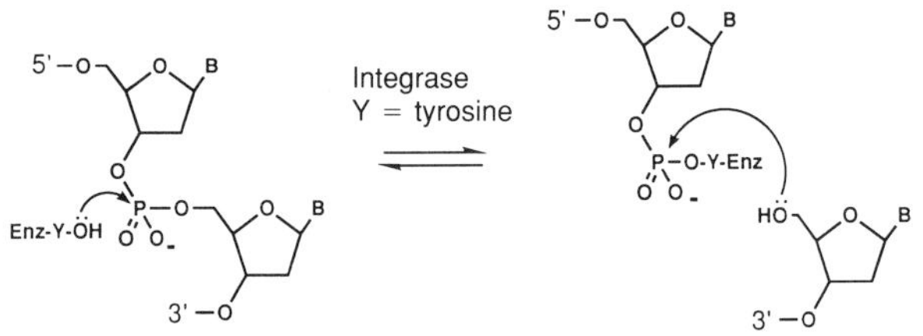

Figure 5. The integrase family catalysis transesterification reaction.

equivalent specific positions within the two participating recombination sites. This pair of complementary transesterifications is followed by a nucleophilic acid attack of the two released 5' OH groups on the phosphotyrosines of the partner duplex, to mediate the second pair of transesterification and complete the first pair of strand exchanges. The first pair of strand exchanges occurs at the junction of the recombinase-binding site and the central/overlap region (Fig. 4) in those systems in which the point of strand exchange has been determined. This gener-ates a Holliday junction in which two of the four strands have been exchanged (Fig. 6). Molecules containing the first pair of strand exchanges are resolved into complete recombinants by a second pair of strand exchanges some 7–8 bp away. This second pair of exchanges requires branch migration of the initial Holliday junction from the site of the first exchanges to that of the second exchange. It is likely that isomerization of the Holliday junction occurs during branch migration, which is most easily visualized occurring as a 250–290° rotation of the helices in the four-fold symmetric form of the junction. Note that this process exchanges the relative positions of the recombinase monomers (Fig. 6), so that recombinase molecules are conveniently positioned to mediate each pair of strand exchanges. We predict that in XerC/D-mediated recombination, each pair of strand exchanges is mediated by XerC and XerD alternately. Recent experiments with the FLP recombinase have suggested that the recombinase monomer providing the tyrosine nucleophile attacks a phosphodiester bond on the partner duplex

(Chen *et al.*, 1992). Two variations of this model, as applied to XerC/D-mediated recombination are shown in Fig. 6. In one (the *trans-diagonal* model), one would predict that one molecule of XerC and one of XerD would participate in each of the four strand exchanges. In the other (the *trans-vertical* model) two molecules of XerC would catalyse each of the two strand exchanges at one position, while two molecules of XerD would catalyse each of the two other strand exchanges. The reason for the XerC system using two separate recombinases remains unclear, since most other systems function perfectly well with one. It could reflect the need in this system to control the activity and specificity of the reaction in a way that we do not yet understand; it also appears to provide a mechanism to ensure that only correctly aligned sites are, though alternative mechanisms are used for systems with just a single recombinase.

6. The functions of XerC/D-mediated site-specific recombination in bacterial chromosome partition

We believe it likely that a major function of XerC/D-mediated site-specific recombination at *dif*, is to ensure that chromosome dimers, arising by an odd number of homologous recombination exchanges between sister chromosomes during or after replication, are converted to monomers prior to partition at cell division. Obviously a single chromosome dimer cannot be partitioned to two daughter cells. The properties of *xerC*, *xerD*, and *dif* mutants are consistent with such a role (aberrant nucleoid segregation and cell division) as is the observation that the mutant phenotype of *xerC* cells is suppressed in cells deficient in homologous recombination (i.e. *xerC recA* mutants). It would also be satisfyingly economical if the role of XerC/D in chromosome partition were identical to that of XerC/D in plasmid monomerization. However, whereas XerC/D recombination at *cer* and other natural plasmid sites, shows the required resolution selectivity and has the necessary determinants (accessory sequences and accessory proteins), the recombination properties at *dif* (no selectivity, no obvious accessory sequences or proteins) do not readily provide an explanation of how recombination at *dif* could effectively convert chromosomal dimers to monomer. Moreover, it is difficult to imagine how resolution selectivity using a topological filter could differentiate synapses with *dif* sites on separate monomers from *dif* sites 4.7 Mb apart and separated by some 20 000 supercoils on the same molecule. In Fig. 7 we present two models which could explain how *dif* could function in monomer chromosome partition despite the lack of selectivity in its recombination.

In the first model, Xer-mediated recombination at *dif* occurs repeatedly once the chromosomal *dif* site has been duplicated: therefore 50% of the time the chromosome will be in the recombinant (dimer) configuration and 50% of the time in the non-recombinant (monomeric) configuration. The presumed motive

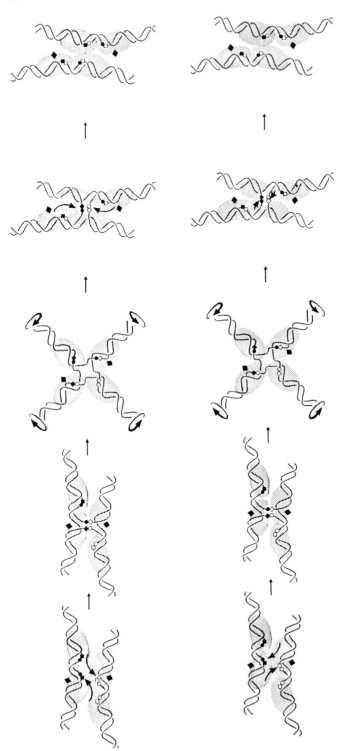

Figure 6. Possible strand exchange mechanisms for XerC/D and other integrase family recombinases. Cartoon of possible strand exchange mechanism for integrase family recombinases (adapted from Stark *et al.*, 1989b and from Chen *et al.*, 1992). The diagrams are intended to suggest only the relative movements of the recombinase subunits and the DNA, and possible local symmetries. When considering XerC/D-mediated recombination, XerC monomers are distinguished from the XerD monomers by the small solid square of the edge of the XerC molecule. The two models presented in the top and bottom panels are derived from the experiments and models of Chen *et al.* (1992) with FLP. The major conclusions of these experiments were: (i) a functional active site can involve contributions from two recombinase monomers, one providing the active site tyrosine and the other, the three additional conserved active site residues (R, H, R) (the *fractional*

active site hypothesis); (ii) the active site tyrosine of a monomer bound to one duplex attacks a phosphodiester bond of the partner duplex. This recombinase monomer and a recombinase molecule bound to the partner duplex each contribute a *fractional active site* during a single strand exchange. Note that the experiments of Chen *et al.* do not directly address the question of whether wild-type recombinase can act as a nucleophile in *cis*, and whether such a nucleophile needs to cooperate with another molecule in a single strand exchange reaction. *Top panel, trans-diagonal* model favoured by Chen *et al.* (1992), but with sites arranged in the antiparallel configuration. The first pair of (top) strand exchanges is initiated by nucleophilic attack of the active site tyrosines of two symmetrically positioned XerD molecules on phosphodiester bonds of the opposite duplex partner. Strand exchange is adjacent to where two XerC molecules are bound and involves an active site contribution from them. After branch migration and isomerization, the positions of the recombinase molecules are exchanged, and the DNA is positioned for the second pair of strand exchanges mediated by the active site tyrosines of the two XerC molecules. Again XerD molecules bound adjacent to the point of strand exchange can make a fractional active site contribution. In this model, one XerC molecule and one XerD molecule contribute to a single strand exchange. Four of the six possible pairs of recombinase-recombinase interactions are used in the processes of cooperative binding of XerC/D to a *core* site and the following strand exchange (there are no mandatory XerC–XerC interactions and XerD–XerD interactions). *Bottom panel, trans-vertical* model with sites arranged antiparallel. The first pair of (top) strand exchanges is initiated by nucleophilic attack from the tyrosines of XerC molecules on phosphodiester bonds of the partner duplex. Here the two XerC molecules contribute to a single active site (the second bound adjacent to the point of strand exchange). After branch migration and isomerization, the positions of the molecules are exchanged and the DNA is again positioned for the second pair of strand exchanges, mediated by a pair of XerD molecules that make nucleophilic attacks on 3'-phosphates of the partner duplex. In this case XerC/D interactions are required for the cooperative binding to each of the recombination sites, but are not intimately involved in catalysis. All six of the possible pairwise combination of recombinase interactions are utilised.

Model 1

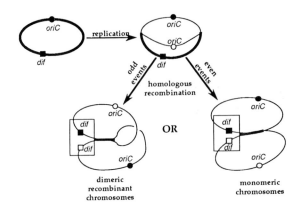

OR

dimeric
recombinant
chromosomes

monomeric
chromosomes

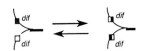

Repeated Xer-mediated site-specific recombination, after *dif* replication would lead to the chromosomes being in the monomeric configuration 50% of the time, when they can be partitioned to daughter cells.

unreplicated double stranded DNA ▬▬▬

replicated double stranded DNA ───

Model 2

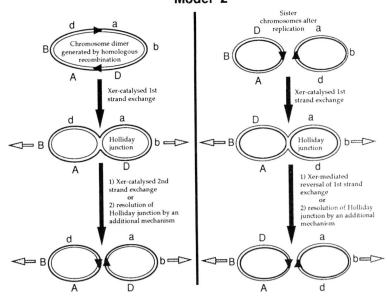

The choice of resolution mechanism must be imposed by external factors, eg., the putative partitioning mechanism for separating sister chromosomes.

Top strand ▬▬▬

Bottom strand ───

Indicates the chromosomes under motive force during partition

force that separates the daughter chromosome nucleoids will therefore be effective in separating the chromosomes during periods that they are in the monomeric state. Model 2 is based on an unpublished observation (R McCulloch) that Xer-mediated recombination *in vivo* can generate a substantial level of Holliday junction intermediates that have undergone just a single specific strand exchange. The formation of these is XerC/D dependent and has been observed mainly with natural plasmid sites that show resolution selectivity. If such Holliday intermediates are formed at *dif* within the chromosome they could avoid the need for rapid/frequent exchanges at *dif*. This is a consequence of the symmetry of the 'uncrossed' conformation of a Holliday junction (Fig. 5). The formation of such a structure in either recombinant (dimeric) or a non-recombinant (monomeric) chromosome would yield DNA molecules that when acted on by the 'machine' that moves the replicated chromosome to the daughter cells would yield monomers as a consequence of conformational constraints on the Holliday junction, though the resolution would involve different pairs of strand exchanges in the two cases (Fig. 6). Since we suspect that XerC and XerD catalyse different pairs of strand exchanges, this model would lead to resolution by either XerC or XerD depending on whether recombination has occurred elsewhere or not. Alternatively, resolution could result from the activities of a different enzyme or enzymes (RuvC?, topoisomerase?). If Holliday intermediates at *dif* function in chromosome segregation, then they could be seen as serving the same type of role as recombinational chiasmata in eukaryote meiosis, since crossing over is required for ensuring that homologues are partitioned to separate cells in many cell types. Recombination may therefore be a common prelude to normal chromosome partition. In bacteria, we believe that the sort of system described here is ubiquitous; we have demonstrated that both *xerC-* and *xerD*-like genes are widely distributed in bacteria (G Blakely unpublished). Moreover, we know of no natural plasmids found in Gram-negative bacteria that lack site-specific recombination systems that could function in plasmid partition. Despite our belief that XerC/D site-specific recombination at *dif* has a major role in ensuring that chromosomal dimers, arising through homologous recombination, are monomeric at partition, it is possible that this recombination has additional, more subtle, roles in the partition process. As our understanding of this system grows, its precise functions should become apparent.

Figure 7. Two models that could explain how *dif* functions in chromosome partition. The models are explained in the text.

Acknowledgements

We thank Dr Sue Lovett for the provision of strains, plasmids and unpublished information. The manuscript was prepared by Margaret White. The research was supported by the Medical Research Council.

References

Argos P, Landy A, Abremski K, Egan JB, Haggard-Ljungquist E, Hoess RH, Kahn ML, Kalionis B, Narayana SVL, Pierson LS, Sternberg N, Leong JM. (1986) The integrase family of site-specific recombinases: regional similarities and global diversity. *EMBO J.* **5:** 433-440.

Arthur A, Sherratt DJ. (1979) Dissection of the transposition process: a transposon-encoded site-specific recombination system. *Mol. Gen. Genet.* **175:** 267-274.

Austin S, Ziese M, Sternberg N. (1981) A novel role for site-specific recombination in maintenance of bacterial replicons. *Cell,* **25:** 729-736.

Begg KJ, Donachie WD. (1991) Experiments on chromosome separation and positioning in *Escherichia coli. New Biol.* **3:** 475-486.

Blakely G, Colloms S, May G, Burke M, Sherratt DJ. (1991) *Escherichia coli* XerC recombinase is required for chromosomal segregation at cell division. *New Biol.* **3:** 789-798.

Chen J-W, Lee J, Jayaram M. (1992) DNA cleavage in trans by the active site tyrosine during Flp recombination: switching protein partners before exchanging strands. *Cell,* **69:** 647-658.

Colloms SD, Sykora P, Szatmari G, Sherratt DJ. (1990) Recombination at ColE1 *cer* requires the *Escherichia coli xerC* gene product, a member of the Lambda integrase family of site-specific recombinases. *J. Bacteriol.* **172:** 6973-6980.

De Boer PAJ, Cook WP, Rothfield LI. (1990) Bacterial cell division. *Annu. Rev. Genet.* **24:** 249-274.

Donachie WD, Begg KJ. (1989) Cell length, nucleoid separation, and cell division of rod-shaped and spherical cells of *Escherichia coli. J. Bacteriol.* **171:** 4633-4639.

Futcher AB. (1986) Copy number amplification of the 2 μm circle plasmid of *Saccharomyces cerevisiae. J. Theor. Biol.* **119:** 197-204.

Hatfull GF, Grindley NDF. (1983) Resolvases and DNA invertases: a family of enzymes active in site-specific recombination. In: *Genetic Recombination* (Kucholapati R, Smith GR, eds). Washington DC: American Society for Microbiology Press. pp 357-396.

Hiraga S, Niki H, Ogura T, Ichinose C, Mori H, Ezaki B, Jaffe A. (1989) Chromosome partitioning in *Escherichia coli*: novel mutants producing anucleate cells. *J. Bacteriol.* **171:** 1496-1505.

Hiraga S, Ogura Teru, Niki H, Ichinose C, Mori H. (1990) Positioning of replicated chromosomes in *Escherichia coli. J. Bacteriol.* **172:** 31-39.

Johnson RC. (1991) Mechanism of site-specific DNA inversion in bacteria. *Curr. Opin. Genet. Dev.* **1:** 404-411.

Kanaar R, Cozzarelli NR. (1992) Role of supercoiled DNA structure in DNA transactions. *Curr. Opin. Structural Biol.* **2:** 369-379.

Kato J, Nishimura Y, Imamura R, Niki H, Hiraga D, Suzuki H. (1990) New topoisomerase essential for chromosome segregation in *E. coli. Cell,* **63:** 393-404.

Kuempel PL, Henson JM, Dircks L, Tecklenburg M, Lim DF. (1991) *dif,* A *recA*-independent recombination site in the terminus region of the chromosome of *Escherichia coli. New Biol.* **3:** 799-811.

Landy A. (1989) Dynamic, structural and regulatory aspects of lambda site-specific recombination. *Annu. Rev. Biochem.* **58:** 913-949.

Lovett S, Kolodner RD. (1991) Nucleotide sequence of the *Escherichia coli* recJ chromosomal region and construction of RecJ-overexpression plasmids. *J. Bacteriol.* **173:** 353-364.

Murphy E. (1989) Transposable elements in Gram-positive bacteria. In: *Mobile DNA* (Berg DE, Howe MM, eds). Washington DC: American Society for Microbiology Press. pp 269-288.

Niki H, Jaffe A, Imamura R, Ogura T, Hiraga S. (1991) The new gene *mukB* codes for a 177 kd protein with coiled-coil domains involved in chromosome partitioning of *E. coli. EMBO J.* **10:** 183-193.

Sadowski P. (1986) Site-specific recombinases: changing partners and doing the twist. *J. Bacteriol.* **165:** 341-347.

Schmid MB. (1990) A locus affecting nucleoid segregation in *Salmonella typhimurium. J. Bacteriol.* **172:** 5416-5424.

Schmidt FRJ, Nucken GJ, Henschke RB. (1989) Structure and function of hotspots providing signals for site-directed specific recombination and gene expression in Tn21 transposons. *Mol. Microbiol.* **3:** 1545-1555.

Sherratt DJ. (1989) Tn3 and related transposable elements: site-specific recombination and transposition. In: *Mobile DNA* (Berg DE, Howe MH, eds). Washington DC: American Society for Microbiology Press. pp 163-184.

Stark WM, Sherratt DJ, Boocock MR. (1989a) Site-specific recombination by Tn3 resolvase: topological changes in the forward and reverse reactions. *Cell,* **58:** 779-790.

Stark WM, Boocock MR, Sherratt DJ. (1989b) Site-specific recombination by Tn3 resolvase. *Trends Genet.* **5:** 304-309.

Stark WM, Boocock MR, Sherratt DJ. (1992) Catalysis by site-specific recombinases. *Trends Genet.* **8:** 432-439.

Stirling CJ, Stewart G, Sherratt DJ. (1988a) Multicopy plasmid stability in *Escherichia coli* requires host-encoded functions that lead to plasmid site-specific recombination. *Mol. Gen. Genet.* **214:** 80-84.

Stirling CJ, Szatmari G, Stewart G, Smith MCM, Sherratt DJ. (1988b) The arginine repressor is essential for plasmid-stabilising site-specific recombination at the ColE1 *cer* locus. *EMBO J.* **7:** 4389-4395.

Stirling CJ, Colloms SD, Collins JF, Szatmari G, Sherratt DJ. (1989) *xerB,* and *Escherichia coli* gene required for plasmid ColE1 site-specific recombination, is identical to *pepA,* encoding aminopeptidase A, a protein with substantial similarity to bovine lens leucine aminopeptidase. *EMBO J.* **8:** 1623-1627.

Summers DK. (1989) Derivatives of ColE1 *cer* show altered topological specificity in site-specific recombination. *EMBO J.* **8:** 309-315.

Summers DK, Sherratt DJ. (1984) Multimerisation of high copy number plasmids causes instability: ColE1 encodes a determinant for plasmid monomerisation and stability. *Cell,* **36:** 1097-1103.

Chapter 4

The *E. coli* genome project: towards the first megabase

Fredrick R Blattner, Donna L Daniels, Valerie D Burland, Guy Plunkett III and S Chuang

1. Introduction

There are many reasons to sequence the genome of *Escherichia coli* K-12. This bacterial genome has been thoroughly studied and although 1200 to 1500 genes are known, there are nearly 3000 more to discover by sequencing. *E. coli* is also the predominant bacterium for genetic engineering. Knowing more about this organism will be useful to the industrial and commercial sector and will lead to possibilities for engineering this genome to improve its efficiency in industrial applications. *E. coli* will also provide very interesting evolutionary comparisons. Some genes discovered in *E. coli* will have homologues in human, yeast and other systems of study. The complete gene bank of *E. coli* will provide a major set with which to compare other genomes. Some *E. coli* strains are important pathogens in human and animal disease. Finally, there is the simple philosophical point that to obtain the complete blueprint of life will be a milestone intellectual achievement. With a complete list of all the genes and all of the control elements of a life form, it will be possible to analyse the function of an entire living cell.

The complete sequence analysis of the 4 720 000 basepairs of the *E. coli* genome provides a technical challenge and acts as a prototype for the sequencing of larger genomes – yeast, mammalian, plant and animal. The *E. coli* genome is 1000 times smaller than the human genome but the number of genes expected is only 10 to 100-fold fewer. Thus the genome of *E. coli* provides a cost-effective source of genetic information compared to other genomes.

For the sequencing project, we chose, in consultation with Dr Barbara Bachman, strain MG1655, a wild-type strain of *E. coli* K-12. MG1655 is directly related to the original K12, having only been cured of bacteriophage lambda by a non-mutagenic technique. The K-12 strain of *E. coli* was isolated in 1922 from a patient and has been used for research ever since. A large number of mutations and modifications have been introduced into the genome through experimental manipulation of the organism in research laboratories around the world since the first skeletal circular map was constructed in the 1950s.

The *E. coli* genome is circular. Positions around the circle – map coordinates – are given in minutes (0–100), a convention arising from the early discovery that the chromosome can be transferred linearly between two cells at a constant rate, which takes 100 minutes at 37°C. Conventionally, the 100–0 join (where the origin of transfer, at Hfr, is located) is drawn at the top of the circle and numbering is clockwise. Minutes are surprisingly close to proportional physical distances (Figs 1 and 2).

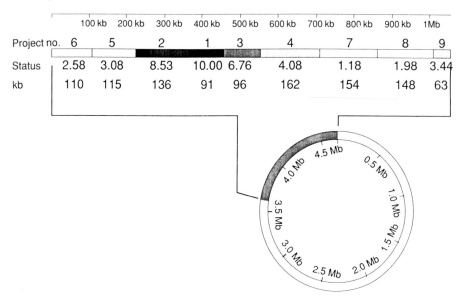

	100 kb	200 kb	300 kb	400 kb	500 kb	600 kb	700 kb	800 kb	900 kb	1Mb
Project no.	6	5	2	1	3	4		7	8	9
Status	2.58	3.08	8.53	10.00	6.76	4.08		1.18	1.98	3.44
kb	110	115	136	91	96	162		154	148	63

Figure 1. Progress towards sequencing the first megabase (Mb) of *E. coli*. The circular genome is diagrammed in the lower right. The megabase starting to the left of the conventional '0' has been divided into 9 projects; their sizes and status of sequencing (on a 0–10 scale) are indicated at the top of the figure. The definition of each status is given in Table 5.

Sequencing efforts are underway in several laboratories which will lead to the sequencing of the entire genome. We will concentrate towards the left of the 100 minute–zero joint (Fig. 1) while Dr Mitzibuki, Dr Xung and others will concentrate to the right of that area. Dr George Church in Boston will initially concentrate his work at the bottom of the map. Through this collaboration, the entire genome will be sequenced.

Figure 1 shows our plan and progress (September 15, 1992) toward the first megabase. We are focusing on the area between 80 minutes and 100 minutes which has been divided into nine segments of approximately 100 kb each. The second segment chosen for sequencing contains the origin of replication, it is a well studied area and would therefore enable us to compare our sequence with those already found in the GenBank and EMBL databases. Regardless of the outcome we feel it is necessary to repeat the sequencing of all the regions

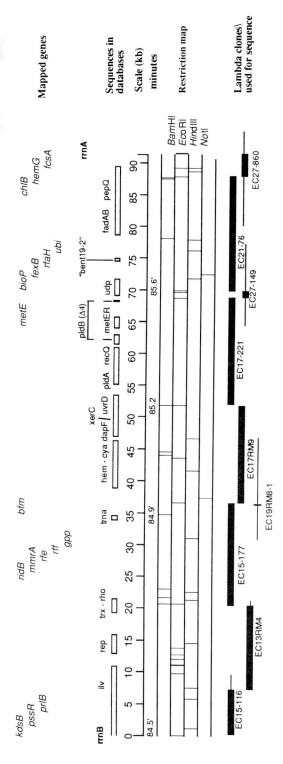

Figure 2. The 91 kb segment of *E. coli* that has been completely sequenced and analysed (Daniels *et al.*, 1992).

previously analysed. Progress is shown on a scale from 1 to 10 in which 10 represents a complete and annotated sequence, 5 represents a sequence that is essentially in a single piece but not annotated or proofread and 1 means that the shotgun library for the sequence has been created but data-gathering has not begun. The first 91 kb segment is at status 10 and was published in *Science* (Daniels *et al.*, 1992). The two adjoining segments are at status 8.5 and 6.5. Both are well on their way toward completion. The sum of segment 1, 2 and 3 is 323 kb.

2. Completed sequence

The segment of *E. coli* that has been sequenced and published is depicted in Fig. 2. As seen at the bottom we completely sequenced seven lambda clones and two additional short segments of other lambda clones to cover the overlap. We are using the lambda clone set isolated by Daniels (1990). The policy of our sequencing effort is to sequence every base at least four times. When two lambda clones appear to join up exactly but do not overlap, we sequence across the joint either in another lambda phage or from the genome itself.

Within the first segment, about 84.2% of the genomic area is coding region (encodes proteins) and we have discovered 82 open reading frames which are candidates to be genes (Table 1). Of these 48 were assigned to genes, while the

Table 1. Contents of the first 91.4 kb segment of the *E. coli* genome that has been completely sequenced and analysed

	Number	Span (kb)	%
Unassigned ORFs	34		
ORFs assigned to genes	48		
Total ORFs	82	76.9	84.2
Non-protein coding regions		14.5	15.8
tRNA genes	8	0.64	0.7
16S rRNA genes	1	1.5	1.6
23S rRNA genes	5′ end	0.8	0.8
Intergenic regions		11.6	12.7

functions of 34 are unknown. There were eight tRNA genes, one 16S rRNA gene, half of a 23S rRNA gene, and a number of other features which will be described. The intergenic regions include the spaces between these open reading frames as well as regions that appear to contain no identifiable genes which we have dubbed 'grey holes'.

Figure 3 shows the second segment we have sequenced, from 11 lambda clones, which is a length of about 136 kb, from 81.9 to 84.7 minutes. Table 2 shows the preliminary analysis of the open reading frames identified in segment 2. Ninety-seven open reading frames were found in 112 kb of this segment so far

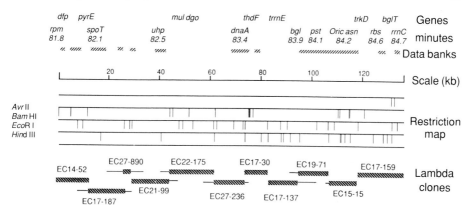

Figure 3. Project 2: current status and results of analysis of the second 136 kb segment of *E. coli* to be sequenced.

analysed, of which 43 were assigned to known genes. There was one each of a tRNA, 16S–23S and 5S rRNA. At the moment, we are not in a position to define the status of the segment regarding grey holes because the annotation has not been completed.

Subsections of segment 2 are illustrated in Fig. 4. Two regions of the lambda clones have been annotated and are ready for merger into a final sequence. The first is clone EC15-15 which contains the most sequenced area of *E. coli* and the origin of replication, *oriC* (Fig. 4a). As shown from the arrows in the figure some regions of this segment have been sequenced in the literature up to seven times, and in this lambda clone there are only 200 basepairs that have not previously been determined at least once in another laboratory. We have painstakingly resolved any discrepancies with all previous determinations of this sequence by

Table 2. Preliminary contents of the second 136 kb to be sequenced

Segment 2 – Preliminary	
112 kb	
ORFs	
Identified	57
Unknown	40
Total	97
RNA Genes	
tRNA	1
16S	1
23S	1
5S	1
Total	4

(a)   **EC 15-15**

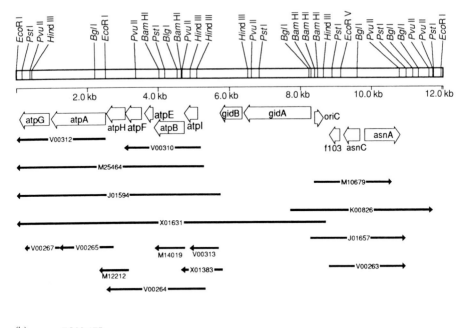

(b)   **EC22-175**

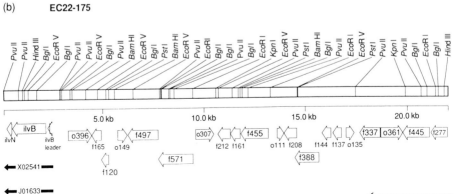

Figure 4. Detailed restriction and gene maps of segments sequences in project 2.

contacting the authors, and by carrying out our own analysis to determine what we consider to be the correct sequence where overlapping has occurred.

Figure 4b shows a second lambda clone from segment 2, EC22-175. In this case, we have two determinations of a region at one end and the remainder of the clone contains over a dozen new genes, most of which have not been identified. This is typical of the patchiness of the determination of the *E. coli* sequence by the random sequencing by different laboratories around the world which tend to focus on areas of known interest.

In summary, the quarter million basepairs of *E. coli* we have sequenced contain 105 known and 88 unknown genes. We were surprised at the frequency of genes that cannot be identified. We found a high frequency of *lysR* family regulator genes: the *lysR* family is a regulatory gene that is frequently arranged in the chromosome so that the regulatory protein is coded in one direction and adjacent to it (and diverging from the same promoter region) is a gene, one of several possible genes controlled by that system. The *lysR* family is characterized by a motif in the protein probably corresponding to a DNA binding domain. We have found these to be located 50–30 kb apart, indicating that there may be as many as 100–200 per genome, if the region sequenced is typical. Thus, the *lysR* family might constitute a substantial fraction of the regulatory apparatus of the genome.

3. Transcriptional units

In the course of defining transcriptional units we discovered that the computer programs recognize too many 'hits' to be useful. We had to develop a procedure for proposing transcriptional units that pruned this number down. The method we used was entirely empirical and consisted primarily of removing promoters from the list that were within proposed coding regions or which were very near to other stronger promoters. The *E. coli* genome obviously contains many sequences that, according to the computer, ought to be good promoters. It remains to be seen whether these consensus signals actually function or whether they are in some way suppressed either by the translational mechanism or by some aspect of chromosomal organization or structure.

An interesting discovery is the 'grey hole', an area in which there are no open reading frames longer than perhaps 75 or 100 amino acids. We found that approximately 2–3% of the genome consists of these areas which are 500–1000 basepairs in length. Although it is possible that small genes could be coded in some of these regions, it also appears likely that the majority of the grey holes do not contain protein coding regions and must be there for another reason. They could be remnants of evolution or remnants of recombination events in which the chromosome was spliced to other pieces of DNA. Alternatively, grey holes may serve structural purposes in chromosomal organization.

We were able to discover two genes that are defective. These are the arylsulphatase gene, *atsA*, and the *ilvG* genes. These are genes that contain terminator mutations which produce shortened peptides in *E. coli* K-12 MG1655. The altered genes are known by comparison with other organisms such as *Salmonella* or *Klebsiella* in which the genes are functional. We also found one instance of a probable evolutionary re-arrangement: the arysulphatase operon in *Klebsiella* consists of two genes that are oriented in the same direction, whereas in *E. coli* the two genes point toward each other with a REP element located in between. REP elements are repeated sequences that appear in the genome about every 20 kb. Another family, ERIC, has also been identified in our sequence. DNA curving sites were also found using the method of Trifonov: we located a substantial number of regions of the DNA that are predicted to deviate more than 75 degrees in a 100 bp window. These are spaced on the order of 5 kb apart in the genome.

4. Chi sites

Another extremely interesting observation has to do with the distribution of chi sites. Chi sites are asymmetric sequences in DNA which are probably binding sites for the recBC nuclease and which are important in the initiation of recombination (Smith, 1987). It was pointed out to us by Andy Taylor and Gerry Smith of the University of Washington that the chi sites in the segment 1 sequence appear to be highly asymmetrically located, with over 80% being on one strand. In the sequenced region, chi sites are arranged in a divergent manner from the origin of replication as shown in Fig. 5. The chi site is a recombinational hot spot and its function may be related to the direction of replication over it.

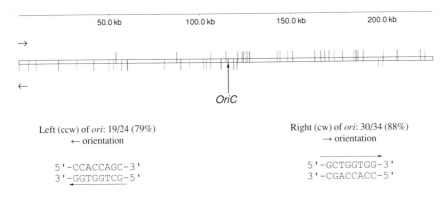

Figure 5. The assymmetric distribution of chi sites with respect to the origin of replication (cw = clockwise; ccw = counterclockwise).

5. Open reading frame (ORF) analysis and sequence errors

In the analysis of an unidentified open reading frame, ORF 233, we found a strong homology with similarity indices between 37.5 and 58.2 with an open reading frame in the *lux* operon of several marine luminescent bacteria (Table 3). We have concluded that this ORF is *ubiB*, coding for flavin reductase. We postulate that the function of *luxG* in bioluminescence is production of flavin, a substrate of the

Table 3. The similarities of the *luxG* and *ubiB* genes from marine luminescent bacteria and *E. coli*. *luxG* is a strong candidate to encode flavin reductase in the marine luminescent bacteria

---

- *E. coli* gene *ubiB* (also known as *fre* or *fadI*) encodes flavin reductase.
- No gene identified for flavin reductase in luminescent marine bacteria.
- UbiB amino acid sequence displays extensive homology to deduced product of *luxG* (ORF in *lux* operon of all marine luminescent bacteria):

Similarity indices for pair-wise alignments of UBIB and LuxG proteins

	Pph	Vfi	Vha
Eco	40.9	40.2	37.7
Vha	39.3	41.0	
Vfi	58.2		

Eco: *Escherichia coli* UbiB
Vha: *Vibrio harveyi* LuxG
Vfi: *Vibrio fischeri* LuxG
Pph: *Photobacterium phosphoreum* LuxG (partial sequence)

∴ Propose *luxG* is gene encoding the flavin reductase of these bacteria.

---

bioluminescent reaction. This is one case in which a gene has been determined by computer analysis of homologues in other organisms.

Figure 6 shows the arrangement of the *ats* operon in *E. coli* compared with *Klebsiella aerogenes*. The re-arrangement, with the REP element in between, may be correlated with the affective *atsB* phenotype of *E. coli*. The largest and

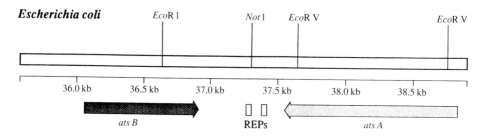

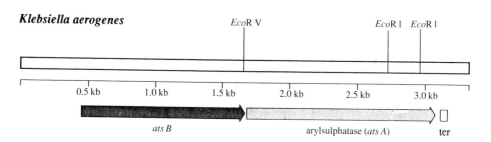

Figure 6. The arrangement of the *ats* operon in *E. coli* compared with that in *Klebsiella aerogenes*.

most complex REP element we found is located between *luxG* and *fadA*. This consists of six cases of a triple direct repeat of the inverted repeat that comprises the basic REP element.

We have also discovered several errors in previous data. One sequence, ECPLDB, contains a 4894 basepair deletion downstream of the gene. We also located a rogue sequence, bent 19, joining a piece from 86 minutes to one from 95 minutes. Only one of those two pieces actually contains a bend. Several instances were found in which, as have been noted elsewhere, vector sequences are presented in the database, attached to the genes sequenced. ECOUDP and ECODEOD3 have 25 and 96 bases of lambda at the 3′ end respectively.

In the case of the *rbsR* gene, Table 4, we found a considerable sequence

Table 4. A comparison between sequence data obtained in the genome sequence project with previously published results

	Sequence differences in rbsR, the ribose operon repressor
DNA	
UW	TTGCCCGCCTGGCG**GGGG**TTTCTA
GenBank	TTGCCCGCCTGGCG**C**———TTTCTA
(M13169)	^20 ^30
PROTEIN: first 15 aas of the DNA binding domain	
UW	MATMKDVARLA**GV**ST
Conserved	MAT KDVA **A**GVST
GenBank	MATMKDVARLA–LST
(M13619)	^1 ^10

Ref: **Mauzy CA and Hermodson M.** (1992) *Protein Science*, **1**: 843–849.

variation between our data and the GenBank entry, which altered a single amino acid and inserted another. This was apparently caused by a difficulty in reading compressed sequencing gels. Our revision eliminates an anomaly noted in the DNA binding domain of the RbsR protein by the authors.

The most troublesome common type of sequence difference is the single insertion or deletion of a basepair leading to a reading frame shift in published sequences termed 'indel'. Such errors can have significant impact on the overall structure of a protein as it appears in the database. The *pepQ* peptidase, the *rephel* (rep helicase) and the *fdhE* (formate dehydrogenase) genes all had changes which result in a region in the middle of the protein having a shifted and thus a completely erroneous amino acid sequence. Sometimes the indel affects the end of the protein by a change in the termination codon arising from the phase shift, shown in Fig. 7.

Finally, in the case of *thdF* (Fig. 8) we discovered a massive difficulty with the GenBank entry as shown in this figure. The published sequence contains not only missing areas but areas that are reverse complemented, areas that are in the wrong

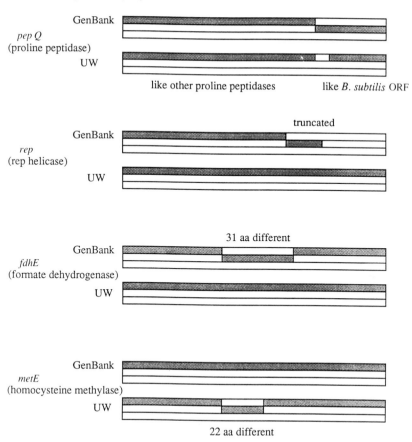

Figure 7. Errors in GenBank DNA sequences that give rise to frameshift errors on analysis (aa = amino acids).

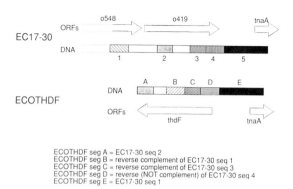

Figure 8. A comparison of the sequences of EC17-30 and ECOTHDF suggests that errors occurred during sequence data assembly in the latter.

position and two areas in which the sequence is presented in GenBank in a reversed but not complemented form. This entry clearly was scrambled as the result of some major error either in the database or the sequencing laboratory.

We do not present the sequencing errors in order to disparage other laboratories. Sequencing is extremely difficult and errors are unavoidable with current technology. In fact, we are already aware of an error in our determination of the *metE* gene. The point that we would like to make is that the data in the database do need to be confirmed and re-evaluated to a consistent level of accuracy. In our view it is not adequate to sequence only regions in between GenBank entries.

6. Genome function

We are determining global transcription response in the genome in order to help in interpretation and ultimate understanding of *E. coli* genome function. To do this experiment, a spot blot containing overlapping lambda clones from the Kohara or the Daniels set is used to measure the genomic origin of RNA preparations. RNA preparations are made from *E. coli* growing in log phase and compared with a similar culture grown under a physiologically altered condition. This can be a physical stimulus such as heat shock, a genetic stimulus such as a mutation in a controlled gene, or a physiologically altered situation such as growth in the mouse gut. The two RNA preparations are converted into radiolabelled DNA by reverse transcriptase and is used as a probe on the spot blot. After hybridization with the control and experimental samples, ratios are taken to determine the induction of RNA in a given genomic interval as a result of the stimulus applied. These samples are quantified with a ^{32}P radioactive imaging machine which counts them directly into the computer. The positions of genes that are strongly affected in their transcription by the particular stimulus can then be located. Figure 9 shows the results for heat shock. The inside of the circular diagram shows the heat shock genes which we discovered by this method. On the outside are the genes for heat shock that had been mapped previously. The method approximately doubles the number of known genes that respond to heat stimulus. As we progress with our project we want to make further refinements of this analysis so that each spot corresponds to a single gene. At this point it will be possible to determine the response quantitatively and globally.

7. Sequencing technology

We would also like to consider briefly some of the methodological aspects of our approach to sequencing. Most of the genome projects have decided that the most

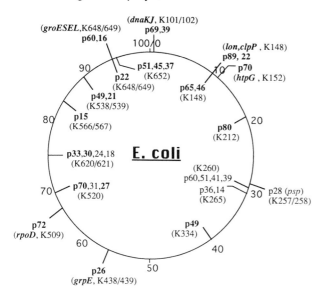

Figure 9. The genetic map of *E. coli* heat shock genes. Outside: previously known genes. Inside: previously uncharacterized genes. Bold type: *rpoH* dependent. The analysis has approximately doubled the number of known genes.

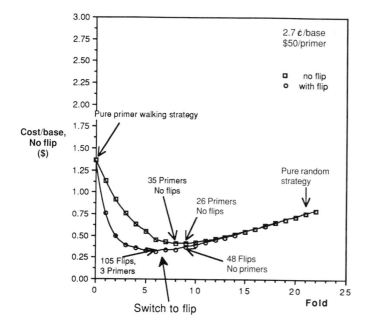

Figure 10. Costs of various sequencing strategies. Left: using a pure primer walking strategy; right: using entirely random strategy (see text).

cost effective and rapid approach to sequencing is through an initial random sequencing combined with a directed strategy to close gaps at the end. Figure 10 illustrates calculations involved in assessing the point at which the random phase should end and the directed phase should begin. For the purposes of this analysis we show a purely directed strategy on the left of the figure: here we would make a primer approximately every 100 basepairs and extend it approximately 400 base-pairs to get four-fold coverage. We assumed that the cost of one primer is $50 and the cost of sequencing one base is 3.7 cents. If there is no random component in the strategy the cost is about $1.50 a base for this particular numerical example. The purely random approach calls for about 20-fold over sequencing in order to achieve four-fold coverage of every base in the genome. This is represented at the far right side where we have a 20-fold random coverage. This is also quite expensive. The intermediate positions form a curve with a minimum cost of sequencing at around 8- or 9-fold random.

The Janus vector and the flipping strategy

We have found a method for decreasing the cost of sequencing while increasing the rate of closure which is depicted in the lower curve. This strategy consists of three steps. First, a random phase generates random sequence and contigs, then the approach flips to sequence the other strand of selected clones, before a final stage of primer walking to close the gaps at the very end. This flipping strategy is made possible by our new vector known as Janus.

The Janus vector is an M13 vector where the cloning site is flanked by lambda attachment sites which allow one to sequence both strands by growing the M13 phage in a host cell containing Int recombinase. This causes an inversion of the cloning site and the DNA clone there, so the opposite end of the opposite strand can be sequenced. Using the Janus vector we have a strategy that is better than the original strategy in a number of respects. We start by random sequencing which produces a result with a number of contigs but some gaps and some thin areas. We then select a series of Janus clones within this set for flipping both to close the weak regions and to sequence areas of thin depth to a greater depth for accuracy. This Janus vector strategy definitely improves the overall efficiency of sequencing.

The process by which we isolate DNA from the M13 clones for sequencing is semi-manual. We are able to grow the M13 clones made from the shotgun library in microtitre dishes and then use simple steps involving a table top centrifuge for spinning microtitre dishes and multiple pipetters: one individual is able to produce over 300 DNAs per day. There is no need for a robotic device. It is very important that the shotgun libraries are constructed with a completely random method. We therefore have used sonication or passage through a French pressure cell as the method for fragmenting the DNA for random sequencing libraries.

It is extremely useful to use robotic procedures to generate the actual sequencing reactions. In this laboratory we have three robots in which the basic robotic hardware, that of the Gilson Company, is a two-dimensional positioning device with syringe driven pump that is able to pipette and transfer samples from microtitre dish wells. A system devised in our laboratory, in collaboration with DNAStar, is used to implement the sequencing reactions. We process 24 DNAs yielding 96 samples for loading on the gel in approximately 50 minutes. At present we are running three of these robots approximately 8 hours a day in order to create the numbers of samples that are run on the gels. This is a very efficient robotic protocol, very trouble free, and it produces very good results with minimal attendance.

The gel boxes, while of a basically conventional nature, have a few features that greatly improve their efficiency in a mass sequencing operation. We use a very rigorous temperature control consisting of a 3/8 inch aluminium plate with heating elements and precise temperature controller. The gel is clamped against this plate for isothermal operation. The gel is larger both vertically and horizontally than the film that will be exposed. The region of gel above the top of the film is not exposed because the bands are not resolved in that area. The regions to the right and left provide an area so that thermal distortion (smiling) does not affect the active area that is being probed with film. This results in a very precise, uniform, straight lane gel that is easily scanned by automatic scanners.

8. Sequence reading and analysis

The scanner for transmitting the data from the sequence film to the computer consists of a four-channel photoelectric reading device that turns the data on the film into a series of traces very similar to that found in a fluorescence based sequencing machine. The traces are analysed by computer to call a sequence and give a quality score for each base. In addition this program provides the ability to edit traces when errors are detected. The strategy we use, however, calls generally for inputing the data into the computer without editing. After a consensus alignment has been created, we look at areas where it is apparent that an error has been made due to the lack of agreement of multiple determinations. The data of the traces can be called to screen on the Macintosh software from DNAStar that is used to assemble the sequence for review. The scanner can also be used to locate the data on the film for direct analysis.

The assembly program contains a number of features that aid in the accuracy and convenience of assembly. The most important of these is the trapezoidal weighting function which causes the consensus to be constructed with preference given to the part of the gel where the data are the most accurate. This prevents consensus degradation and causes the assembly to proceed with fewer errors. A second major feature is the prepass process in which every sequence is compared by a very rapid algorithm with every other, to put sequences into the project that match well first, leaving the more dubious data to the end where they will not degrade the consensus.

9. Sequence analysis

The software that we have developed for identifying genes includes the ability to locate open reading frames and to apply either the Gribskov or Staden scores to identify reading frames that are the most likely to be transcribed on the basis of their codon preference value.

Another program called Findit is used to do BLAST searches against the entire protein database in order to locate the regions of our sequence that correspond to those in the database. This program, developed by Jude Shavlik and his collaborators in the Computer Science Department at the University of Wisconsin, is graphical and allows the operator to study the relationships depicted with other genes to determine whether the homology is sensible with a region of the genome of *E. coli*.

10. Progress

The final section of this paper is a discussion of the current rate of sequence progress. We use an assembly line operation which is monitored by a 10–level project status report: as each lambda clone is sequenced it passes from 1 to 10. Status of larger segments is obtained by length weighted average of the status of each of the elements of the segment. Thus at any given moment there is a certain total length of sequence in the genome that is assembled, a smaller set that has been edited, a still smaller set that has been bio-checked against the databases, and a smaller subset that has been annotated. Finally, the smallest subset has been published. The definition of each level is shown in Table 5 and briefly in the key to Fig. 11 (see p. 174).

Figure 11 shows a bar graph illustrating the progress since the beginning of 1992. We use a gray scale illustrated at the top that depicts the segments of DNA that have achieved the status indicated. Status 5 represents assembled data, meaning that the data have been assembled into one or very few contigs. At this time we have approximately half a million basepairs at that status. Approximately a quarter of a million is annotated, and 90 000 have been published. From the slopes of these lines it is possible to make estimates as to the overall rate of our sequencing operation. At the moment we are operating at the demonstrated rate of a quarter million basepairs a year and this has been increasing. We anticipate we will be able to sequence and fully analyse a megabase per year of *E. coli* DNA as our process is improved.

Table 5. The status levels used to coordinate and analyse progress in the large-scale sequencing project

Project status level	Includes
(10) Analysed, published	Splicing adjacent lambda clones. Note of sequence similarities; gene/function identification; proposed transcriptional units, promoters, and terminators; repeated sequence elements; other features. Online and Genbank/EMBL submission. Publication.
(9) Annotated	Annotate ORFs. Compare with published sequences, annotate correlations and differences.
(8) Bio-checked	Bio-check with 'Geneplot' and 'FIND-IT' to identify ORFs and locate frameshift errors. Re-proofread (scientific staff).
(7) Provisional	Films proofread in ambiguous regions (editing staff).
(6) Edited	Edit alignments on screen. Obtain more data for thin or ambiguous areas.
(5) Assembled	Assembly yields complete sequence in 1–4 contigs.
(4) Initial assembly	Raw sequence data prepassed and assembled.
(3) Data gathering	Raw data collection in progress.
(2) Shotgun made	Subclone DNA into M13 vector.
(1) Project chosen	Minimal set clone chosen, DNA preparation made, restriction map verified.

Note

This is paper No. 3344 from the University of Wisconsin Laboratory of Genetics.

References

Daniels DL. (1990) Constructing encyclopedias of genomes. In: *The Bacterial Chromosome* (Drlica K, Riley M, eds). Washington DC: American Society of Microbiology Press. pp 43-52.

Daniels DL, Plunkett G III, Burland V, Blattner FR. (1992) Analysis of the *Escherichia coli* genome: DNA sequence of the region from 84.5 to 86.5 minutes. *Science,* **257:** 771-778.

Smith GR. (1987) Mechanism and control of homologous recombination in *Escherichia coli. Annu. Rev. Genet.* **21:** 179-201.

Chapter 5

The equivalent of mitosis in bacteria

Moselio Schaechter and Ulrik von Freiesleben

1. Introduction

The bacterial nucleoid is a highly lobular intracytoplasmic body, generally located towards the centre of the cell and not bound by a membrane. Within this region, the DNA is found at a local concentration of approximately 2–5% (w/v). The reason why this long molecule (4.7×10^6 basepairs in *Escherichia coli*, see Blattner *et al.*, Chapter 4, this volume) is compacted and physically limited to the nucleoid region is not well understood. It is known that *in vivo* the DNA of *E. coli* is negatively supercoiled into some 50 individual domains. Nucleoids isolated by gentle breakage of cells in the presence of divalent cations have superhelicity and dimension similar to those seen intracellularly. There are no indications that the nucleoid changes in gross physical characteristics during the cell cycle.

Replication takes place from a distinct site on the covalently linked circular DNA, the origin, *oriC*, and proceeds bidirectionally towards a terminus. The *E. coli* origin is rich in sequence repeats and possesses 11 sites for methylation by the main DNA methyltransferase of this organism, the Dam enzyme. Initiation of DNA replication is under the influence of a protein that is highly conserved among many bacteria, DnaA. Once initiated, DNA replication takes place at a nearly constant rate in moderately fast and fast growing *E. coli*, until it reaches a terminus. Doubling of the chromosome takes 40 minutes at 37°C, which requires that in cultures growing faster than this time, initiation of chromosome replication takes places before the end of the previous round. If the organisms grow faster than that, new rounds of replication will initiate before the ongoing replication is finished, giving rise to a so-called multifork replication pattern.

As yet, we have not learned a great deal about the process of chromosome partition in bacteria. About all we know with certainty is that bacterial chromosomes segregate quite precisely. Anucleated bacteria, which would be expected to arise if partition were unregulated, are formed very rarely after cell division. A detailed treatment of chromosome segregation in *E. coli* can be found in reviews by Helmstetter and Leonard (1987), Leonard and Helmstetter (1989), Smith and Zyskind (1989) and Zyskind and Smith (1992). A recent review by Hiraga (1992) emphasizes the genetics of the process. A somewhat formal comparison between bacterial chromosome segregation and mitosis in higher cells can be found in a review by Schaechter (1990). Woldringh and Nanninga (1985) review the structure of the nucleoid, and Brock (1988) gives a historical account.

We have not taken full advantage of the ease of genetic manipulation in bacteria to define the structural and regulatory elements involved in segregation. An obvious reason for this uncharacteristic lack of progress is that prokaryotic cells are small. Thus, to study chromosome segregation in bacteria is to jump directly into molecular details, without the benefit of a comfortable cytological sojourn. In addition, bacteria have fewer parts and do not have the luxury of totally separate machinery for the equivalent of mitosis. Rather, this process makes use of cellular and molecular elements that also have other functions. The best examples are the cell envelopes, widely believed to participate in the prokaryotic equivalent of mitosis. Clearly, the mitotic function of the bacterial membrane and wall is difficult to separate from the many other physiological activities of these busy structures.

We will focus on three aspects of the bacterial equivalent of mitosis:
(1)   How is chromosome segregation timed?
(2)   How are newly replicated sister chromosomes decatenated?
(3)   How does chromosome movement take place?

2. The timing of chromosome segregation

Comparison with eukaryotic mitosis

Bacteria divide rapidly, often faster than the time it takes to replicate a chromosome. This means that at the time of segregation, chromosomes may still be in the process of replication. We know that this dynamic situation holds for the two best studied species, *E. coli* and *Bacillus subtilis*, and it is probably true for many others as well.

In principle, it is possible that chromosome segregation in bacteria follows the same sequence of events as in higher cells: chromosomes first replicate, then they segregate. It is at least as likely that chromosome segregation does not wait for the chromosome to be completely replicated, but that partitioning begins before that. We favour an extreme view, that segregation is tightly coupled to replication and that chromosome partitioning begins soon after the onset of replication. If so, dividing the cell cycle into the phases used to describe eukaryotic cells (G1, S, G2, M) is cumbersome if not irrelevant. A different notation adopted since the classical findings of Helmstetter and Cooper (1968) conveniently divides the cell cycle into a period, C, required for DNA replication, and a period, D, from termination of replication to cell division. Accordingly, we think that segregation starts at the beginning of the C period and lasts for that entire period.

Parenthetically, the analogy with classical mitosis works better for plasmids than for the bacterial chromosome. Thus, plasmid replication occupies a short fraction of the cell cycle only, an 'S-phase' (Yarmolinsky and Sternberg, 1988). It has been recently shown that the S-phase of certain plasmids occupies a particular time in the cell cycle, different from the time when chromosome replication starts (Keasling et al., 1991). Some plasmids are known to have a 'centromere', that is, they have a sequence (not their replicative origin!) that works in cis to ensure precise partition when cloned into plasmids that otherwise do not partition precisely (e.g. Hiraga, 1992).

The centromere function of plasmid sequences invites us to speculate on the existence of such sequences in the bacterial chromosome. We know of no sequence from the E. coli chromosome (including the replicative origin or the terminus) that stabilizes otherwise unstable plasmids. This suggests that chromosome segregation requires more than the single constituent needed for plasmid partition. The fact that the replicative origin itself does not stabilize plasmids has led some to doubt that it plays an essential role in chromosome segregation (Niki et al., 1991). A more conservative conclusion would be that the origin is not sufficient to account for segregation, but it may nevertheless play a subtle role. Another attractive possibility is that the terminus may have centromere function. However, deletions of the terminus region do not lead to abnormalities in chromosome segregation (Henson and Kuempel, 1985), making this an unlikely possibility.

The regulation of initiation of DNA replication in E. coli

Before delving into how initiation of chromosome segregation may be closely linked to replication, let us consider aspects of the regulation of initiation of replication, a subject that has received considerable attention. Replication of the bacterial chromosome starts when a regulatory protein, DnaA, reaches a certain concentration (see Zyskind and Smith, 1992). This protein works in a complex manner: it probably binds to four DNA repeats of the origin region, called DNA boxes, then participates in opening the DNA at nearby AT-rich sites. Once the DNA is opened, several other proteins can bind to make an early initiation complex. Several of these and subsequent steps have been elucidated with an *in vitro* system (Bramhill and Kornberg, 1988; Sekimizu et al., 1988). DnaA is not reused and must be made anew for each new round of replication. The reason for this is not really known, but several notions have been put forth to explain it, including that only newly synthesized DnaA has the proper folding characteristics for activity (Katayama and Nagata, 1991), that 'spent' DnaA becomes sequestered in the membrane (Hwang et al., 1990), or that it is selectively cleaved. These various possibilities, considered broadly, may not be mutually exclusive. A factor with possibly similar properties to DnaA has been described in eukaryotes (see DePamphilis, Chapter 6, this volume).

The role of DNA methylation and DNA–membrane interactions in chromosome segregation

Our model states that segregation begins shortly after the start of chromosomal replication, when newly made DNA binds to a component of the cell envelope. We base this notion on the finding that *newly replicated* origin DNA, the first part of the chromosome to be synthesized, has a unique ability to bind to a membrane component. We do not think that this step is essential for the process of chromosome partitioning, but have reason to believe that it is concerned with its precise timing.

The specific affinity of newly made origin DNA for membranes is due to its unique state of methylation. For a substantial period after replication, newly made origin DNA is hemimethylated, that is, the new DNA strand has not had a chance of becoming methylated by the major DNA methyltransferase of *E. coli*, the Dam enzyme. This enzyme methylates the adenine of the sequence GATC, which is found in unusual abundance in the origin of *E. coli* and other enteric bacteria (typically, 11 GATC in 245 bp; Zyskind *et al.*, 1983). We have shown that *in vitro* hemimethylated origin DNA possesses an unusual affinity for a membrane preparation (Ogden *et al.*, 1988). Further work has shown that this binding takes place mainly with a specific fraction that is separable on sucrose density gradients (Chakraborti *et al.*, 1992). The cytological location of this fraction is not yet known. We have recently found a 28 kDa protein with special affinity for hemi-methylated origin DNA and are currently studying it in our laboratory.

There is an apparent problem with invoking the state of DNA methylation in chromosome segregation. Surprisingly, null mutants in Dam methylase are viable and capable of growing at nearly normal rates. However, these mutants have a defect in the timing of initiation. Normal *E. coli* growing at rapid rates have more than one chromosome per cell but the initiation of their replication is highly synchronous (Skarstad *et al.*, 1986). Dam mutants, on the other hand, lack such synchrony and thus are altered in the precision with which initiation is timed (Bakker and Smith, 1989; Boye and Løbner-Olesen, 1990). We expect that, under these conditions, the onset of segregation in sister chromosomes will also be poorly timed. Accordingly, the state of methylation of the DNA may not be relevant to the actual process of segregation of the chromosomes (which can take place in the absence of methylation) but may be involved in precisely timing the initiation of segregation (Bakker and Smith, 1989; Boye and Løbner-Olesen, 1990). Coupling the timing of segregation directly to that of initiation of replication makes this an economical notion that requires no new elements. Note that segregation may begin less than a second after initiation of replication because *E. coli* at 37°C makes about 800 nucleotides of DNA per second.

As pointed out by Bakker and Smith (1989), the majority of bacteria are not known to have a system comparable to that of the Dam methyltransferase. They may either have an alternative mechanism to accurately time initiation, or lack a way of precisely timing the process. On general grounds, we consider the latter possibility unlikely.

Methylation of the newly synthesized origin DNA has been shown to be significantly delayed when compared with the average time required to methylate GATCs elsewhere on the chromosome (Campbell and Kleckner, 1990; Ogden *et al.*, 1988). The mechanism responsible for this delay is not known but at least two possibilities have been suggested. One is a direct consequence of the low intracellular concentration of Dam methyltransferase as well as some of its kinetic properties (Bergerat and Løbner-Olesen, 1992). Consequently, regions with higher GATC content may be methylated more slowly than single GATC sites. The other possibility is that when hemimethylated origin DNA is bound to the membrane, the GATC sequence becomes unavailable to the enzyme. This requires that such sequestration be relieved at some point, possibly by a cell-cycle dependent event (Schaechter *et al.*, 1991).

Preventing methylation of the origin may be a way to inhibit premature initiation of a further round of replication. Whereas hemimethylated DNA can be replicated in an *in vitro* system (Boye, 1991), it fails to be replicated *in vivo* (Hughes *et al.*, 1984; Messer *et al.*, 1985; Russell and Zinder, 1987; Smith *et al.*, 1985), perhaps because of attachment to the membrane. It is of interest that the addition of a membrane fraction to the *in vitro* system also prevents replication (Landoulsi *et al.*, 1989)

What does the delay in methylation of origin DNA have to do with chromosome segregation?

The delay in methylation of origin DNA is about 10 minutes long, a considerably shorter time than the 40 minutes required for chromosome replication at 37°C. We have suggested that this time may be sufficient to give the daughter chromosomes their sense of direction and that 10 minutes may be sufficient to ensure that each becomes destined to occupy one of the cell halves (Schaechter, 1990). However, this may well be an oversimplification because, in addition to attachment at the origin, the bacterial chromosome also appears to be bound to the membrane at random sites along the DNA. The number of these attachment sites has been estimated as being between 20 and 80 per nucleoid (for a review, see Leibowitz and Schaechter, 1975). The function of these non-specific attachments is not known, but nothing at present precludes their participation in chromosome partition, directly or indirectly.

Multiple copies of oriC *do not perturb chromosomal segregation*

A surprising fact is that an *E. coli* cell is not disturbed by carrying a large number of copies (10–20) of minichromosomes, that is, plasmids with the chromosomal replicative origin (Leonard *et al.*, 1990). It is not yet known if high copy number minichromosomes act like other medium-to-high copy number plasmids, segregating at random and not requiring a partition system. Alternatively, it is possible that the chromosomal segregation system is also used here and that the number of *oriC* specific membrane binding sites increases in response to the presence of minichromosomes. This raises the possibility that membrane binding sites are regulated by the number of *oriC* copies – in other words, are made on demand.

3. Decatenation of sister chromosomes – the last requirement before segregation

Resolution of replicative catenates

As the result of replication, daughter chromosomes form a catenate that is incapable of separation without first being topologically resolved. How does the catenate open to yield two separable chromosomes? Several topoisomerases capable of this resolution are known in *E. coli* (for a recent review see Wang, 1991), allowing for the participation of several mechanisms. The catenated chromosomes may consist either of intertwined fully replicated duplex molecules or they could be a pair of unfinished, thus gapped, molecules linked by an as yet unreplicated duplex. Intertwined completed molecules may be resolved by a type II DNA topoisomerase, such as DNA gyrase. The unlinking of unfinished molecules, on the other hand, can be carried out by either type I or type II topoisomerases. In *Salmonella typhimurium* and *E. coli*, both DNA gyrase (Steck and Drlica, 1984) and a recently discovered locus (*parC* encoding a subunit of topoisomerase IV in *E. coli*, Kato *et al.*, 1990, and *clmF* in *S. typhimurium*, Luttinger *et al.*, 1991; Schmidt, 1990), have been implicated in the resolution of catenated chromosomes. The type I topoisomerases, topoisomerases I and III, have been shown to efficiently unlink gapped plasmid DNA *in vitro* (DiGate and Marians, 1989; Minden and Marians, 1986).

Speaking in support of the role of topoisomerases in decatenation is the fact that several laboratories have isolated temperature sensitive mutants of *E. coli* and *S. typhimurium* that exhibit a phenotype called Par for *partition*, implying that they are defective in chromosome segregation. At the non-permissive temperature, these mutants continue to synthesize DNA and to grow without dividing, thus they make filaments. The interesting thing about these mutants is that the DNA remains localized in the centre of the filament, instead of partitioning into discrete nucleoids, as in the case of cell division mutants. One of the genes involved, *parA*, codes for a subunit of DNA gyrase, GyrB, whereas another, *parD*, codes for GyrA (Hussain *et al.*, 1987a, b; Kato *et al.*, 1989). The existence and role of other topoisomerases has been discussed by Kato *et al.* (1990) for *E. coli* and Schmidt (1990) for *S. typhimurium*. The participation of gyrase in segregation may not be straightforward because a mutant in GyrB has been found to suppress a defect in a protein required for cell division, FtsZ without affecting DNA topology (Ruberti *et al.*, 1991). Another mutant with the Par phenotype, *parB*, is defective in *dnaG*, the gene coding for DNA primase and is slowed down in DNA replication at the non-permissive temperature (Norris *et al.*, 1986).

Resolution by recombination

Recently, site specific recombination has been implicated in the decatenation process. This is suggested by the finding that a RecA-dependent homologous exchange takes place with great frequency in the region of termination of chromosome replication (Louarn *et al.*, 1991). Mutants deficient in RecA protein have defects in partitioning (Zyskind *et al.*, 1992). Site-specific recombinases, XerC and XerD appear to have a role in chromosome segregation because mutants in these enzymes are also unable to partition correctly. XerC and XerD bind specifically to a site near the chromosomal terminus called *dif*, which functions to resolve dimeric chromosomes produced by sister chromatid exchange (Blakeley *et al.*, 1991; Kuempel *et al.*, 1991). This topic is discussed in more detail by Sherratt *et al.* (Chapter 3, this volume).

It is not clear how well the various phenotypes of mutants indicate that the defect is in chromosome segregation. As is discussed below, mutations in other genes, such as that coding for a histone like protein HU, also result in anucleated cells.

4. The process of chromosome movement

How do daughter chromosomes separate from one another?

Bacteria do not have a visible system of microtubules or other fibrillar apparatus involved in chromosome movement. There is actual movement, as opposed to just placing the chromosomes at both sides of the cell's septum, because just before separation the microscopically visible nucleoids can be seen at positions corresponding to one quarter and three quarters of the cell length. Although under the microscope they do not appear to be in contact with the envelope, such contact cannot be excluded. The bacterial nucleoid is irregularly lobular in shape and not nearly as compact *in vivo* as was once thought (Kellenberger, 1990): thin loops of DNA extending to the periphery of the cell would not be visible in thin sections. Nonetheless, the bulk of the DNA does move laterally and, at least in the case of *B. subtilis*, appears to do so rather rapidly (Sargent, 1974).

A number of models have been proposed for an active mechanism of chromosome movement. The simplest and most economical was the first one, the replicon hypothesis (Jacob *et al.*, 1963). It stated that the newly replicated DNA attaches to the membrane on both sides of the incipient septum. All that is needed is for the membrane to grow by deposition of new material at this site so that, by the time the cell divides, the two chromosomes not only have finished replication, but have also been pushed to the midline of the incipient progeny cells. Unfortunately for Occam's razor, it turns out that the membrane is not synthesized in such a semiconservative mode. Rather, it has been shown that the membrane of *E. coli* is made by insertion of new material at many sites (Woldringh *et al.*, 1987). Also, the replicon model does not suggest a mechanism for the replicating progeny chromosomes to attach to the membrane at either side of the septum.

Mutants defective in partitioning

What little is known about the machinery that may be involved in chromosome partitioning is derived from mutant studies. In a particular phenotype, the mutation permits cell division but no segregation, thus one cell contains two copies of the genome, the other is anucleated but of normal size (see Hiraga, 1992, for a review). These mutants must be distinguished from minicell formers, where division takes place abnormally near one end of the cell and which are defective in regulation of septum location rather than chromosome partitioning. One of the mutants isolated by Niki *et al.* (1991) makes normal-sized anucleated cells with a frequency of 10–20% of all cells in a growing population. Similar phenotypes have also been observed in other mutants of *E. coli* (Jaffé *et al.*, 1988). The gene, *mukB*, identified by Niki *et al.*, codes for a particularly interesting 177 kDa protein (MukB) which has an N-terminal globular head probably involved in nucleotide binding, two α-helical domains, and a large carboxy terminal globular domain. The α-helical portions may form a coiled coil, recalling that of myosin heavy chain and other force generating proteins. MukB binds DNA but without any known specificity. The authors suggest that MukB may be involved in the movement of the chromosome, possibly by making filaments of polymerized protein (Hiraga *et al.*, 1990). Hiraga *et al.* (1989) found that a mutation in the gene for a membrane protein, *MukA* (or *Tol C*) also results in anucleated normal-size cells. This raises the possibility that several proteins may be involved in chromosome movement in bacteria. The connection between these proteins and the membrane is as yet unknown.

Do physico-chemical properties of nucleoids play a role in segregation?

Recent genetic progress is certainly exciting, but certain difficulties intrinsic in the study of intracellular structures in bacteria must be kept in mind. The nucleoid, not being separated from the cytosol by a membrane, retains its intracellular shape by virtue of physical forces. We can think that the nucleoid–cytosol compartments represent an aqueous phase separation system. *In vitro*, bacterial nucleoids remain compact as long as the concentration of counterions is sufficiently high, but it is not known if this is the reason why they appear condensed *in vivo*. The possibility remains that chromosome movement is not an active process, but a consequence of properties of this particular aqueous phase separation system. Perhaps the simplest explanation for chromosome movement is that once the chromosomes have replicated and resolved, their observed intracellular localization in the centre of the cell is physically the more stable one. Relevant to such a passive mechanism may be the role of DNA binding proteins such as the histone-like HU. This protein is coded for by two genes and a double mutant lacking both genes produces a large number of anucleated cells (Wada *et al.*, 1988), which has been taken as an indication of a defect in chromosome segregation.

A totally passive mechanism for partitioning may not be sufficient because it has been reported (based on inhibition of protein synthesis) that chromosome movement requires protein synthesis (Donachie and Begg, 1989; Hiraga *et al.*, 1990). However, these data may be difficult to interpret because it has been known for many years that inhibition of protein synthesis results in marked re-arrangements of the shape of nucleoids (Kellenberger, 1990), leading even to the fusion of individual nucleoids within a cell (Dworsky and Schaechter, 1973; Schaechter and Laing, 1961).

The role of the cell wall in segregation

The driving element in segregation may not be the membrane alone but may require the participation of the murein cell wall. The membrane may not be sufficiently rigid to allow segregation without the help of a stiff structure such as the cell wall. Keeping the problem with membrane segregation in mind, we may ask how the cell wall segregates. A widely held view is that the major component of the wall of rod-shaped bacteria – its cylindrical portion – is synthesized at many sites and is therefore not conserved upon growth. There are data to suggest that this is not the case for the wall at the poles. Thus, it has been shown in *B. subtilis* that the murein of the poles is conserved (e.g. Koch, 1985). For *E. coli*, there are no definitive data at present, although there are suggestions that here too the pole wall may be conserved (Burman *et al.*, 1983; Cooper, 1988; Woldringh *et al.*, 1987). Thus, attachment to the murein at the poles could serve to segregate daughter chromosomes. In *E. coli* the murein is located between the cytoplasmic membrane and an outer membrane. Zones of adhesion between the outer and the inner membrane are scattered throughout the cell surface (Bayer, 1979). Because these membrane regions ('Bayer patches') must traverse the cell wall, their topology is influenced by the topology of the cell wall. In addition to the Bayer patches, there is a spatially defined structure, the periseptal annulus, a ring-shaped zone of attachment of these two membranes that flanks the division septum (MacAlister *et al.*, 1983). There is no direct evidence for the attachment of DNA to zones of membrane adhesion, but their existence allows one to speculate that the chromosome may bind to them.

References

Bakker A, Smith DW. (1989) Methylation of GATC sites is required for precise timing between rounds of DNA replication in *Escherichia coli. J. Bacteriol.* **171:** 5738-5742.

Bayer ME. (1979) The fusion sites between outer membrane and cytoplasmic membrane of bacteria: their role in membrane assembly and virus infection. In: *Bacterial Outer Membrane* (Inouye M, ed). New York: John Wiley and Sons. pp 167-202.

Bergerat A, Løbner-Olesen A. (1992) How does Dam methylase of *E. coli* leave a portion of DNA temporarily hemimethylated on GATC sites in the replication fork? *FEBS Lett.* in press.

Blakeley G, Colloms S, May G, Burke M, Sherratt D. (1991) *Escherichia coli* XerC recombinase is required for chromosomal segregation at cell division. *New Biol.* **3:** 789-798.

Boye E. (1991) The hemimethylated replication origin of *Escherichia coli* can be initiated *in vitro. J. Bacteriol.* **173:** 4537-4539.

Boye E, Løbner-Olesen A. (1990) The role of dam methyltransferase in the control of DNA replication in *E. coli. Cell,* **62:** 981-989.

Bramhill D, Kornberg A. (1988) Duplex opening by dnaA protein at novel sequences in initiation of replication at the origin of the *E. coli* chromosome. *Cell,* **52:** 743-755.

Brock TD. (1988) The bacterial nucleus: a history. *Microbiol. Rev.* **52:** 397-411.

Burman LG, Raichler J, Park JT. (1983) Evidence for multisite growth of *Escherichia coli* murein involving concomitant endopeptidase and transpeptidase activities. *J. Bacteriol.* **156:** 386-392.

Campbell JL, Kleckner N. (1990) *E. coli oriC* and the *dnaA* gene promoter are sequestered from dam methyl transferase following the passage of the chromosomal replication fork. *Cell,* **62:** 967-979.

Chakraborti A, Gunji S, Shakibai N, Cubeddu J, Rothfield L. (1992) Characterization of the *E. coli* membrane domain responsible for binding *oriC* DNA. *J. Bacteriol.* **174:** 7202-7206.

Cooper S. (1988) Rate and topography of cell wall synthesis during the division cycle of *Salmonella typhimurium. J. Bacteriol.* **170:** 422-430.

DiGate RJ, Marians KJ. (1989) Molecular cloning and DNA sequence analysis of *Escherichia coli topB*, the gene encoding topoisomerase III. *J. Biol. Chem.* **264:** 17924-17930.

Donachie WD, Begg KJ. (1989) Chromosome partition in *Escherichia coli* requires post-replication protein synthesis. *J. Bacteriol.* **171:** 5405-5409.

Dworsky P, Schaechter M. (1973) Effect of rifampicin on the structure and membrane attachment of the nucleoid of *Escherichia coli. J. Bacteriol.* **116:** 1364-1374.

Helmstetter CE, Cooper S. (1968) DNA synthesis during the division cycle of rapidly growing *Escherichia coli* B/r. *J. Mol. Biol.* **31:** 507-518

Helmstetter CE, Leonard AC. (1987) Mechanism for chromosome and minichromosome segregation in *Escherichia coli. J. Mol. Biol.* **197:** 195-204.

Henson JM, Kuempel PL. (1985) Deletion of the terminus region (340 kilobase pairs of DNA) from the chromosome of *Escherichia coli*. *Proc. Natl Acad. Sci. USA*, **82:** 3766-3770.

Hiraga S. (1992) Chromosome and plasmid partition in *Escherichia coli*. *Annu. Rev. Biochem.* **61:** 283-306.

Hiraga S, Niki H, Ogura T, Ichinose C, Mori H, Ezaki B, Jaffé A. (1989) Chromosome partitioning in *Escherichia coli*: novel mutants producing anucleate cells. *J. Bacteriol.* **171:** 1496-1505.

Hiraga S, Ogura T, Niki H, Ichinose C, Mori H. (1990) Positioning of replicated chromosomes in *Escherichia coli*. *J. Bacteriol.* **172:** 31-39.

Hughes P, Squali-Houssaini F, Forterre P, Kohiyama M. (1984) In vitro replication of dam methylated and nonmethylated *oriC* plasmid. *J. Mol. Biol.* **176:** 155-159.

Hussain K, Begg KJ, Salmond GP, Donachie WD. (1987a) *ParD*: a new gene coding for a protein required for chromosome partitioning and septum localization in *Escherichia coli*. *Mol. Microbiol.* **1:** 73-81.

Hussain K, Elliott EJ, Salmond GP. (1987b) The *parD* mutant of *Escherichia coli* also carries a *gyrAam* mutation. The complete sequence of *gyrA*. *Mol. Microbiol.* **1:** 259-273.

Hwang DS, Crooke E, Kornberg A. (1990) Aggregated dnaA protein is dissociated and activated for DNA replication by phospholipase or dnaK protein. *J. Biol. Chem.* **265:** 19244-19248.

Jacob F, Brenner S, Cuzin F. (1963) On the regulation of DNA replication in bacteria. *Cold Spring Harb. Symp. Quant. Biol.* **28:** 339-348.

Jaffé A, D'Ari R, Hiraga S. (1988) Minicell-forming mutants of *Escherichia coli*: Production of minicells and anucleate rods. *J. Bacteriol.* **170:** 3094-3101.

Katayama T, Nagata T. (1991) Initiation of chromosomal DNA replication which is stimulated without oversupply of DnaA protein in *Escherichia coli*. *Mol. Gen. Genet.* **226:** 491-502.

Kato J, Nishimura Y, and Suzuki H. (1989) *Escherichia coli parA* is an allele of the *gyrB* gene. *Mol. Gen. Genet.* **217:** 178-181.

Kato J, Nishimura Y, Imamura R, Niki H, Hiraga S, Suzuki H. (1990) New topoisomerase essential for chromosome segregation in *E. coli*. *Cell*, **63:** 393-404.

Keasling J, Palsson BO, Cooper S. (1991) Cell-cycle-specific F plasmid replication: regulation by cell size control of initiation. *J. Bacteriol.* **173:** 2673-2680.

Kellenberger E. (1990) Intracellular organization of the bacterial chromosome. In: *The Bacterial Chromosome* (Drlica K, Riley M, eds). Washington DC: American Society for Microbiology Press. pp 173-186.

Koch A. (1985) How bacteria divide in spite of internal hydrostatic pressure. *Can. J. Microbiol.* **31:** 1071-1084.

Kuempel PL, Henson JM, Dircks L, Tecklenburg M, Lim DF. (1991) *dif*, a *recA*-independent recombination site in the terminus region of the chromosome of *Escherichia coli*. *New Biol.* **3:** 799-811.

Landoulsi A, Hughes P, Kern R, Kohiyama M. (1989) *dam* methylation and the initiation of DNA replication on *oriC* plasmids. *Mol. Gen. Genet.* **216:** 217-223.

Leibowitz PJ, Schaechter M. (1975) The attachment of the bacterial chromosome to the cell membrane. *Int. Rev. Cytol.* **41:** 1-28

Leonard AC, Helmstetter C. (1989) Replication and segregation control of *Escherichia coli* chromosomes. In: *Chromosomes: Eukaryotic, Prokaryotic, and Viral*, vol. 2 (Adolph KW, ed). Boca Raton: CRC Press.

Leonard AC, Theisen PW, Helmstetter CE. (1990) Replication timing and copy number control of *oriC* plasmids. In: *The Bacterial Chromosome* (Drlica K, Riley M, eds). Washington DC: American Society for Microbiology Press. pp 279-286.

Louarn J-M, Louarn J, Francois V, Patte J. (1991) Analysis and possible role of hyperrecombination in the termination region of the *Escherichia coli* chromosome. *J. Bacteriol.* **173:** 5097-5104.

Luttinger AL, Springer AL, Schmidt MB. (1991) A cluster of genes that affect nucleoid segregation in *Salmonella typhimurium*. *New Biol.* **3:** 687-697.

MacAlister TJ, MacDonald B, Rothfield LI. (1983) The periseptal annulus: an organelle associated with cell division in gram-negative bacteria. *Proc. Natl Acad. Sci. USA*, **80:** 1372-1376.

Messer W, Bellekes U, Lother H. (1985) Effect of *dam* methylation on the activity of the *E. coli* replication origin, *oriC*. *EMBO J.* **4:** 1327-1332.

Minden JS, Marians KJ. (1986) *Escherichia coli* topoisomerase I can segregate replicating pBR322 daughter DNA molecules *in vitro. J. Biol. Chem.* **261:** 11906-11917.

Niki H, Jaffé A, Imamura R, Ogura T, Hiraga S. (1991) The new gene *mukB* codes for a 177 kd protein with coiled-coil domains involved in chromosome partitioning of *E. coli. EMBO J.* **10:** 183-193.

Norris V, Alliotte T, Jaffé A, D'Ari R. (1986) DNA replication termination in *Escherichia coli parB* (a *dnaG* allele), *parA*, *gyrB* mutants affected in DNA distribution. *J. Bacteriol.* **168:** 494-504.

Ogden G, Pratt MJ, Schaechter M. (1988) The replicative origin of the *Escherichia coli* chromosome binds to cell membranes only when hemimethylated. *Cell*, **54:** 127-135.

Ruberti I, Crescenzi F, Paolozzi L, Ghelardini P. (1991) A class of *gyrB* mutants, substantially unaffected in DNA topology, suppresses the *Escherichia coli* K12 *ftsZ84* mutation. *Mol. Microbiol.* **5:** 1065-1072.

Russell DW, Zinder ND. (1987) Hemimethylation prevents DNA replication in *E. coli. Cell*, **50:** 1071-1079.

Sargent MG. (1974) Nuclear segregation in *Bacillus subtilis. Nature*, **250:** 252-254.

Schaechter M. (1990) The bacterial equivalent of mitosis. In: *The Bacterial Chromosome* (Drlica K, Riley M, eds). Washington DC: American Society for Microbiology Press. pp 313-322.

Schaechter M, Lang VO. (1961) Direct observation of fusion of bacterial nuclei. *J. Bacteriol.* **81:** 667-668.

Schaechter M, Polaczek P, Gallegos R. (1991) Membrane attachment and DNA bending at the origin of the *Escherichia coli* chromosome. *Res. Microbiol.* **142:** 151-154.

Schmidt MB. (1990) A locus affecting nucleoid segregation in *Salmonella typhimurium. J. Bacteriol.* **172:** 5416-5424.

Sekimizu K, Bramhill D, Kornberg A. (1988) Sequential early stages in the *in vitro* initiation of replication at the origin of the *Escherichia coli* chromosome. *J. Biol. Chem.* **263:** 7124-7130.

Skarstad K, Boye E, Steen HB. (1986) Timing of initiation of chromosome replication in individual *Escherichia coli* cells. *EMBO J.* **5:** 1711-1717.

Smith DW, Garland AM, Herman G, Enns RE, Baker T, Zyskind JW. (1985) Importance of state of methylation of *oriC* GATC sites in initiation of DNA replication in *Escherichia coli. EMBO J.* **4:** 1319-1326.

Smith DW, Zyskind JW. (1989) The chromosomal DNA replication origin, *oriC*, in bacteria. In: *Chromosomes: Eukaryotic, Prokaryotic, and Viral*, vol. 2 (Adolph KW, ed). Boca Raton: CRC Press.

Steck TR, Drlica K. (1984) Bacterial chromosome segregation: evidence for DNA gyrase involvement in decatenation. *Cell*, **36:** 1081-1088.

Wada M, Kano Y, Ogawa T, Okazaki T, Imamoto F. (1988) Construction and characterization of the deletion mutant of *hupA* and *hupB* genes in *Escherichia coli. J. Mol. Biol.* **204:** 581-591.

Wang JC. (1991) DNA topoisomerases: why so many? *J. Biol. Chem.* **266:** 6659-6662.

Woldringh CL, Nanninga N. (1985) In: *Molecular Cytology of Escherichia coli.* (Nanninga N, Woldringh CL, eds). London: Academic Press.

Woldringh CL, Hus P, Pas E, Brakenhoff GJ, Nanninga N. (1987) Topography of peptidoglycan synthesis during elongation and polar cap formation in a cell division mutant of *Escherichia coli* MC4100. *J. Gen. Microbiol.* **133:** 575-586.

Yarmolinsky MB, Sternberg N. (1988) In: *The Bacteriophages,* vol. 1 (Calendar R, ed). New York: Plenum Publishing.

Zyskind JW, Smith DW. (1992) DNA replication, the bacterial cell cycle, and cell growth. *Cell*, **69:** 5-8.

Zyskind JW, Cleary JM, Brusilow WSA, Harding NE, Smith DW. (1983) Chromosomal replication origin from the marine bacterium *Vibrio harveyi* functions in *Escherichia coli*: *oriC* consensus sequence. *Proc. Natl Acad. Sci. USA*, **80:** 1164-1168.

Zyskind JW, Svitil AL, Stine WB, Biery MC, Smith DW. (1992) RecA protein of *Escherichia coli* and chromosome partitioning. *Mol. Microbiol.* **6:** 2525-2537.

Chapter 6

Eukaryotic origins of DNA replication

Melvin L DePamphilis

1. Introduction

All eukaryotic genomes can be divided into two groups. 'Simple' genomes are represented by lower eukaryotes such as protozoa (*Tetrahymena*), yeast, and slime mold (*Physarum*), by animal viruses, and by the mitochondria of higher eukaryotes. 'Complex' genomes are represented by the metazoa. (Metazoa include all multicellular animals except sponges and simple multicellular parasites that inhabit various invertebrates.) In simple genomes, origins of replication consist of well characterized sequences that interact with specific proteins (reviewed in Challberg and Kelly, 1989; Clayton, 1991; Fangman and Brewer, 1991; Kornberg and Baker, 1992; Stillman, 1989; Tsurimoto *et al.*, 1990; Wang, 1991), while in complex genomes, the nature of replication origins is not yet clear (reviewed in Benbow *et al.*, 1985; Held and Heintz, 1992; Orr-Weaver, 1991; Umek *et al.*, 1989; Schaechter and von Freiesleben, Chapter 5, this volume). This review attempts to illuminate origins of replication in complex genomes by first identifying the essential features of origins in simple eukaryotic genomes, and then asking whether or not the same features are present in complex genomes.

There is both good news and bad news. The good news is that there is light at the end of the tunnel. Origins of replication in simple genomes share a common anatomy: an origin recognition sequence (ORS), a DNA unwinding element (DUE), and one or more binding sites for specific transcription factors. Four different mechanisms have been recognized by which these transcription factors facilitate origin activity, and this increased appreciation of the similarities between replication and transcription suggests that cells might regulate DNA replication by linking it to gene expression. In complex genomes, new technology has permitted mapping of initiation sites for DNA replication in single copy sequences in the cellular chromosomes of metazoa (Vassilev and DePamphilis, 1992). The results demonstrate that DNA replication also begins at specific sites (origin of bidirectional replication, OBR) using the same replication fork mechanism favoured by simple genomes. This strengthens the hypothesis that unique *cis-*

acting sequences are responsible for determining where replication begins in all genomes.

The bad news is that the tunnel is longer than we thought. The origin of bidirectional replication lies within a larger initiation zone recognized by the appearance of replication bubbles randomly distributed throughout its DNA. Moreover, DNA replication appears to be independent of DNA sequence when DNA is injected into non-mammalian eggs, or plasmids are transfected into differentiated cells, and the frequency of initiation events in natural chromosomes can change during animal development, suggesting that replication can begin at a large number of different DNA sequences. Perhaps initiation of replication in complex genomes utilizes a novel principle not yet observed in simpler genomes. While *cis*-acting sequences apparently determine where replication begins, chromatin structure or nuclear organization may establish a primary initiation site among many potential initiation sites within the initiation zone.

2. Origins of DNA replication in simple genomes

Functional and genetic origins are coincident

Origins of DNA replication are recognized in two ways: (1) The 'genetic origin' (*ori*) is defined by *cis*-acting mutations in a DNA sequence; (2) 'functional origin' is the actual site where DNA replication begins. In simple genomes, these two origins are coincident. Therefore, one would expect that sites where replication begins in complex genomes will be associated closely with a specific *cis*-acting origin sequence.

Replication most frequently occurs by the replication fork mechanism (Fig. 1). Binding of an origin recognition protein is followed by DNA unwinding and then RNA-primed DNA synthesis (Kornberg and Baker, 1992; Wang, 1991). As the two forks advance, DNA is synthesized continuously on the 'forward arm' but discontinuously on the 'retrograde arm' via repeated synthesis and joining of short RNA-primed nascent DNA chains commonly referred to as Okazaki fragments. This is a consequence of the facts that DNA polymerases synthesize only in the 5' to 3' direction, and that the two DNA strands are antiparallel. Various methods have shown that genetic and functional origins are coincident in the linear adenovirus (Ad) genome and a variety of prokaryotic origins (Kornberg and Baker, 1992), but the most precise method for mapping functional origins that exist within a chromosome is to identify the transition from discontinuous to continuous DNA synthesis that must occur on each template strand if initiation of replication occurs by the replication fork mechanism. This defines the origin of bidirectional replication (OBR, Fig. 1). The OBR has been identified at the resolution of single nucleotides in SV40 (simian virus 40; Guo *et al.*, 1991; Hay and DePamphilis, 1982), PyV (polyomavirus; Hendrickson *et al.*, 1987a), mtDNA

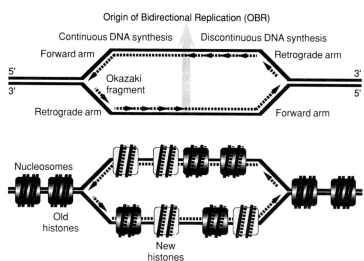

Figure 1. Replication fork model with random segregation of prefork histone octamers (for more detail, see Cusick *et al.*, 1989).

(mitochondrial DNA; Clayton, 1991), *Escherichia coli* (Kohara *et al.*, 1985), and bacteriophage λ (Yoda *et al.*, 1988). In each case, a sharply defined OBR was found adjacent to or within *ori*.

Another approach has been the analysis of DNA replication structures such as forks and bubbles based on their mobility under various conditions of two-dimensional (2-D) gel electrophoresis (Fangman and Brewer, 1991; Zhu *et al.*, 1992). These methods place the OBR at or very near the dyad symmetry origin recognition element (ORE) of EBV (Epstein-Barr virus) *oriP* (Gahn and Schild-kraut, 1989), within 0.2 kb of the BPV (bovine papilloma virus) *ori* (Ustav *et al.*, 1991; Yang and Botchan, 1990), and within ±0.1–0.4 kb of a yeast autonomously replicating sequence (ARS) that permits yeast plasmids to replicate once per S-phase. Initiation of replication in yeast (*Saccharomyces cerevisiae*) chromosomes coincides with ARS elements (±0.3 kb), revealing that, at least in *S. cerevisiae*, genetic and functional origins are coincident. The same appears to be true in the non-transcribed spacer regions of ribosomal DNA genes in the protozoan, yeast and slime mold where initiation sites have been localized to specific sites (≤1 kb) and these loci also contain *cis*-acting sequences that promote replication (Daniel and Johnson, 1989; Larson *et al.*, 1986; Linskens and Huberman, 1988; Yao and Yao, 1989). Replication bubbles also have been mapped to a ~1 kb locus proximal to the *profilin-P* gene in the slime mold *Physarum* (Benard and Pierron, 1992), but analysis of *cis*-acting sequences is not yet available.

Basic components of an origin of replication

In simple genomes, *ori* consist of a core component and one or more auxiliary components. *Ori*-core is the minimal essential *cis*-acting sequence required to initiate DNA replication under all conditions; it is analogous to a transcription promoter. *Ori*-core consists of an ORE, a DNA unwinding element (DUE), and an A/T-rich element in which one strand is T-rich and one strand is A-rich. Spacing, orientation and arrangement of these three core components is usually critical for *ori* function. The A/T-rich element is part of ORE in some origins (Ad, yeast), while in others it appears to facilitate the DUE activity (SV40, PyV), and in still others it may serve as the DUE element itself (herpes simplex virus, HSV).

Origin recognition element (ORE). ORE is the DNA binding site for one or more 'origin recognition proteins' that are required for initiation of replication (Fig. 2, Table 1). These proteins participate in replication either directly by unwinding the DNA (helicase activity) or indirectly through their association with replication proteins. This association could serve simply to guide replication proteins to *ori*, or it could modify their activity. Examples of helicases are SV40 large tumour antigen (T-ag) (Borowiec *et al.*, 1990; Gutierrez *et al.*, 1990), PyV T-ag (Lorimer *et al.*, 1991, 1992; Seki *et al.*, 1990) and HSV UL9 protein (Fierer and Challberg, 1992). SV40 T-ag has been shown to unwind specifically the SV40 *ori*-core. Two examples illustrate association with replication proteins. Ad preterminal protein-dCMP forms a complex with Ad DNA polymerase that binds specifically to Ad

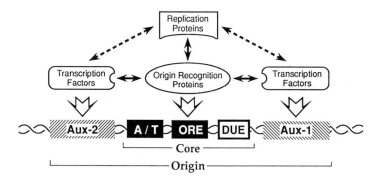

Figure 2. Anatomy of a eukaryotic origin of DNA replication (*ori*). Origin auxiliary component *aux-1* flanks the DUE element of *ori*-core, while *aux-2* flanks the A/T-rich element. Transcription factors that activate *aux-1* or *aux-2* are listed, and their roles in replication suggested in the text (see appropriate numbered sections). The A/T-rich element is part of the ORE in some *ori*-cores (Ad, yeast) but distinct in others (SV40, PyV). Origin recognition proteins that bind to the natural sequences are listed. The DUE has been identified in three origins, and is coincident with the OBR in two of them.

Table 1. Proteins that activate eukaryotic origins of DNA replication

Origin	Origin recognition protein(s)	Transcription factors	
		Aux-2	Aux-1
SV40	T-ag	Sp1, T-ag	T-ag
PyV	T-ag	AP1	T-ag
BPV	E1	E2	
Ad2	[pre TP:Ad DNA pol]	NF1, OCT1	
EBV *oriP*	EBNA1	EBNA1	
HSV *oriS*	UL9	several candidates	
mtDNA *oriH*	RNase MRP	mtRNA polymerase	
Yeast ARS	ACBP?, ORC?	ABF1	ABF1

ori-core (Chen *et al.*, 1990; Temperley *et al.*, 1991), and T-ag binds the enzyme required to initiate SV40 and PyV DNA synthesis, DNA polymerase-α, (Dornreiter *et al.*, 1990; Gannon and Lane, 1990). The latter association must be important because initiation of SV40 or PyV DNA replication requires both the cognate viral T-ag and DNA primase–DNA polymerase-α from a permissive host cell (Eki *et al.*, 1991). An origin recognition complex of 6 to 8 polypeptides has been isolated from yeast that binds to both strands of the ARS1 ORE *in vitro* (elements A+B1, Bell and Stillman, 1992), and genomic footprinting of ARS1 *in situ* suggests that this complex also interacts with ARS1 *in vivo* (Diffley and Cocker, 1992). This may be a cellular equivalent of viral replication complexes.

DNA unwinding element (DUE). The DUE is an easily unwound DNA region. Although it is not a unique DNA sequence like an ORE, a DUE is determined by base stacking interactions, and therefore is not simply a function of AT content, but depends on nucleotide sequence (Natale *et al.*, 1992). A DUE is recognized by its hypersensitivity to single-strand specific nucleases and reactive reagents, its insensitivity to point mutations, and its sensitivity to larger sequence changes that affect its thermodynamic characteristics. DUE was first demonstrated as a component of yeast (Umek and Kowalski, 1988) and bacterial (Kowalski and Eddy, 1989) origins by Kowalski and co-workers who suggested the DUE as the site where DNA synthesis begins. In yeast origins, the DUE lies within the ~100 bp flanking the 3'-end of the T-rich strand of the ORE that is essential for ARS activity. The inhibitory effect of GC-rich linker scanner mutations together with the absence of an effect of point mutations suggest that the yeast ARS B2 element mapped by Marahrens and Stillman (1992) functions as a DUE. In SV40 and PyV origins, the DUE predicted by thermodynamic analysis of DNA sequence (D Kowalski, personal communication) is coincident with the site where T-ag dependent DNA unwinding begins (Borowiec and Hurwitz, 1988; Lorimer, 1992), which is coincident with the OBR (DePamphilis *et al.*, 1988; Guo *et al.*, 1991; Hay and DePamphilis, 1982; Hendrickson *et al.*, 1987b). Therefore, the DUE likely determines where DNA unwinding begins.

Where does DNA synthesis begin? Origins are not symmetrical arrangements of elements, either in terms of sequence or function. They tend to be highly polarized with initiation of DNA synthesis occurring only on one strand of *ori*-core (e.g. Ad, mtDNA, SV40, PyV). Thus, in SV40 and PyV, RNA-primed DNA synthesis initiation events appear only on the early mRNA template strand of [A/T-ORE-DUE] (DePamphilis *et al.*, 1988; Guo *et al.*, 1991; Hay and DePamphilis, 1982; Hay *et al.*, 1984; Hendrickson *et al.*, 1987a, b). Therefore, initiation of replication is unlikely to be a symmetrical event with two forks moving simultaneously out of the ORE as indicated in some models (Borowiec *et al.*, 1990; Tsurimoto *et al.*, 1990). More likely is that forward arm synthesis begins on one strand of *ori*-core, and then progresses beyond *ori*-core before DNA synthesis is initiated in the opposite direction on the complementary strand (initiation zone model, DePamphilis *et al.*, 1988).

Auxiliary components. Ori-auxiliary components consist of transcription factor binding sites that facilitate *ori*-core activity (Fig. 2). These transcriptional elements may determine when replication begins. The evidence is now compelling that *ori*-auxiliary components must bind specific transcription factors (Table 1) in order to facilitate origin activity. The reason for this presumably involves specific protein–protein interactions between transcription factors and origin recognition proteins, as well as replication proteins such as DNA polymerase. Transcription factors that activate one origin do not necessarily activate another, and the ability of a transcription factor to stimulate a promoter does not necessarily reflect its ability to stimulate an origin. Since *ori*-auxiliary sequences do not alter the mechanism of replication and are dispensable under some conditions (Guo *et al.*, 1991; Martínez-Salas *et al.*, 1988; Prives *et al.*, 1987; Temperley and Hay, 1991), they are analogous to transcription enhancers.

Four different mechanisms have been identifed by which transcription factors facilitate origin activity. In the first mechanism, transcription factors initiate transcription through *ori*-core. The only eukaryotic example is mitochondrial DNA where some of the transcripts from an upstream promoter are cleaved by a site-specific endoribonuclease (RNase MRP for mitochondrial RNA processing). These 3' ends are then used by mitochondrial DNA polymerase as RNA primers for DNA synthesis (Clayton, 1991).

In the second mechanism, transcription factors facilitate the assembly of an initiation complex. The best example of this mechanism is the Ad2 origin where transcription factor NF1 binds to Ad DNA polymerase and facilitates binding of subsaturating concentrations of the Ad2 preterminal protein/Ad DNA polymerase complex (pTP-pol) to the Ad ORE (Mul and van der Vliet, 1992). A second example is the enhancer-specific BPV E2 protein which appears to facilitate E1 binding to the BPV ORE (Ustav *et al.*, 1991; Yang *et al.*, 1991).

The third mechanism is for transcription factors to facilitate the activity of an initiation complex. Once a replication complex has formed, transcription factors may stimulate its ability to stimulate subsequent steps in replication such as DNA unwinding or initiation of DNA synthesis. For example, SV40 *ori*-auxiliary sequences apparently stimulate DNA unwinding (Gutierrez *et al.*, 1990). This may occur because as T-ag unwinds the DNA at *ori*-core, T-ag converts its sequence-specific high affinity DNA binding site into single-stranded DNA that has a non-specific low affinity for T-ag. Therefore, during the transition from DNA unwinding to initiation of DNA synthesis, the moderate affinity of T-ag for *aux-1* and *aux-2* sequences, or its ability to interact with other proteins that bind to these sequences, may promote DNA unwinding by stabilizing the initial unwound intermediate (Gutierrez *et al.*, 1990).

The activity of SV40 *ori*-auxiliary components requires binding of transcription factors whose specific activation domains can interact with the T-ag initiation complex bound to *ori*-core. SV40 *aux-1* consists of a strong T-ag DNA binding site (Guo *et al.*, 1991); T-ag is a transcription factor (Zhu *et al.*, 1991) as well as a replication protein. The SV40 enhancer stimulates SV40 *ori*-core only in the absence of the natural *aux-2* sequence (Guo *et al.*, 1989). The natural *aux-2* is part of the early gene promoter and contains three Sp1 and three T-ag binding sites. A synthetic oligonucleotide that only bound Sp1 provided ~75% of *aux-2* activity, while a sequence that only bound T-ag provided ~20% of *aux-2* activity (Guo and DePamphilis, 1992). Thus, the combined effects of these two proteins could account for SV40 *aux-2* activity *in vivo*. Tandem AP1 or NF1 binding sites could completely substitute for SV40 *aux-2* (Cheng and Kelly, 1989; Guo and DePamphilis, 1992; Hoang *et al.*, 1992), but other transcription factors, including the c-*jun* activation domain, could not. These replicationally inactive transcription factors did, however, activate appropriate promoters or enhancers in the same cells. Therefore, *aux-2* activity depends on binding proteins whose specific activation domains can interact with the T-ag initiation complex bound to *ori*-core.

The fourth mechanism is for transcription factors to prevent repression caused by chromatin structure. Chromatin structure acts as a general repressor of promoter and origin activity by interfering with binding of initiation factors to promoters and origins of replication (Felsenfeld, 1992). This mechanism has been suggested for SV40 *aux-2* because prebinding NF1 (Cheng and Kelly, 1989) or GAL4:VP16 (Cheng *et al.*, 1992) to a synthetic *aux-2* sequence composed of appropriate DNA binding sites prevented the subsequent assembly of nucleosomes from interfering with DNA replication when these chromatin templates were incubated in appropriate cell extracts. NF1 had no effect when DNA templates underwent replication in the absence of nucleosome assembly. However, while NF1 DNA binding sites also stimulated SV40 *ori*-core activity *in vivo*; GAL4:VP16 DNA binding sites, in the presence of GAL4:VP16 protein, did not stimulate SV40 DNA replication *in vivo*, although it was highly active in transcription (Guo and DePamphilis, 1992; Hoang *et al.*, 1992). Therefore, the significance of

GAL4:VP16 action *in vitro* is questionable. Furthermore, simply prebinding T-ag to *ori* prevents nucleosomes from repressing *ori* activity (Ishimi, 1992; J Sogo, personal communication). Thus, the critical question of whether proteins bound to *aux-2* facilitate T-ag in preventing repression during nucleosome assembly has not been answered.

Nevertheless, the ability of enhancers to prevent repression by chromatin structure appears to be a valid model for *aux-2* activity in PyV, as well as for the effect of enhancers on promoters. The *aux-1* of PyV, like that of its close relative SV40, is a strong T-ag binding site (Weichbraun *et al.*, 1989) and presumably serves the same role as in SV40. PyV *aux-2* consists of several functionally redundant enhancer elements, one of which is an AP1 binding site that strongly stimulates *ori*-core. PyV *aux-2* function, like that of SV40, requires binding of transcription factors with specific activation domains. Synthetic AP1 binding sites alone could completely substitute for PyV *aux-2* (Guo and DePamphilis, 1992; Murakami *et al.*, 1991), although different members of the AP1 protein family appear to activate *ori* in different cells. In general, several different transcription activation domains stimulated PyV *ori*-core activity 10 to 30-fold ($\leq 30\%$ of the activity produced by the PyV enhancer), while DNA binding domains alone had no effect (Baru *et al.*, 1991; Bennett-Cook and Hassell, 1991; Guo and DePamphilis, 1992; Murakami *et al.*, 1991; Nilsson *et al.*, 1991; Wasylyk *et al.*, 1990). Therefore, PyV *aux-2*, like that of SV40, requires binding of proteins with specific activation domains, presumably to interact with T-ag bound to *ori*-core.

PyV *ori*-core's need for an enhancer appears to depend on formation of a repressive chromatin structure. The PyV enhancer (*aux-2*) is not required to activate PyV *ori*-core in cytoplasmic extracts (Prives *et al.*, 1987), conditions where chromatin assembly does not occur (Gruss *et al.*, 1990). Enhancers also are not required to activate either the PyV *ori*-core or a promoter when the DNA is injected into the paternal pronucleus of mouse eggs (Martínez-Salas *et al.*, 1988, 1989; Wiekowski *et al.*, 1991), conditions where repression by chromatin structure is absent (S Majumder, M Wiekowski and M DePamphilis, unpublished results). Enhancers have been shown to modify chromatin structure and may prevent chromatin formation over an origin or promoter by forming a loop of DNA in which the enhancer is linked to the origin or promoter through association of enhancer-bound proteins with either origin recognition proteins (DePamphilis *et al.*, 1988) or the RNA polymerase initiation complex (Felsenfeld, 1992). This association would redistribute nucleosomes in the linked region and prevent new nucleosomes from forming. Such a DNA loop has been shown to occur with purified EBV DNA and EBNA1 protein which binds both to the EBV enhancer and to the EBV *ori*-core (Middleton and Sugden, 1992). This suggests that an [enhancer:EBNA1:EBNA1:ORE] complex is required for replication.

3. Origins of replication

In simple genomes, origins of replication consist of 18 to about 200 bp of DNA that can be divided into required sequences (core components) and auxiliary sequences. DNA replication begins when an origin recognition protein binds to the ORE that is part of *ori*-core. This protein initiates DNA unwinding, either by virtue of its own helicase activity (e.g. SV40, PyV) or by its association with another replication protein (e.g. Ad). The site of DNA unwinding is determined by the DUE which corresponds to the OBR (e.g. SV40, PyV). *Ori*-auxiliary sequences bind transcription factors that facilitate *ori*-core activity. The requirement for specific protein activation domains to facilitate Ad, SV40, PyV and EBV origins strongly suggests direct interaction with proteins that bind to *ori*-core, and evidence for T-ag:DNA polymerase-α, NF1:Ad DNA polymerase and EBNA1:EBNA1 complexes has been reported. The extent to which these transcription factors are needed may simply reflect the relative abilities of origin recognition proteins to bind ORE and initiate DNA unwinding or DNA synthesis. For example, Ad4 DNA polymerase is better than Ad2 DNA polymerase at binding to the ORE and therefore does not need help from NF1 protein. Similarly, SV40 T-ag binds more strongly to its ORE than PyV T-ag binds to its ORE (Scheller and Prives, 1985), and therefore may be less sensitive to interference from chromatin structure. Since transcription factors can differ significantly in their ability to bind to DNA organized into nucleosomes (Taylor *et al.*, 1991), the same would be expected of replication factors. Enhancers may provide a specific mechanism for preventing repression by chromatin structure (e.g. PyV, EBV, yeast), while other *ori*-auxiliary sequences (e.g. SV40 and PyV *aux-1*, Ad2 NF3 binding site) may provide a mechanism for promoting DNA unwinding. This would account for the consistently strong effect of *aux-1 in vitro* and *in vivo*, and for the ability to dispense with enhancers to activate origins or promoters when chromatin structure is altered or absent.

4. Origins of DNA replication in complex genomes

Replication structures appear throughout a broad initiation zone

The most surprising feature of DNA replication in complex genomes that differs from simple genomes is the appearance of replication bubbles distributed throughout a large DNA region. Electron microscopy of metazoan DNA from cells in S-phase sometimes reveals clusters of small replication bubbles (0.1 to 0.5 kb microbubbles) spread over 3 to 30 kb (Micheli *et al.*, 1982). Although the significance of these structures has never been established, their appearance throughout the non-transcribed spacer region of rDNA that serves as an initiation zone for replication in *Xenopus* larvae (Bozzoni *et al.*, 1981) suggests that replication is initiated at many DNA sites within a large (~10 kb) initiation zone. This concept has been reinforced by analyses of DNA replication structures such as forks and bubbles based on their mobility under various conditions of 2-D gel electrophoresis (Fangman and Brewer, 1991; Zhu *et al.*, 1992).

The availability of a Chinese hamster ovary (CHO) cell line containing ~1000 tandem, integrated copies of the dihydrofolate reductase (*DHFR*) gene region (CHO C400 cells) provided an opportunity to observe DNA replication at a unique locus in mammalian chromosomes. Application of 2-D gel methods to CHO C400 cells that were either proliferating exponentially or synchronized at their G1/S-border revealed trace amounts of replication bubbles distributed throughout a 28 kb initiation zone and replication forks facing in both directions (Vaughn *et al.*, 1990). The fraction of bubbles relative to forks has been increased by enrichment for DNA with single-stranded character that is associated with nuclear matrix (Dijkwel *et al.*, 1991), and by synchronization of cells in G1 with mimosine (Dijkwel and Hamlin, 1992). This region has now been extended to 55 kb between the *DHFR* gene and a second gene designated *2BE2121* (Dijkwel and Hamlin, 1992; P Dijkwel, personal communication; Fig. 3, 2-D gel electrophoresis). Replication bubbles were not detected in either gene, and replication forks in these genes appear to migrate only away from the initiation zone.

A somewhat different result was reported using the same method but at a locus that undergoes programmed amplification during animal development. The *Drosophila* chorion gene region is amplified specifically more than 60-fold in follicle cells during oogenesis, and this amplification is regulated by five *cis*-acting sequence elements located within ~8 kb of DNA (Orr-Weaver, 1991). While replication bubbles were again distributed throughout the amplified region, the authors concluded that ~80% of the initiation events occurred within a 1 kb region of DNA encompassing the *cis*-acting amplification enhancer region-d, closest to the amplification control element (Delidakis and Kafatos, 1989; Heck and Spradling, 1990). These results suggest that DNA replication in metazoan chromosomes occurs at many DNA sites distributed throughout an initiation zone of 8 to 55 kb, but that some initiation sites are highly preferred over others. Moreover, the primary initiation site corresponds to a component of the genetically defined origin.

5. Most replication events begin within a small DNA locus

The most extensive analysis of mammalian DNA replication to date has been carried out in the 300 kb region around the *DHFR* gene of CHO cells. CHO C400 cells made it possible to identify the earliest DNA fragments that could be radiolabelled with DNA precursors in the amplified *DHFR* gene region when synchronized cells were released into S-phase. This approach identified three different initiation loci (*ori-α*, *ori-β* and *ori-γ*; Fig. 3). In the *ori-β* region, several studies have localized the earliest labelled DNA to a 1.8 kb fragment (Fig. 3, earliest labelled fragment). The same results were obtained regardless of whether radiolabelling was carried out in whole cells or cell lysates. When nascent DNA was trapped at its site of origin by pretreating cells with psoralen to introduce

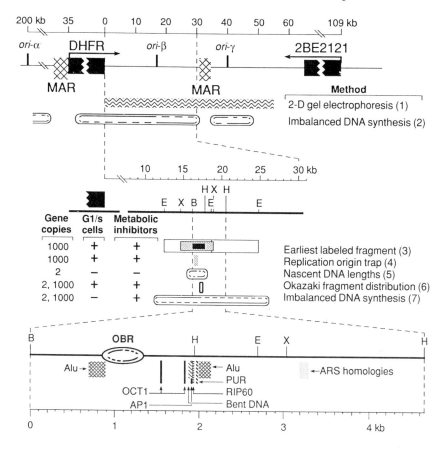

Figure 3. Initiation of DNA replication in the hamster *DHFR* gene region. Three initiation zones have been identified. *Ori-γ* exists about 22 kb downstream from *ori-β*, and *ori-α* may exist upstream of *ori-β* in a CHO cell line that contains an additional several hundred kb of amplified *DHFR* locus (262). The *DHFR* and *2BE2121* genes, matrix attachment regions (MAR), Alu sequences, yeast ARS sequences, bent DNA and binding sites for RIP60, PUR, Oct1 and AP1 proteins are indicated (see text for details). Restriction sites are *Eco*RI (E), *Xba*I (X), *Bam*HI (B) and *Hind*III (H). Methods used to map DNA replication initiation sites are described in (1) Vaughn *et al.*, 1990; Dijkwel and Hamlin, 1992; Dijkwel *et al.*, 1991; P Dijkwel, personal communication; (2) Handeli *et al.*, 1989; (3) (light shading) Heintz and Hamlin, 1982; Burhans *et al.*, 1986a; Heintz and Stillman, 1988; (medium shading) Burhans *et al.*, 1986a, b; (filled rectangle) Leu and Hamlin, 1989; (4) Anachkova and Hamlin, 1989; (5) Vassilev *et al.*, 1990; (6) Burhans *et al.*, 1990; and (7) Burhans *et al.*, 1991; Handeli *et al.*, 1989. Width of bar indicates outer limits of initiation site.

DNA crosslinks, replication bubbles were confined to a 0.5 kb DNA sequence in the region defined by the earliest labelled DNA fragments (Fig. 3, replication origin trap).

Interpretation of these studies was complicated by the fact that they required cells containing highly amplified sequences and synchronized at the beginning of S-phase by amino acid starvation and DNA synthesis inhibitors. Thus, the appearance of a specific origin of replication may have been either unique to amplified DNA sequences or an artefact of experimental conditions. Both possibilities were eliminated by a mapping technique that relates the lengths of nascent DNA strands, identified by incorporation of labelled precursors, to the site where they are initiated (Vassilev and Johnson, 1989). Nascent DNA strands are fractionated according to their length and then the smallest strands containing a unique sequence marker are identified. If these chains initiate replication bidirectionally, the origin is found at their centre. If they replicate unidirectionally, sequence markers at different genomic locations will give inconsistent results. The polymerase chain reaction (PCR) method is used to amplify the marker sequence in order to observe single-copy sequences in cells growing exponentially in the absence of metabolic inhibitors. This method revealed that initiation in the *DHFR* *ori-β* region of CHO cells begins within a 3 kb segment of the early labelled DNA fragment (Fig. 3, nascent DNA length).

This method has identified initiation sites at other genomic locations as well. Craig Chinault (Baylor Medical School, personal communication) has localized a specific initiation zone near the mouse adenosine deaminase (*ADA*) gene to a 2 kb site within a 4 kb fragment that promotes autonomous replication in plasmids carrying the EBV nuclear retention sequence (Krysan *et al.*, 1989). This initiation zone is active both as a single copy sequence at its normal chromosome location and in amplified DNA that replicates extrachromosomally. Vassilev and Johnson (1990) localized initiation events to a 2.5 kb site upstream of the human c-*myc* gene, consistent with results from a different method that measures the direction of fork movement using a nuclear run-off approach (McWhinney and Leffak, 1990). Sequences within this c-*myc* initiation zone have been reported to promote autonomous replication of plasmid DNA in human cells (Iguchi-Ariga *et al.*, 1987; McWhinney and Leffak, 1990).

6. Replication forks are initiated at a specific site within the initiation zone

Taken together, the studies described above argue strongly that most replication events in metazoan chromosomes begin at a specific site within the larger initiation zone defined by the presence of replication bubbles. In support of this hypothesis, two independent methods have revealed that DNA replication in mammalian chromosomes occurs by the replication fork mechanism and that most replication forks emanate from a specific OBR. In the replication fork mechanism (Fig. 1), an OBR is marked by a transition between continuous and discontinuous DNA synthesis on each template. These transitions have been located in two ways. First, the distribution of Okazaki fragments was determined by annealing them to separated strands of unique restriction fragments. Okazaki fragments were identified on the basis of their size (100-150 nucleotides; Burhans et al., 1990), the presence of an RNA primer (Burhans et al., 1991), their rapid joining to long nascent DNA strands, and the fact that they annealed predominantly, if not exclusively, to DNA templates representing the retrograde arms of replication forks (Burhans et al., 1990). This DNA was purified and then hybridized to cloned single-stranded DNA templates representing various unique DNA segments in the DHFR gene region. At least 80% of the replication forks within the 27 kb examined (ori-β region) emanated from an OBR located within a 0.45 kb region 17 kb downstream from the DHFR gene (Fig. 3, Okazaki fragment distribution). The same was true for the amplified DHFR gene region in CHO C400 cells. This method has also been used to identify an OBR within a ~5 kb region located ~10 kb downstream of the CAD gene promoter, and a second OBR identified within a ~7 kb region located ~30 kb upstream of the ADA gene in mouse cells (Carroll et al., 1993; G Wahl, personal communication). These two OBRs were found in both single chromosomal copies and amplified extrachromosomal copies of the genes.

An OBR can also be identified by a second method that involves enrichment of long nascent DNA chains on forward arms by eliminating DNA chains synthesized on retrograde arms. Originally, cells were treated with emetine, a general inhibitor of protein synthesis, in order to block histone synthesis in the belief that histone octamers in front of replication forks would segregate exclusively to the forward arms, leaving nascent DNA on retrograde arms unprotected and therefore sensitive to non-specific endonucleases such as micrococcal nuclease (Handeli et al., 1989). If this were true, then an OBR could be recognized by the transition from nuclease protected to nuclease sensitive nascent DNA on each template strand, analogous to measuring the transition from continuous to discontinuous DNA synthesis. However, it was later found that enrichment of long nascent DNA strands on forward arms results from imbalanced DNA synthesis due to preferential inhibition of Okazaki fragment synthesis at replication forks (Burhans et al., 1991). The same is true with cycloheximide (J Sogo, ETH Zürich, unpublished data). In the absence of protein synthesis, prefork histone octamers are distributed to both arms of replication forks (Fig. 1) in circular DNA replicating in whole cells (Burhans et al., 1987; Cusick et al., 1984; Sogo et al., 1986), in cell extracts (Krude and Knippers, 1991) or with purified mammalian replication proteins (Sugasawa et al., 1992).

When this method was applied to the DHFR gene region in CHO cells and in CHO C400 cells, replication forks downstream of the DHFR gene emanated from an OBR somewhere within a 14 kb region that encompassed the initiation sites identified in previous studies of *ori*-β (Fig. 3, imbalanced DNA synthesis). Nascent DNA on forward arms of replication forks was enriched 5 to 7-fold over nascent DNA on retrograde arms. This bias was observed on both sides of the OBR identified by the distribution of Okazaki fragments, consistent with at least 85% of replication forks within this region emanating from OBR-β. Replication forks were also observed coming from *ori*-γ and *ori*-α. The results from these two origin mapping methods clearly demonstrate that, in metazoan chromosomes, replication is initiated at specific sites using the same replication fork mechanism found in simple genomes.

7. Composition of mammalian origins of bidirectional replication

The *DHFR ori*-β region of CHO cells has several features that might facilitate origin activity (Fig. 3). *Ori*-β is flanked by two matrix attachment regions (Dijkwel and Hamlin, 1988; Foreman and Hamlin, 1989), commonly associated with newly replicated DNA (Cook, 1991), and, in some cases, origins of cellular replication (Carri *et al.*, 1986). The OBR is flanked also by two Alu repeats, sequences associated with DNA amplification (Beitel *et al.*, 1991, McArthur *et al.*, 1991) and autonomously replicating plasmid DNA (Johnson and Jelinek, 1986). A segment of bent DNA is also present as are several close matches to the yeast ARS consensus sequence that is required for ARS activity. However, mutations in the bent DNA region of yeast ARS1 that reduce plasmid replication efficiency (Snyder *et al.*, 1986), act by interfering with binding of yeast transcription factor ABF-1, not by eliminating DNA bending (Marahrens and Stillman, 1992). A report that the 3'-flanking region of the ARS1 core component can be replaced with synthetic bent DNA (Williams *et al.*, 1988) may be explained by simply replacing one DUE element with another.

Clustered within a 500 bp region adjacent to OBR-β are binding sites for at least four proteins that may function in initiation of DNA replication (Held and Heintz, 1992). RIP60 is a protein that co-purifies with a DNA helicase activity (Dailey *et al.*, 1990). PUR is a protein that binds to a specific purine-rich sequence found within a region of bent DNA present in both the human c-*myc* and hamster *DHFR ori*-β initiation zones (Bergemann and Johnson, 1992). PUR is analogous to a yeast ARS consensus binding protein (Hofmann and Gasser, 1991) in that both proteins bind specific regions of single-stranded DNA. Transcription factor OCT1 can facilitate initiation at the Ad2 replication origin (Verrijzer *et al.*, 1990), and AP1 can facilitate initiation at either the SV40 or PyV origins (Guo and DePamphilis, 1992). Whether or not any of these proteins plays a role in *ori*-β activity remains to be demonstrated.

8. How does a metazoan initiation zone function?

The preceding studies confine initiation in metazoan chromosomes to specific loci of 10 to 100 kb that may be restricted to intergenic regions, but analyses of the hamster *DHFR* gene region produced a paradox. On the one hand, analyses of DNA structures reveal replication bubbles distributed over distances as great as 55 kb with equivalent numbers of replication forks facing in both directions, consistent with a nearly random selection of initiation sites throughout a large initiation zone. On the other hand, analyses of newly synthesized DNA reveal (1) that DNA synthesis begins within a 0.5 to 3 kb site, and (2) that 80–90% of Okazaki fragments originate from only one of the DNA strands, and this strand changes polarity as it crosses the OBR (~0.4 kb), as expected from the replication fork mechanism (Fig. 1). How can these apparently contradictory results be reconciled?

Before this paradox was evident, Benbow *et al.* (1985) proposed a strand separation model (Fig. 4) in which replication begins by unwinding DNA at many

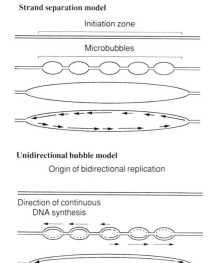

Strand separation model

Initiation zone

Microbubbles

Unidirectional bubble model
Origin of bidirectional replication

Direction of continuous
DNA synthesis

Figure 4. 'Strand separation' and 'unidirectional bubble' models for replication of metazoan chromosomes (see text for details).

sites throughout an initiation zone that varies in size from ~5 kb to perhaps 100 kb. These preferred initiation sites may be analogous to the easily unwound DNA sequences found in *S. cerevisiae* origins (Umek and Kowalski, 1990). DNA unwinding continues in the absence of DNA synthesis to generate extensive single-stranded DNA regions. RNA-primed DNA synthesis is then initiated at many sites on the exposed templates. This model accounts for the appearance of replication bubbles throughout the initiation zone, the general difficulty in finding replication forks in metazoan chromosomes by electron microscopy (Benbow *et al.*, 1985), and the appearance of extensive amounts of single-stranded DNA in

Xenopus cleavage embryos and other cells (Gaudette and Benbow, 1986). However, this model does not utilize the replication fork mechanism (Fig. 1). DNA primase–DNA polymerase-α, the enzyme responsible for initiation of DNA synthesis in chromosomes of eukaryotic cells and small nuclear viruses (Wang, 1991), initiates synthesis at many sites on single-stranded DNA templates (Vishwanatha *et al.*, 1986; Yamaguchi *et al.*, 1985). Although some pyrimidine rich sites are strongly preferred (Davey and Faust, 1990), the strand separation model would still be expected to produce a mixture of short and long RNA-primed nascent DNA chains distributed throughout both DNA templates.

Assuming that all of the experimental results are valid, there are essentially three ways to resolve the *DHFR ori-β* paradox. Linskens and Huberman (1990) proposed a unidirectional bubble model (Fig. 4) in which at least one bidirection-ally replicating bubble always forms at the OBR, while additional replication bubbles form at other randomly located sites within the initiation zone and expand unidirectionally away from the OBR. Thus, the frequency of newly synthesized DNA strands would be greatest at the site of the OBR, and the direction of fork movement in bubbles on one side of the OBR would be opposite to the direction of fork movement in bubbles on the other side of the OBR. An alternative model, suggested by the Jesuit's dictum that many are called but few are chosen, offers two other solutions (Fig. 5, Jesuit model). Either all initiation events produce

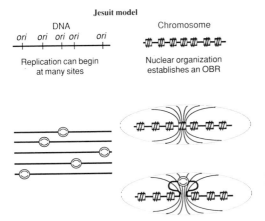

Figure 5. 'Jesuit' model for replication of metazoan chromosomes (see text for details).

productive bidirectional replication events but most of them occur at a single primary initiation site, the OBR, or else initiation events that occur outside the OBR are aborted.

Both models predict that more bubbles would appear at the OBR than at other sites in the initiation zone, although the Jesuit model predicts a greater difference than the unidirectional bubble model. However, the unidirectional bubble model is difficult to rationalize in terms of a mechanism that would establish this unique pattern of unidirectional fork movement. Moreover, it fails to account for reports that the length of nascent DNA strands increases proportionally with distance from the OBR (McWhinney and Leffak, 1990; Vassilev *et al.*, 1990; Vassilev and Johnson, 1990; C Chinault, personal communication). On the other hand, the Jesuit Model assumes that 2-D gel analysis of metazoan genomes is not sufficiently quantitative to distinguish a primary initiation site (OBR) from secondary sites, or smaller aborted bubbles from larger productive ones. This notion stems from the fact that the fraction of bubble structures observed in metazoan cells is generally low relative to studies in yeast cells, indicating that replication bubbles are more readily lost or broken into fork structures during preparation of metazoan DNA. In addition, 2-D gel patterns of metazoan genomes are generally more complex than those from simpler genomes, increasing the difficulty of interpreting DNA mobility patterns (Dijkwel *et al.*, 1991; Vassilev and DePamphilis, 1992).

While all three of these models push the bounds of credibility (*deus ex machina*) in order to provide plausible solutions, they do illuminate the two most critical questions: Do complex genomes utilize specific *cis*-acting *ori* sequences of the type described for simple genomes, and what distinguishes initiation events that mark the OBR from initiation events outside the OBR?

9. Metazoan cells can initiate replication at many DNA sites

It is clear that replication can occur at many DNA sites because virtually any DNA injected into the cytoplasm of fertilized eggs from frogs (*Xenopus laevis*, Endean and Smithies, 1989; Harland and Laskey, 1980; Marini *et al.*, 1988; Mechali and Kearsey, 1984), or added to extracts of *Xenopus* eggs (Blow and Laskey, 1986) will replicate semiconservatively, once per cell division cycle. In fact, analysis of circular replicating DNA molecules reveals that initiation occurs at one randomly chosen site in each molecule, regardless of the eukaryotic DNA sequence present (Hyrien and Méchali, 1992; Mahbubani *et al.*, 1992). The same appears to be true for eggs from sea urchin (McMahon *et al.*, 1985; McTiernan and Stambrook, 1984) and zebra fish (Stuart *et al.*, 1988). The ability to initiate replication at many sites may be unique to embryos undergoing rapid cleavage events. 2-D gel analysis reveals replication bubbles distributed throughout the tandem repeats of histone genes in early *Drosophila* embryos (Shinomiya and Ina, 1991), but replication bubbles are 5 to 10-fold more frequent in the early embryos of *Drosophila* and newt than in differentiated cells from the same animal (Blumenthal *et al.*, 1974; Callan, 1974). Therefore, while many sequences can function as origins of replication, some become preferred over others when the rate of cell division slows and cell differentiation begins.

This preference for certain DNA sequences can also reveal itself by promoting plasmid replication in mammalian cells. Unfortunately, sequences that can promote autonomous plasmid replication in metazoan cells, comparable to ARS sequences that fulfil this function in yeast, have proven difficult to identify and reports of such sequences (reviewed in Landry and Zannis-Hadjopoulos, 1991;

McWhinney and Leffak, 1990; Pearson *et al.*, 1991; Umek *et al.*, 1989) have not been confirmed (Burhans *et al.*, 1990; Gilbert and Cohen, 1989; Lüscher and Eisenman, 1990). Plasmids carrying the *DHFR ori*-β region are not preferentially replicated in transient assays (Burhans *et al.*, 1990; DePamphilis *et al.*, 1988). Perhaps the most promising approach utilizes plasmids containing the EBV EBNA1 dependent enhancer region, a sequence that retains plasmids within the nuclei of human cells through many cell divisions (Krysan *et al.*, 1989; J Yates and C Chinault, personal communication). These plasmids replicate once per S-phase (Haase and Calos, 1991), and the isolated human sequences alone can promote transient replication in some cell lines. Unfortunately, it is not clear whether replication depends primarily on DNA size or DNA sequence. Although replication is best with fragments of human DNA ≥10 kb, human DNA is at least 4 to 10-fold better than similar size fragments of bacterial DNA, and variations of 4 to 7-fold exist even within human DNA fragments of similar size (Heinzel *et al.*, 1991). Moreover, this approach has been used to isolate a replication origin near the mouse *ADA* gene that maps within a 2 kb locus (C Chinault, personal communication), and large human DNA segments can extend the host range of EBV-derived plasmids to rodent cells (J Yates, personal communication). While these results suggest that specific DNA sequences are involved, analysis of some plasmids by the 2-D gel method suggests that initiation occurs randomly throughout both human and bacterial sequences (Krysan and Calos, 1991).

The ability to initiate replication at many DNA sites is not inconsistent with the appearance of specific chromosomal initiation sites *in vivo*. Clearly, some *cis*-acting sequence information (*ori*?) must determine the site of an OBR, otherwise it would not be retained during cell division. The fact that the same OBR was found both in cells containing single copy sequences and in cells containing many copies of the sequence confirms that each copy of DNA contains whatever information is necessary to establish the same OBR. Otherwise, initiation events would occur at a new site in each DNA copy and thus preclude observing replication fork polarity (Burhans *et al.*, 1990). In fact, Handeli *et al.* (1989) reported that a 14 kb region containing OBR-β could be transplanted to other chromosomal sites and still function as an origin of replication. Therefore, either plasmid DNA replication under the conditions described above does not accurately reflect DNA replication in cellular chromosomes, or it reveals that metazoan cells are promiscuous in their choice of initiation sites, but that once a replication site is established, it remains throughout all subsequent rounds of replication. Thus, initiation of replication in plasmid DNA could appear random with respect to the whole population, but site specific with respect to the progeny of a single molecule.

10. Replication of metazoan DNA requires chromatin structure and nuclear organization

Primary initiation sites are most likely distinguished from secondary sites by differences in chromatin structure or nuclear organization (Fig. 5). *DHFR ori-β* does not promote autonomous replication in plasmids (Burhans *et al.*, 1990), suggesting that metazoan origins of DNA replication may function only in the context of a native chromosome. Chromatin structure clearly can determine origin accessibility, since nucleosome positioning affects the activity of yeast ARS in plasmids (Simpson, 1990), and both yeast and *Drosophila* origins of replication exhibit strong position effects when relocated to other chromosomal sites (Ferguson and Fangman, 1992; Karpen and Spradling, 1990). In mammals, a DNA 'locus control region' affects both chromatin structure and timing of replication (Forrester *et al.*, 1990). However, chromatin organization may simply be a prelude to nuclear organization. Initiation of DNA replication in *Xenopus* eggs occurs by a sequential pathway: DNA is first assembled into chromatin, then condensed into spheres and modified to allow binding of lamins and nuclear membrane vesicles that are finally fused to form an inner and an outer nuclear membrane containing nuclear pores (Forbes *et al.*, 1983; Newport, 1987; Sheehan *et al.*, 1988; Shiokawa *et al.*, 1986; Wilson and Newport, 1988). DNA replication in these pseudo-nuclei (Cox and Laskey, 1991), like DNA replication in normal mammalian nuclei (Fox *et al.*, 1991; Kill *et al.*, 1991), occurs at many discrete sites throughout the nucleus. Replication efficiency is increased greatly if injected DNA is preincubated in meiotically arrested eggs to allow chromatin assembly prior to activating the egg to begin DNA replication (Sanchez *et al.*, 1992; Wangh, 1989). This may be a prerequisite for nuclear envelope formation (Pfaller *et al.*, 1991). In extracts from *Xenopus* eggs, formation of nuclei must precede each round of DNA replication; the more efficiently the extract makes nuclei, the more efficiently it replicates DNA (Blow and Laskey, 1986; Blow and Sleeman, 1990; Hutchison *et al.*, 1988; Newport, 1987; Sheehan *et al.*, 1988). Lamin LIII, the only lamin in *Xenopus* embryonic nuclei, is not required for nuclear membrane assembly, but is essential for initiating DNA replication (Meier *et al.*, 1991). These results strongly suggest that initiation of cellular DNA replication requires attachment to some component of nuclear structure. The long-standing observation that newly replicated DNA is associated with nucleoskeleton may reflect association of nuclear structure with replication forks that form at the OBR (Carri *et al.*, 1986; Cook, 1991).

11. Summary and perspective

In summary, the investigation of metazoan origins of replication is akin to blind men describing an elephant. Each one describes that part he happens to touch, but no one perceives the whole animal. What seems clear is that most DNA replication events in differentiated metazoan cells are initiated predominantly at highly preferred sites (~10:1) about 0.4 to 3 kb in length that are genetically inheritable and therefore determined by some, as yet unidentified, *cis*-acting sequence (*ori*?). As with simpler systems, DNA synthesis occurs by the replication fork mechanism, and nucleosome segregation is distributive (Fig. 1). These primary initiation sites are embedded within a larger initiation zone (~8 to 55 kb) defined by the presence of replication bubbles. What remains unclear is the mechanism by which replication is initiated. Based on simpler genomes, one would expect the OBR to be coincident with an *ori* sequence, but one could infer from the *Drosophila* chorion gene locus that ORE (contained within the amplification control element) and DUE (contained with each amplification enhancer region) are widely separated. Moreover, a metazoan *ori* probably requires chromatin structure or nuclear organization in addition to DNA sequence in order to establish an OBR. Definitive answers will require genetic and biochemical manipulation of metazoan initiation zones similar to those carried out in simpler systems, but these analyses will require new methods for isolating functional origins from their natural habitat.

The possibility that origins of replication in metazoan chromosomes are more complex than simpler systems has several intriguing implications. First, the stability of OBR sites within a preformed nucleus would discourage foreign DNA from establishing itself as an episome, because foreign DNA must compete with resident DNA for initiation sites. Viruses are designed to escape this restriction. For example, plasmid DNA injected into the nuclei of mouse embryos does not replicate unless it contains a functional viral origin of replication and is provided with the viral initiation factors (Martínez-Salas *et al.*, 1988). Once viral replication is established in mammalian cells, however, some plasmid molecules can compete successfully for initiation sites in the nucleus and thus establish an episome that replicates, on average, once per S-phase (Chittenden *et al.*, 1991; Kusano *et al.*, 1987).

A second implication is that metazoan origins may differ from those of fungi and simpler organisms because nuclei are disassembled during mitosis in metazoan cells, but remain largely intact during mitosis in fungi (Heath, 1982). This could explain why yeast origin activity does not require a nuclear matrix attachment region (Amati *et al.*, 1990), whereas, in mammalian cells, establishment of extrachromosomal DNA is strongly facilitated by a nuclear retention activity such as found in the EBV genome (Haase and Calos, 1991). This nuclear retention activity is provided by sequences that lie within a strong nuclear matrix attachment region (Jankelevich *et al.*, 1992).

Finally, the concept of an origin as a sequence plus its nuclear attachment component imparts at least one advantage to metazoa. Metazoan cells could initiate replication at thousands of different origins without ever replicating the same region twice, because the act of replication would disassemble the origin structure and thus inactivate the origin. Nuclei from G1-phase cells can initiate one round of replication in *Xenopus* egg extracts, but nuclei from cells that have completed replication such as sperm (Blow and Laskey, 1988), erythrocytes (Coppock *et al.*, 1989) and G2-phase HeLa cells (Leno *et al.*, 1992) must have their nuclear envelope disrupted before a *Xenopus* egg extract can initiate one round of DNA replication. A second round of replication requires disrupting the nuclei either by mitosis (Hutchison *et al.*, 1988) or agents that permeabilize nuclear membranes (Blow and Laskey, 1988). Similar observations have been reported for extracts of *Drosophila* embryos (Crevel and Cotterill, 1991). These results are generally interpreted in terms of a licensing factor (Blow and Laskey, 1988) that establishes initiation sites in chromatin, and must be replenished from the cytoplasm after each S-phase. This bears analogy with *E. coli* where re-initiation of DNA replication is governed by association of the hemimethylated origin sequence with a cell membrane component (Malki *et al.*, 1992).

References

Amati BB, Pick L, Laroche T, Gasser S. (1990) Nuclear scaffold attachment stimulates, but is not essential for ARS activity in *Saccharomyces cerevisiae*: analysis of the *Drosophila* ftz SAR. *EMBO J.* **9**: 4007-4016.

Anachkova B, Hamlin JL. (1989) Replication in the amplified dihydrofolate reductase domain in CHO cells may initiate at two distinct sites, one of which is a repetitive sequence element. *Mol. Cell. Biol.* **9**: 532-540.

Baru M, Shlissel M, Manor H. (1991) The yeast GAL4 protein transactivates the polyoma-virus origin of DNA replication in mouse cells. *J. Virol.* **65**: 3496-3503.

Beitel LK, McArthur JG, Stanners CP. (1991) Sequence requirements for the stimulation of gene amplification by a mammalian genomic element. *Gene*, **102**: 149-156.

Bell SP, Stillman B. (1992) ATP-dependent recognition of eukaryotic origins of DNA replication by a multiprotein complex. *Nature*, **357**: 128-134.

Benard M, Pierron G. (1992) Mapping of a *Physarum* chromosomal origin of replication tightly linked to a developmentally-regulated profilin gene. *Nucl. Acids Res.* **20**: 3309-3315.

Benbow RM, Gaudette MF, Hines PJ, Shioda M. (1985) Initiation of DNA replication in eukaryotes. In: *Control of Animal Cell Proliferation*, vol. 1 (Boynton AL, Leffert HL, eds). New York: Academic Press. pp 449-483.

Bennett-Cook ER, Hassell JA. (1991) Activation of polyomavirus DNA replication by yeast GAL4 is dependent on its transcriptional activation domains. *EMBO J.* **10**: 959-969.

Bergemann AD, Johnson EM. (1992) The HeLa Pur factor binds single-stranded DNA at a specific element conserved in gene flanking regions and origins of DNA replication. *Mol. Cell. Biol.* **12**: 1257-1265.

Blow JJ, Laskey RA. (1986) Initiation of DNA replication in nuclei and purified DNA by a cell-free extract of *Xenopus* eggs. *Cell*, **47**: 577-587.

Blow JJ, Laskey RA. (1988) A role for the nuclear envelope in controlling DNA replication within the cell cycle. *Nature*, **332**: 546-548.

Blow JJ, Sleeman AM. (1990) Replication of purified DNA in *Xenopus* egg extract is dependent on nuclear assembly. *J. Cell Sci.* **95**: 383-391.

Blumenthal AB, Kriegstein HJ, Hogness DS. (1974) The units of DNA replication in *Drosophila melanogaster* chromosomes. *Cold Spring Harbor Symp. Quant. Biol.* **38**: 205-223.

Borowiec JA, Hurwitz J. (1988) Localized melting and structural changes in the SV40 origin of replication induced by T-antigen. *EMBO J.* **7**: 3149-3158.

Borowiec JA, Dean FB, Bullock PA, Hurwitz J. (1990) Binding and unwinding – how T-antigen engages the SV40 origin of DNA replication. *Cell*, **60**: 181-184.

Bozzoni I, Baldari CT, Amaldi F, Buongiorno-Nardelli M. (1981) Replication of ribosomal DNA in *Xenopus laevis*. *Eur. J. Biochem.* **118**: 585-590.

Burhans WC, Selegue JE, Heintz NH. (1986a) Isolation of the origin of replication associated with the amplified Chinese hamster dihydrofolate reductase domain. *Proc. Natl Acad. Sci. USA*, **83**: 7790-7794.

Burhans WC, Selegue JE, Heintz NH. (1986b) Replication intermediates formed during initiation of DNA synthesis in methotrexate-resistant CHO C400 cells are enriched for sequences derived from a specific, amplified restriction fragment. *Biochemistry* **25**: 441-449.

Burhans WC, Vassilev LT, Caddle MS, Heintz NH, DePamphilis ML. (1990) Identification of an origin of bidirectional DNA replication in mammalian chromosomes. *Cell*, **62**: 955 —965.

Burhans WC, Vassilev LT, Wu J, Sogo JM, Nallaseth F, DePamphilis ML. (1991) Emetine allows identification of origins of mammalian DNA replication by imbalanced DNA synthesis, not through conservative nucleosome segregation. *EMBO J.* **10**: 4351-4360.

Callan HG. (1974) DNA replication in the chromosomes of eukaryotes. *Cold Spring Harbor Symp. Quant. Biol.* **38**: 195-203.

Carri MT, Micheli G, Graziano E, Pace T, Buongiorno-Nardelli M. (1986) The relationship between chromosomal origins of replication and supercoiled loop domains in the eukaryotic genome. *Exp. Cell Res.* **164**: 426-436.

Carroll SM, DeRose ML, Nonet GH, Kelly RE, Wahl GM. (1993) Localization of a bidirectional DNA replication origin in the wild type and in episomally amplified murine adenosine deaminase loci. *Mol. Cell. Biol. in press*.

Challberg MD, Kelly TJ. (1989) Animal virus DNA replication. *Annu. Rev. Biochem.* **58**: 671-717.

Chen M, Mermod N, Horwitz M. (1990) Protein-protein interactions between adenovirus DNA polymerase and nuclear factor I mediate formation of the DNA replication preinitiation complex. *J. Biol. Chem.* **265**: 18634-18642.

Cheng L, Kelly TJ. (1989) Transcriptional activator nuclear factor I stimulates the replication of SV40 minichromosomes *in vivo* and *in vitro*. *Cell*, **59**: 541-551.

Cheng L, Workman JL, Kingston RE, Kelly TJ. (1992) Regulation of DNA replication in vitro by the transcriptional activation domain of GAL4-VP16. *Proc. Natl Acad. Sci. USA*, **89**: 589-593.

Chittenden T, Frey A, Levine AJ. (1991) Regulated replication of an episomal SV40 origin plasmid in COS7 cells. *J. Virol.* **65**: 5944-5951.

Clayton DA. (1991) Replication and transcription of vertebrate mitochondrial DNA. *Annu. Rev. Cell Biol.* **7**: 453-478.

Cook PR. (1991) The nucleoskeleton and the topology of replication. *Cell*, **66**: 627-635.

Coppock DL, Lue RA, Wangh LJ. (1989) Replication of *Xenopus* erythrocyte nuclei in a homologous egg extract requires prior proteolytic treatment. *Dev. Biol.* **131**: 102-110.

Cox LS, Laskey RA. (1991) DNA replication occurs at discrete sites in pseudonuclei assembled from purified DNA *in vitro*. *Cell*, **66**: 271-275.

Crevel G, Cotterill S. (1991) DNA replication in cell-free extracts from *Drosophila melanogaster*. *EMBO J.* **10**: 4361-4369.

Cusick ME, DePamphilis ML, Wassarman PM. (1984) Dispersive segregation of nucleosomes during replication of SV40 chromosomes, J. *Mol. Biol.* **178**: 249-271.

Cusick ME, Wassarman PM, DePamphilis ML. (1989) In: *Methods in Enzymology – Nucleosome*, vol. 170 (Wassarman PM, Kornberg RD, eds). San Diego: Academic Press. pp 290-316.

Dailey L, Caddle MS, Heintz N, Heintz NH. (1990) Purification of mammalian proteins with origin-specific DNA binding and ATP-dependent DNA helicase activities. *Mol. Cell. Biol.* **10**: 6225-6235.

Daniel DC, Johnson EM. (1989) Selective initiation of replication at origin sequences of the rDNA molecule of *Physarum polycephalum* using synchronous plasmodial extracts. *Nucl. Acids Res.* **17**: 8343-8362.

Davey SK, Faust EA. (1990) Murine DNA polymerase-α-primase initiates RNA-primed DNA synthesis preferentially upstream of a 3'-CC(C/A)-5' motif. *J. Biol. Chem.* **265**: 3611-3614.

Delidakis C, Kafatos FC. (1989) Amplification enhancers and replication origins in the autosomal chorion gene cluster of *Drosophila. EMBO J.* **8**: 891-901.

DePamphilis ML, Martínez-Salas E, Cupo DY, Hendrickson EA, Fritze CE, Folk WR, Heine U. (1988) Initiation of polyomavirus and SV40 DNA replication and the requirements for DNA replication during mammalian development. In: *Cancer Cells*, vol. 6 (Stillman B, Kelly T, eds). New York: Cold Spring Harbor Press. pp 165-175.

Diffley JFX, Cocker JH. (1992) Protein-DNA interactions at a yeast replication origin. *Nature*, **357**: 169-172.

Dijkwel PA, Hamlin JL. (1988) Matrix attachment regions are positioned near replication initiation sites, genes and an interamplicon junction in the amplified dihydrofolate reductase domain of CHO cells. *Mol. Cell. Biol.* **8**: 5398-5409.

Dijkwel PA, Hamlin JL. (1992) Replication in the *DHFR* locus initiates only during the first hour of the S-period in CHO cells synchronized with mimosine. *Mol. Cell. Biol.* **12**: 3715-3722.

Dijkwel PA, Vaughn JP, Hamlin JL. (1991) Mapping of replication initiation sites in mammalian genomes by two-dimensional gel analysis: stabilization and enrichment of replication intermediates by isolation on the nuclear matrix. *Mol. Cell. Biol.* **11**: 3850-3859.

Dornreiter I, Höss A, Arthur AK, Fanning E. (1990) SV40 T-antigen binds directly to the large subunit of purified DNA polymerase-α. *EMBO J.* **9**: 3329-3336.

Eki T, Enomoto T, Masutani C, Miyajima A, Takada R, Murakami Y, Ohno T, Hanaoka F, Ui M. (1991) Mouse DNA primase plays the principal role in determination of permissiveness for polyomavirus DNA replication. *J. Virol.* **65**: 4874-4881.

Endean DJ, Smithies O. (1989) Replication of plasmid DNA in fertilized *Xenopus* eggs is sensitive to both the topology and size of the injected template. *Chromosoma* **97**: 307-314.

Fangman WL, Brewer BJ. (1991) Activation of replication origins within yeast chromosomes. *Annu. Rev. Cell Biol.* **7**: 375-402.

Felsenfeld G. (1992) Chromatin as an essential part of the transcriptional mechanism. *Nature*, **355**: 219-224.

Ferguson BM, Fangman WL. (1992) A position effect on the time of replication origin activation in yeast. *Cell*, **68**: 333-339.

Fierer DS, Challberg MD. (1992) Purification and characterization of UL9, the HSV origin binding protein. *J. Virol.* **66**: 3986-3995.

Forbes DJ, Kirschner MW, Newport JW. (1983) Spontaneous formation of nucleus-like structures around bacteriophage DNA microinjected into *Xenopus* eggs. *Cell*, **34**: 13-23.

Foreman PK, Hamlin JL. (1989) Identification and characterization of a gene that is coamplified with dihydrofolate reductase in a methotrexate-resistant CHO cell line. *Mol. Cell. Biol.* **9**: 1137-1147.

Forrester WC, Epner E, Driscoll MC, Enver T, Brice M, Papayannopoulou T, Groudine M. (1990) A deletion of the human β-globin locus activation region causes a major alteration in chromatin structure and replication across the entire β-globin locus. *Genes Dev.* **4**: 1637-1649.

Fox MH, Arndt-Jovin DJ, Jovin TM, Baumann PH, Robert-Nicoud M. (1991) Spatial and temporal distribution of DNA replication sites localized by immunofluorescence and confocal microscopy in mouse fibroblasts. *J. Cell Sci.* **99**: 247-253.

Gahn TA, Schildkraut CL. (1989) The Epstein-Barr virus origin of plasmid replication, *oriP*, contains both the initiation and termination sites of DNA replication. *Cell*, **58**: 527-535.

Gannon JV, Lane DP. (1990) Interactions between SV40 T-antigen and DNA polymerase. *New Biol.* **2**: 84-92.

Gaudette MF, Benbow RM. (1986) Replication forks are underrepresented in chromsomal DNA of *Xenopus laevis* embryos. *Proc. Natl Acad. Sci. USA*, **83**: 5953-5957.

Gilbert D, Cohen SN. (1989) Autonomous replication in mouse cells: a correction. *Cell* **56**: 143-144.

Gruss C, Gutierrez C, Burhans WC, DePamphilis ML, Koller Th, Sogo JM. (1990) Nucleosome assembly in mammalian cell extracts before and after DNA replication. *EMBO J.* **9**: 2911-2922.

Guo Z-S, DePamphilis ML. (1992) Specific transcription factors stimulate both simian virus 40 and polyomavirus origins of DNA replication. *Mol. Cell. Biol.* **12**: 2514-2524.

Guo Z-S, Gutierrez C, Heine U, Sogo JM, DePamphilis ML. (1989) Origin auxiliary sequences can facilitate initiation of simian virus 40 DNA replication *in vitro* as they do *in vivo*. *Mol. Cell. Biol.* **9**: 3593-3602.

Guo Z-S, Heine U, DePamphilis ML. (1991) T-antigen binding to site I facilitates initiation of SV40 DNA replication but does not affect bidirectionality. *Nucl. Acids Res.* **19**: 7081-7088.

Gutierrez C, Guo ZS, Roberts JM, DePamphilis ML. (1990) Simian virus 40 origin auxiliary sequences weakly facilitate T-antigen binding, but strongly facilitate DNA unwinding. *Mol. Cell. Biol.* **10**: 1719-1728.

Haase SB, Calos MP. (1991) Replication control of autonomously replicating human sequences. *Nucl. Acids Res.* **19**: 5053-5058.

Handeli S, Klar A, Meuth M, Cedar H. (1989) Mapping replication units in animal cells. *Cell*, **57**: 909-920.

Harland RM, Laskey RA. (1980) Regulated replication of DNA microinjected into eggs of *Xenopus laevis. Cell*, **21**: 761-771.

Hay RT, DePamphilis ML. (1982) Initiation of simian virus 40 DNA replication *in vivo*: location and structure of 5'-ends of DNA synthesized in the *ori*-region. *Cell*, **28**: 767-779.

Hay RT, Hendrickson EA, DePamphilis ML. (1984) Sequence specificity for the initiation of RNA primed-SV40 DNA synthesis *in vivo*. *J. Mol. Biol.* **175**: 131-157.

Heath B. (1982) Variant mitoses in lower eukaryoties: indicators of the evolution of mitosis? *Int. Rev. Cytol.* **64**: 1-80.

Heck MMS, Spradling AC. (1990) Multiple replication origins are used during *Drosophila* chorion gene amplification. *J. Cell Biol.* **4**: 903-914.

Heintz NH, Hamlin JL. (1982) An amplified chromosomal sequence that includes the gene for dihydrofolate reductase initiates replication within specific restriction fragments. *Proc. Natl Acad. Sci. USA*, **79**: 4083-4087.

Heintz NH, Stillman B. (1988) Nuclear synthesis *in vitro* is mediated via stable replication forks assembled in a temporarily specific fashion *in vivo*. *Mol. Cell. Biol.* **8**: 1923-1931.

Heinzel SS, Krysan PJ, Tran CT, Calos MP. (1991) Autonomous DNA replication in human cells is affected by the size and source of DNA. *Mol. Cell. Biol.* **11**: 2263-2272.

Held PG, Heintz NH. (1992) Eukaryotic replication origins. *Biochim. Biophys. Acta.* **1130**: 235-246.

Hendrickson EA, Fritze CE, Folk WR, DePamphilis ML. (1987a) The origin of bidirectional DNA replication in polyoma virus. *EMBO J.* **6**: 2011-2018.

Hendrickson EA, Fritze CE, Folk WR, DePamphilis ML. (1987b) Polyoma virus DNA replication is semi-discontinuous. *Nucl. Acids Res.* **15**: 6369-6385.

Hoang AT, Weidong W, Gralla JD. (1992) Replication activation potential of selected RNA polymerase II promoter elements at the SV40 origin. *Mol. Cell. Biol.* **12**: 3087-3093.

Hofmann JFX, Gasser SM. (1991) Identification and purification of a protein that binds the yeast ARS consensus sequence. *Cell*, **64**: 951-960.

Hutchison J, Cox R, Ford CC. (1988) The control of DNA replication in a cell-free extract that recapitulates a basic cell cycle *in vitro*. *Development* **103**: 553-566.

Hyrien O, Méchali M. (1992) Plasmid replication in *Xenopus* eggs and egg extracts: A 2D gel electrophoretic analysis. *Nucl. Acids Res.* **20**: 1463-1469.

Iguchi-Ariga SMM, Itani T, Kiji Y, Ariga H. (1987) Possible function of the c-*myc* product: Promotion of cellular DNA replication. *EMBO J.* **6**: 2365-2371.

Ishimi Y. (1992) Preincubation of T antigen with DNA overcomes repression of SV40 DNA replication by nucleosome assembly. *J. Biol. Chem.* **267**: 10910-10913.

Jankelevich S, Kolman JL, Bodnar JW, Miller G. (1992) A nuclear matrix attachment region organizes the Epstein-Barr viral plasmid in Raji cells into a single DNA domain. *EMBO J.* **11**: 1165-1176.

Johnson EM, Jelinek WR. (1986) Replication of a plasmid bearing a human Alu-family repeat in monkey COS-7 cells. *Proc. Natl Acad. Sci. USA*, **83**: 4660-4664.

Karpen GH, Spradling AC. (1990) Reduced DNA polytenization of a minichromosome region undergoing position-effect variation in *Drosophila*. *Cell*, **63**: 97-107.

Kill IR, Bridges JM, Campbell KHS, Maldonado-Codina G, Hutchison CJ. (1991) The timing of the formation and usage of replicase clusters in S-phase nuclei of human diploid fibroblasts. *J. Cell Sci.* **100**: 869-876.

Kohara Y, Tohdoh N, Jiang X-W, Okazaki T. (1985) The distribution and properties of RNA primed intiation sites of DNA synthesis at the replication origin of *Escherichia coli* chromosome. *Nucl. Acids Res.* **13**: 6847-6866.

Kornberg A, Baker T. (1992) *DNA Replication.* New York: WH Freeman.

Kowalski D, Eddy MJ. (1989) The DNA unwinding element: a novel cis-acting component that facilitates opening of the *Escherichia coli* replication origin. *EMBO J.* **8**: 4335-4344.

Krude T, Knippers R. (1991) Transfer of nucleosomes from parental to replicated chromatin. *Mol. Cell. Biol.* **11**: 6257-6267.

Krysan PJ, Calos MP. (1991) Replication initiates at multiple locations on an autonomously replicating plasmid in human cells. *Mol. Cell. Biol.* **11**: 1464-1472.

Krysan P, Haase S, Calos M. (1989) Isolation of human sequences that replicate autonomously in human cells. *Mol. Cell. Biol.* **9**: 1026-1033.

Kusano T, Uehara H, Saito H, Segawa K, Oishi M. (1987) Multicopy plasmid with a structure related to the polyoma virus genome. *Proc. Natl Acad. Sci. USA*, **84**: 1789-1793.

Landry S, Zannis-Hadjopoulos M. (1991) Classes of autonomously replicating sequences are found among early-replicating monkey DNA. *Biochim. Biophys. Acta* **1088**: 234-244.

Larson DD, Blackburn EH, Yaeger PC, Orias E. (1986) Control of rDNA replication in *Tetrahymena* involves a *cis*-acting upstream repeat of a promoter element. *Cell*, **47**: 229-240.

Leno GH, Downes CS, Laskey RA. (1992) The nuclear membrane prevents replication of human G2 nuclei, but not G1 nuclei in *Xenopus* egg extract. *Cell*, **69**: 151-158.

Leu T-H, Hamlin JL. (1989) High-resolution mapping of replication fork movement through the amplified dihydrofolate reductase domain in CHO cells by in-gel renaturation analysis. *Mol. Cell. Biol.* **9**: 523-531.

Linskens MHK, Huberman JA. (1988) Organization of replication of ribosomal DNA in *Saccharomyces cerevisiae*. *Mol. Cell. Biol.* **8**: 4927-4935.

Linskens MHK, Huberman JA. (1990) The two faces of higher eukaryotic DNA replication origins. *Cell*, **62**: 845-847.

Lorimer HE. (1992) *Nucleotide dependent interactions of polyomavirus and simian virus 40 large T antigen with viral origin DNA.* Ph. D. Thesis, Columbia University, New York, NY.

Lorimer HE, Wang EH, Prives C. (1991) DNA-binding properties of polyomavirus large T antigen are altered by ATP and other nucleotides. *J. Virol.* **65**: 687-699.

Lüscher B, Eisenman RN. (1990) New light on Myc and Myb. Part I. Myc. *Genes Dev.* **4**: 2025-2035.

Mahbubani HM, Paull T, Elder JK, Blow JJ. (1992) DNA replication initates at multiple sites on plasmid DNA in *Xenopus* egg extracts. *Nucl. Acids Res.* **20**: 1457-1462.

Malki A, Kern R, Kohiyama M, Hughes P. (1992) Inhibition of DNA synthesis at the hemimethylated pBR322 origin of replication by a cell membrane fraction. *Nucl. Acids Res.* **20**: 105-109.

Marahrens Y, Stillman B. (1992) A yeast chromosomal origin of DNA replication defined by multiple functional elements. *Science*, **255**: 817-823.

Marini NJ, Etkin LD, Benbow RM. (1988) Persistence and replication of plasmid DNA microinjected into early embryos of *Xenopus laevis*. *Dev. Biol.* **127**: 421-434.

Martínez-Salas E, Cupo DY, DePamphilis ML. (1988) The need for enhancers is acquired upon formation of a diploid nucleus during early mouse development. *Genes Dev.* **2**: 1115-1126.

Martínez-Salas E, Linney E, Hassell J, DePamphilis ML. (1989) The need for enhancers in gene expression first appears during mouse development with formation of a zygotic nucleus. *Genes Dev.* **3**: 1493-1506.

McArthur JG, Beitel LK, Chamberlain JW, Stanners CP. (1991) Elements which stimulate gene amplification in mammalian cells: Role of recombinogenic sequences/structures and transcriptional activation. *Nucl. Acids Res.* **19**: 2477-2484.

McMahon AP, Flytzanis CN, Hough-Evans BR, Katula KS, Britten RJ, Davidson EH. (1985) Introduction of cloned DNA into sea urchin egg cytoplasm: replication and persistence during embryogenesis. *Dev. Biol.* **108**: 420-430.

McTiernan CF, Stambrook PJ. (1984) Initiation of SV40 DNA replication after micro-injection into *Xenopus* eggs. *Biochim. Biophys. Acta* **782**: 295-303.

McWhinney C, Leffak M. (1990) Autonomous replication of a DNA fragment containing the chromosomal replication origin of the human c-*myc* gene. *Nucl. Acids Res.* **18**: 1233-1242.

Méchali M, Kearsey S. (1984) Lack of specific sequence requirement for DNA replication in *Xenopus* eggs compared with high sequence specificity in yeast. *Cell*, **38**: 55-64.

Meier J, Campbell KHS, Ford CC, Stick R, Hutchison CJ. (1991) The role of lamin LIII in nuclear assembly and DNA replication in cell-free extracts of *Xenopus* eggs. *J. Cell Sci.* **98**: 271-279.

Micheli G, Baldari CT, Carri MT, Cello GD, Buongiorno-Nardelli M. (1982) An electron microscope study of chromosomal DNA replication in different eukaryotic systems. *Exp. Cell Res.* **137**: 127-140.

Middleton T, Sugden B. (1992) EBNA1 can link the enhancer element to the initiator element of the Epstein-Barr virus plasmid origin of DNA replication. *J. Virol.* **66**: 489-495.

Mul M, van der Vliet PC. (1992) Nuclear factor 1 enhances adenovirus DNA replication by increasing the stability of a preinitiation complex. *EMBO J.* **11**: 751-760.

Murakami Y, Satake M, Yamaguchi-Iwai Y, Sakai M, Muramatsu M, Ito Y. (1991) The nuclear proto-oncogenes c-*fos* and c-*jun* as regulators of DNA replication. *Proc. Natl Acad. Sci. USA*, **88**: 3947-3951.

Natale DA, Schubert AE, Kowalski D. (1992) DNA helical stability accounts for mutational defects in a yeast replication origin. *Proc. Natl Acad. Sci. USA*, **89**: 2654-2658.

Newport J. (1987) Nuclear reconstitution *in vitro*: stages of assembly around protein-free DNA. *Cell*, **48**: 205-217.

Nilsson M, Forsberg M, You Z, Westin G, Magnusson G. (1991) Enhancer effect of BPV E2 protein in replication of polyomavirus DNA, *Nucl. Acids Res.* **19**: 7061-7065.

Orr-Weaver TL. (1991) *Drosophila* chorion genes: Cracking the eggshell's secrets. *BioEssays*, **13**: 97-105.

Pearson CE, Frappier L, Zannis-Hadjopoulos M. (1991) Plasmids bearing mammalian DNA-replication origin-enriched (ors) fragments initiate semiconservative replication in a cell-free system. *Biochim. Biophys. Acta* **1090**: 156-166.

Pfaller R, Smythe C, Newport JW. (1991) Assembly/disassembly of the nuclear envelope membrane: cell cycle-dependent binding of nuclear membrane vesicles to chromatin *in vitro*. *Cell*, **65**: 209-217.

Prives C, Murakami Y, Kern FJ, Folk W, Basilico C, Hurwitz J. (1987) DNA sequence requirements for replication of polyomavirus DNA *in vivo* and *in vitro*. *Mol. Cell. Biol.* **7**: 3694-3704.

Sanchez JA, Marek D, Wangh LJ. (1992) The efficiency and timing of plasmid DNA replication in *Xenopus* eggs: correlations to the extent of prior chromatin assembly. *J. Cell Science* **103**: 907-918.

Scheller A, Prives C. (1985) Simian virus 40 and polyomavirus large tumor antigens have different requirements for high-affinity sequence-specific DNA binding. *J. Virol.* **54**: 532-545.

Seki M, Enomoto T, Eki T, Miyahima A, Murakami Y, Hanoaka F, Ui M. (1990) DNA helicase and nucleoside-5′-triphosphatase activities of polyoma virus large tumor antigen. *Biochemistry*, **29**: 1003-1009.

Sheehan MA, Mills AD, Sleeman AM, Laskey RA, Blow JJ. (1988) Steps in the assembly of replication-competent nuclei in a cell-free system from *Xenopus* eggs. *J. Cell Biol.* **106**: 1-12.

Shinomiya T, Ina S. (1991) Analysis of chromosomal replicons in early embryos of *Drosophila melanogaster* by two-dimensional gel electrophoresis. *Nucl. Acids Res.* **19**: 3935-3941.

Shiokawa K, Sameshima M, Tashiro K, Miura T, Nakakura N, Yamana K. (1986) Formation of nucleus-like structures in the cytoplasm of lambda injected fertilized eggs and its partition into blastomeres during early embryogenesis in *Xenopus laevis*. *Dev. Biol.* **116**: 539-542.

Simpson RT. (1990) Nucleosome positioning can affect the function of a cis-acting DNA element *in vivo*. *Nature*, **343**: 387-389.

Snyder M, Buchman AR, Davis RW. (1986) Bent DNA at a yeast autonomously replicating sequence. *Nature*, **324**: 87-89.

Sogo JM, Stahl H, Koller Th, Knippers R. (1986) Structure of replicating SV40 minichromosomes: the replication fork, core histone segregation and terminal structures. *J. Mol. Biol.* **189**: 189-204.

Stillman B. (1989) Initiation of eukaryotic DNA replication *in vitro*. *Annu. Rev. Cell Biol.* **5**: 197-245.

Stuart GW, McMurray JV, Westerfield M. (1988) Replication, integration and stable germ-line transmission of foreign sequences injected into early zebrafish embryos. *Development* **103**: 403-412.

Sugasawa K, Ishimi Y, Eki T, Hurwitz J, Kikuchi A, Hanaoka F. (1992) Nonconservative segregation of parental nucleosomes during SV40 chromosome replication *in vitro*. *Proc. Natl Acad. Sci. USA*, **89**: 1055-1059.

Taylor ICA, Workman JL, Schuetz TJ, Kingston RE. (1991) Facilitated binding of GAL4 and heat shock factor to nucleosomal templates: differential function of DNA binding domains. *Genes Dev.* **5**: 1285-1298.

Temperley SM, Hay RT. (1991) Replication of adenovirus type 4 DNA by a purified fraction from infected cells. *Nucl. Acids Res.* **19**: 3243-3249.

Temperley SM, Burrow CR, Kelly TJ, Hay RT. (1991) Identification of two distinct regions within the adenovirus minimal origin of replication that are required for adenovirus type 4 DNA replication *in vitro*. *J. Virol.* **65**: 5037-5044.

Tsurimoto T, Melendy T, Stillman B. (1990) Sequential initiation of lagging and leading strand synthesis by two different polymerase complexes at the SV40 DNA replication origin. *Nature*, **346**: 534-539.

Umek RM, Kowalski D. (1988) The ease of DNA unwinding as a determinant of initiation of yeast replication origins. *Cell*, **52**: 559-567.

Umek RM, Kowalski D. (1990) Thermal energy suppresses mutational defects in DNA unwinding at a yeast replication origin. *Proc. Natl Acad. Sci. USA*, **87**: 2486-2490.

Umek RM, Maarten HK, Kowalski D, Huberman JA. (1989) New beginnings in studies of eukaryotic DNA replication origins. *Biochim. Biophys. Acta* **1007**: 1-14.

Ustav M, Ustav E, Szymanski P, Stenlund A. (1991) Identification of the origin of replication of bovine papillomavirus and characterization of the viral origin recognition factor E1. *EMBO J.* **10**: 4321-4329.

Vassilev L, Johnson EM. (1989) Mapping initiation sites of DNA replication *in vivo* using polymerase chain reaction amplification of nascent strand segments. *Nucl. Acids Res.* **17**: 7693-7705.

Vassilev LT, Burhans WC, DePamphilis ML. (1990) Mapping an origin of DNA replication at a single copy locus in exponentially proliferating mammalian cells. *Mol. Cell. Biol.* **10**: 4685-4689.

Vassilev L, Johnson EM. (1990) An initiation zone of chromosomal DNA replication located upstream of the c-*myc* gene in proliferating HeLa cells. *Mol. Cell. Biol.* **10**: 4899-4904.

Vassilev LT, DePamphilis ML. (1992) Guide to identification of origins of DNA replication in eukaryotic cell chromosomes. *Crit. Rev. Biochem. Mol. Biol.* **27**: 445-472.

Vaughn JP, Dijkwel PA, Hamlin JL. (1990) Initiation of DNA replication occurs in a broad zone in the *DHFR* gene locus. *Cell*, **61**: 1075-1087.

Verrijzer CP, Kal AJ, der Vliet PCV. (1990) The DNA binding domain (POU domain) of transcription factor OCT-1 suffices for stimulation of DNA replication. *EMBO J.* **9**: 1883-1889.

Vishwanatha JK, Yamaguchi M, DePamphilis ML, Baril EF. (1986) Selection of template initiation sites and lengths of RNA primers synthesized by DNA primase are strongly affected by its organization in a multiprotein DNA polymerase-α complex. *Nucl. Acids Res.* **14**: 7305-7323.

Wang TS-F. (1991) Eukaryotic DNA polymerases. *Annu. Rev. Biochem.* **60**: 513-552.

Wangh LJ. (1989) Injection of *Xenopus* eggs before activation, achieved by control of extracellular factors, improves plasmid DNA replication after activation. *J. Cell Sci.* **93**: 1-8.

Wasylyk C, Schneikert J, Wasylyk B. (1990) Oncogene v-*jun* modulates DNA replication. *Oncogene*, **5**: 1055-1058.

Weichbraun I, Haider G, Wintersberger E. (1989) Optimal replication of plasmids carrying polyomavirus origin regions requires two high-affinity binding sites for large T antigen. *J. Virol.* **63**: 961-964.

Wiekowski M, Miranda M, DePamphilis ML. (1991) Regulation of gene expression in preimplantation mouse embryos: Effects of zygotic gene expression and the first mitosis on promoter and enhancer activities. *Dev. Biol.* **147**: 403-414.

Williams JS, Eckdahl TT, Anderson JN. (1988) Bent DNA functions as a replication enhancer in *Saccharomyces cerevisiae*. *Mol. Cell. Biol.* **8**: 2763-2769.

Wilson KL, Newport J. (1988) A trypsin-sensitive receptor on membrane vesicles is required for nuclear envelope formation *in vitro*. *J. Cell Biol.* **107**: 57-68.

Yamaguchi M, Hendrickson EA, DePamphilis ML. (1985) DNA primase-DNA polymerase-α from simian cells: sequence specificity of initiation sites on simian virus 40 DNA. *Mol. Cell. Biol.* **5**: 1170-1183.

Yang L, Botchan M. (1990) Replication of bovine papillomavirus type 1 DNA initiates within an E2 responsive enhancer element. *J. Virol.* **64**: 5903-5911.

Yang L, Li R, Mohr IJ, Clark R, Botchan MR. (1991) Activation of BPV-1 replication *in vitro* by the transcription factor E2. *Nature*, **353**: 628-632.

Yao MC, Yao CH. (1989) Accurate processing and amplification of cloned germ line copies of ribosomal DNA injected into developing nuclei of *Tetrahymena thermophila*. *Mol. Cell. Biol.* **9**: 1092-1099.

Yoda K-Y, Yasuda H, Jiang X-W, Okazaki T. (1988) RNA-primed initiation sites of DNA replication in the origin region of bacteriophage λ genome. *Nucl. Acids Res.* **16**: 6531-6546.

Zhu J, Newlon CS, Huberman JA. (1992) Localization of a DNA replication origin and termination zone on chromosome III of *Saccharomyces cerevisiae*. *Mol. Cell. Biol.* **12**: 4733-4741.

Zhu J, Rice PW, Chamberlain M, Cole CN. (1991) Mapping the transcriptional transactivation function of SV40 large T-antigen. *J. Virol.* **65**: 2778-2790.

Chapter 7

Substructure of telomere repeat arrays

Eric J Richards, Aurawan Vongs, Michael Walsh, Jin Yang and Steve Chao

1. Telomere functions

At least three different roles have been assigned to the specialized structures at chromosome ends, termed telomeres. A large store of cytological evidence, some dating back to the late nineteenth century, indicates that chromosome ends frequently associate with the nuclear envelope, as well as with each other (Agard and Sedat, 1983; Fussell, 1984; Gruenbaum *et al.*, 1984). It is believed that these interactions play a role in organizing the chromatin within the nucleus and facilitating chromosome pairing initiation in meiosis, although very little is known about the functional consequences or specificity of these interactions. Another role of telomeres was first recognized in the 1930s through the work of Muller (1938) and McClintock (1938, 1939, 1941, 1942). These studies demonstrated that telomeres serve as inert chromosomal caps preventing chromosomal re-arrangements (e.g. translocations, dicentric formation) stemming from the reactive, recombinogenic behaviour of broken chromosome ends (Orr-Weaver *et al.*, 1981). Later, in the 1970s, Olovnikov (1973) and Watson (1972) pointed out that the linear DNA molecules could not be replicated by conventional template-dependent DNA polymerases without becoming slowly whittled away. Telomeres must therefore be replicated by a special mechanism. Work over the past few years has led to the discovery and characterization of a ribonucleoprotein complex, termed telomerase, which can synthesize DNA sequences onto chromosome ends using an internal RNA template (Greider and Blackburn, 1985, 1989; Shippen-Lentz and Blackburn, 1990; Yu *et al.*, 1990). Although telomerase activity has been demonstrated in relatively few organisms to date, including ciliates (Shippen-Lentz and Blackburn, 1989; Zahler and Prescott, 1988) and humans (Morin, 1989), it is widely assumed that telomerase is present in many, if not all eukaryotes. Support for this assumption stems, in part, from the striking conservation of telomere DNA structure.

2. Telomeric DNA structure

Telomeric DNA has been characterized from a wide variety of eukaryotic organisms over the past 15 years (reviewed in Blackburn, 1984; Zakian, 1989). As illustrated in Table 1, telomeric DNA from these organisms conforms to several design features. Most notably, telomeric DNA is composed of tandem arrays of short repeat motifs. The small size of the repeats, generally seven or fewer base pairs, may be a consequence of the limited coding capacity of the relatively short (<200 nucleotides) telomerase RNA (Romero and Blackburn, 1991), but other

Table 1. Examples of telomeric DNA structures. The structural features of telomeric DNA isolated from a variety of organisms are shown and telomere repeat arrays with substructure are noted. Abbreviations: 1° = primary; TRA = telomere repeat array; nd = not determined. References: A–E = (Allshire et al., 1989; Brown et al., 1990; Cheng et al., 1991; Cross et al., 1990; de-Lange et al., 1990); F = (Wilkie et al., 1990); G = (Muller et al., 1991); H and I = (Richards et al., 1992; Richards and Ausubel, 1988); J = (Petracek et al., 1990); K = (Shampay et al., 1984); L and M = (Klobutcher et al., 1981; Pluta et al., 1982); N = (Dawson and Herrick, 1984); O = (Forney and Blackburn, 1988); P = (Blackburn and Gall, 1978); Q = (van der Ploeg et al., 1984).

Organism	Source	1° Repeat(s)	TRA size	Substructure	References
Mammals					
Homo sapiens	Chromosome ends	TTAGGG	~5–15 kb	Yes	A–E
	Healed Chr 16	TTAGGG		No	F
Invertebrates					
Ascaris lumbricoides	Healed somatic telomeres	TTAGGC	2–4 kb	No	G
Plants					
Arabidopsis thaliana	Chromosome ends	TTTAGGG	3–4 kb	Yes	H, I
Algae					
Chlamydomonas reinhardtii	Chromosome ends	TTTTAGGG	0.3–0.35 kb	Yes	J
Fungi					
Saccharomyces cerevisiae	Chromosome ends	TG$_{1-3}$	~0.5 kb		K
Protozoa					
Oxytricha	Macronuclear telomeres	TTTTGGGG	20 bp + 16 nt 3' tail	No	L, M
	Micronuclear telomeres	TTTTGGGG	3–6 kb		N
Paramecium	Chromosome ends	TTT(T/G)GGG	nd		O
Tetrahymena	Macronuclear telomeres	TTGGGG	0.12–0.14 kb	No	P
Trypanosoma	Chromosome ends	TTAGGG	<5.5 kb	Yes	Q

functional constraints may play a role. Although the primary nucleotide sequence of the motifs is not strictly conserved, all contain a short run of G-residues (Blackburn, 1984). In many organisms, the number of repeats at each chromosome end is not fixed, leading to size heterogeneity of the terminal restriction fragments. The orientation of the repeats relative to the chromosome ends is conserved, with the G-rich strand oriented 5'->3' towards the terminus. The structure of the extreme terminus has been difficult to study in many organisms because of the low abundance of telomeric ends. However, the terminus has been characterized in some organisms with small chromosomes and is found to be a short 3' extension (Henderson and Blackburn, 1989; Klobutcher et al., 1981) of the G-rich strand. The 3' overhang can form high-order structures (Henderson et al., 1987; Kang et al., 1992; Sundquist and Klug, 1989; Williamson et al., 1989) through non-Watson–Crick G–G base pairing, but the structure of the extension in vivo is not known.

It should be pointed out that some exceptional telomeric DNA structures have been described which do not conform to the pattern just discussed. For example, no evidence has been found for short tandem repeats at the ends of Drosophila chromosomes. Stabilization of Drosophila chromosome ends can be achieved, at least partially, by a complex set of retroposon-like elements which are found exclusively in heterochromatic regions (Beissmann et al., 1990, 1992). Another example is the telomeres of linear mitochondrial genomes in Tetrahymena which are composed of multiple tandem units of relatively large repeats (31–53 bp) that show no simple-sequence characteristics (Morin and Cech, 1986, 1988). Similarly, the genome of the pathogenic yeast, Candida albicans, contains tandem arrays of a 23 bp repeat located extremely close to the chromosomal termini (Sadhu et al., 1991). Keeping in mind that rare alternative telomere structures exist, we return to a closer consideration of the telomeric repeat arrays found at the ends of most eukaryotic chromosomes.

3. Telomere repeat array substructure

Although the general architecture of telomeric repeats is well conserved throughout the eukaryotic world, the sequence of the repeats can vary within a single telomere. The presence of variant repeats within the telomeric repeat array has been documented by a number of laboratories. Telomere repeat arrays can be categorized into three groups based on the distribution pattern of variants. The first group is composed of arrays of repeats uniform in sequence, with no or extremely few repeat variants. This group includes many of the characterized telomeres, especially recently created chromosome ends such as healed telomeres (i.e. newly acquired at previously non-telomeric sites) and ciliate macronuclear telomeres which are synthesized after each sexual cycle. The second group is formed by telomeres containing telomeric repeats of variable sequence which are interspersed and distributed throughout the array (e.g. *Saccharomyces cerevisiae*, (TG1-3)$_n$ (Shampay *et al.*, 1984); *Paramecium*, (TT(T/G)GGG)$_n$ (Forney and Blackburn, 1988). The origin of the variants repeats is unknown but recent evidence suggests that relaxed template reading or slippage during replication by *Tetrahymena* telomerase can lead to the synthesis of repeats of varying length. This suggestion stems from the results of Yu and Blackburn (Yu and Blackburn, 1991) demonstrating that *Tetrahymena* repeats containing variable numbers of G-residues could be formed directly by the action of a telomerase carrying a mutant template RNA. Alternatively, these authors suggested that variant repeats of uniform length may reflect the action of different telomerase species carrying RNAs with varying template domains. The third class of telomeres is characterized by terminal arrays which exhibit a non-random distribution of variant repeats. A higher concentration of variant repeats at the centromere-proximal edge of the array is the most common pattern observed. This pattern, first described by van der Ploeg *et al.* (1984) for *Trypanosoma* telomeres and Allshire *et al.* (1989) for human telomeres, has been noted subsequently by other investigators working on a variety of organisms (see Table 1). In some cases, the telomere repeat substructure is more complicated than the simple gradient just described. For example, the array can be interrupted by non-telomeric sequences or variant repeats can be clustered to created a mosaic of variant and uniform repeats.

Examples of the third category of telomeres are illustrated in Fig. 1, taken from our work on telomeric DNA structure in the flowering plant, *Arabidopsis thaliana*. Shown are sequences of telomeric repeat arrays carried on three clones, YpAtT1, 5 and 7, isolated by selecting for plant sequences which function to seed telomere formation in the yeast, *S. cerevisiae* (Richards *et al.*, 1992). At least three subdomains can be distinguished in each cloned array: (1) regions composed of long arrays of the previously described *A. thaliana* telomeric repeat, TTTAGGG (uniform repeat region), (2) regions containing repeats of uniform size (7 bp) which contain 1–2 base substitutions (variant repeats), and (3) domains populated largely by related repeats of varying size and sequence (degenerate repeats). There is a spatial order to these subdomains. The most distal regions correspond to uniform repeat domains, flanked by more proximal variant repeat regions which eventually give way to degenerate repeats at the most centromere-proximal edge of the array. This gradient is interrupted in two of the clones yielding a mosaic; YpAtT7 contains a variant repeat block at the most distal end of the cloned insert and the variant repeat domain of YpAtT5 is disrupted by a stretch of ~100 bp of non-telomeric sequence.

As illustrated in Fig. 1, the cloned arrays are not complete due to processing of the native 3–4 kb *A. thaliana* telomere repeat arrays in the yeast host. Similar processing has been noted in all cases where heterologous telomeres are propa-gated in yeast (e.g. Brown, 1989). The question then arises as to whether the substructure seen in Fig. 1 is an artefact or an accurate representation of the native

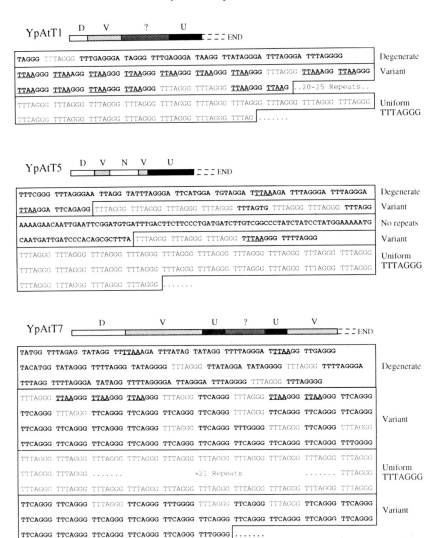

Figure 1. Structure and DNA sequence of three *A. thaliana* telomeric clones. The isolation of the clones has been previously described (Richards *et al.*, 1992) and the primary DNA sequences of the YpAtT1 and YpAtT7 telomere repeat arrays were included in that report. The additional DNA sequence from the YpAtT5 telomere repeat array was determined by standard methods. Subdomains within the sequences are boxed and labelled at right. Maps above the corresponding sequence diagram the size and arrangement of the subdomains (shaded boxes: D = degenerate repeats; V = variant repeats; N = no telomere repeats; ? = sequencing gaps; U = uniform TTTAGGG repeats). Sequences are oriented 5′ to 3′ towards the terminus, and the incomplete nature of the arrays is indicated by the dashed lines extending to the 'end'. *Mse*I (TTAA) sites are underlined and variant repeats are shown in bold.

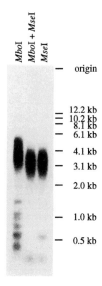

Genomic DNA: Columbia

Probe: (TTTAGGG)$_4$

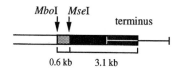

Figure 2. *Mse*I sites are restricted to the centromere-proximal edge of the telomere repeat arrays. A Southern blot of genomic DNA from the *A. thaliana* strain, Columbia, digested with either *Mbo*I, *Mse*I, or both enzymes simultaneously, was hybridized with a telomeric repeat probe. The dispersed bands running in the 2–5 kb range represent telomere repeat arrays, while the more discrete smaller molecular weight signals correspond to cross-hybridizing sequences located at internal chromosome sites (Richards *et al.*, 1991). *Mbo*I cuts at GATC sites within the telomere-associated sequences, trimming the telomere repeat arrays to approximated 3.7 ± 1.2 kb. *Mse*I further trims the arrays to fragments centring on ~3.1 kb in length. The map at the bottom of the figure shows the positions of the most distal *Mse*I and *Mbo*I sites relative to the terminus based on the behaviour of the entire population of telomere repeat arrays in Columbia. Similar results were seen using Landsberg genomic DNA although the telomere repeat arrays in this strain are longer than Columbia arrays (Richards *et al.*, 1992) making small size differences more difficult to detect.

telomeric array. To address this concern we determined the position of *Mse*I sites, (TTAA; underlined in Fig. 1) which mark many of the variant and degenerate repeats, relative to chromosomal termini. If the variant and degenerate domains were scattered throughout the telomere repeat arrays, *Mse*I should digest the arrays to very small fragments. Alternatively, if the variant and degenerate regions were clustered at the centromere-proximal edge of the arrays, consistent with the structure of the cloned inserts, we should find *Mse*I sites restricted to the proximal edge. The results presented in Fig. 2 show that *Mse*I digestion reduces the size of the telomeric repeat arrays by ~600 bp, suggesting that the variant and degenerate repeats are largely restricted to the most proximal 600 bp of the array.

4. Origin of telomere repeat array substructure

How did telomere repeat array substructure arise? It may be significant that the characterized telomeres which display substructure were isolated from telomeres which are relatively 'old', that is chromosomal termini not telomeres created relatively recently from broken or non-telomeric ends. The increasing divergence in repeat sequence at the proximal boundary of the array has been postulated to arise from the accumulation of mutations or telomerase-directed synthesis of variant repeats coupled with an increase in turnover or repair events operating in the distal portions of the array (Allshire *et al.*, 1989). Erasure of repeat mutations by terminal deletion and resynthesis of uniform repeats by telomerase is one probable mechanism and would take place as part of the dynamic length equilibrium occurring at many telomeres. The turnover process may also involve recombination (sister chromatid or interchromosomal exchange, gene conversion) between telomere repeat arrays (Pluta and Zakian, 1989). Similar mechanisms could explain the presence of variant repeat clusters and the patchwork pattern found in some telomere repeat arrays. Formation and local spread of a particular variant, such as the prevalent TTCAGGG motif of YpAtT7, may arise from replication slippage and/or recombination (unequal exchange, gene conversion) (Smith, 1976). Unequal exchange between chromatids, or chromosomes, placing variant regions distal to uniform repeats could also explain the origin of the mosaic patterns, such as that seen in the YpAtT7 telomere repeat array.

5. Are variant repeats part of the telomere?

Should the variants be considered telomeric repeats or simply another type of telomere-associated sequence (TAS)? The TASs, defined as the sequences which lie adjacent to the telomere repeat arrays (Blackburn and Szostak, 1984), are not conserved nor absolutely required for telomere function. Two observations lead us to consider the variant repeats as part of the telomere proper. First, the variant repeat block in YpAtT7 is found at the distal end of the clone and acquired a $(TG1-3)_n$ yeast telomere cap necessary for stable propagation of heterologous telomeres in the yeast host. The variant repeats must have been recognized as a telomere by the yeast telomere replication machinery, and would likely serve as a telomere healing substrate in plant cells as well. More importantly, the genomic telomeric repeat array corresponding to YpAtT1 and YpAtT7 are equal in length (~3.5 kb in strain Landsberg) despite the substantially larger block of variant repeats present in YpAtT7 which extends at least 1 kb towards the chromosomal terminus (Richards *et al.*, 1992). This suggests that the mechanism which measures and regulates telomere length in the plant cell must also recognize the variant repeats as telomeric. We are currently trying to determine how close to the end of the terminal array the variant repeat region extends in the native YpAtT7 telomere.

Not all telomeric sequence variants are functional. Blackburn and co-workers showed that variant repeats synthesized from an altered telomerase template disrupt *Tetrahymena* telomere function and length regulation when present at the distal end of the telomere repeat array (Yu *et al.*, 1990). On the other hand, variant repeats sequestered in proximal regions do not affect *Tetrahymena* telomeres (Yu and Blackburn, 1991). This situation may model the arrangement of variant repeats found at many native telomeres. It should be kept in mind that the naturally occurring variant repeats, such as those described in *A. thaliana*, would be selected against if they interfered with telomere structure and function. Functional constraints may explain why the most common *A. thaliana* telomere repeat variants change only the third or fourth nucleotide of the repeat: TT(C/T/A)(A/G)GGG.

6. Are all telomeres in a cell the same?

While variation in subterminal telomere-associated sequences has been described in many organisms, the structure of the terminal telomere repeat arrays at different chromosome ends within a cell is generally thought to be uniform. However, evidence from *in situ* hybridization experiments suggest that there may be some differences in telomere length between chromosomes in a single cell (Narayanswami *et al.*, 1992). In addition, it is clear from Fig. 1 that the substructure of telomere repeat arrays at different chromosome ends in *A. thaliana* can be quite different.

Not only can the size and arrangements of variant and degenerate repeat regions differ between chromosomal termini, the sequence of the variant repeats can be chromosome-specific. As demonstrated in Fig. 3, an oligonucleotide probe that specifically recognizes blocks of $(TTCAGGG)_n$ detects only a subset of telomeres which hybridize to the primary telomere repeat probe, $(TTTAGGG)_4$, in the *A. thaliana* strain, Landsberg. The number of Landsberg telomeres represented in this subset is not known, but probably represent only one to three of *A. thaliana*'s ten telomeres ($n=5$). The YpAtT7 clone, which contains large numbers of TTCAGGG variants (see Fig. 1), was derived from Landsberg genomic DNA and corresponds to one of the telomeres in the subset. The TTCAGGG repeat

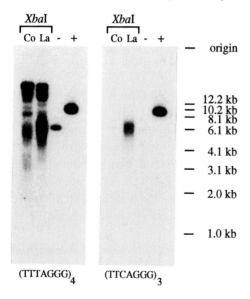

Figure 3. TTCAGGG is a strain-specific and chromosome-specific telomere repeat variant. Duplicate Southern blots were hybridized with end-labelled oligonucleotide probes corresponding to either the primary *A. thaliana* telomeric repeat, TTTAGGG (left panel), or the telomeric repeat variant, TTCAGGG (right panel). The lanes contain: *XbaI*-digested *A. thaliana* genomic DNA from the strain Columbia (Co) or Landsberg erecta (La), or two control DNA samples: pAtT32 corresponding to the YpAtT1 insert which lacks TTCAGGG repeats (–) and pAtT80 corresponding to the YpAtT7 insert which contains numerous TTCAGGG variants (+) (see Fig. 1). The filters were washed at high stringency (0.6 × SSC, 0.2% SDS @50°C) to eliminate cross-hybridization of the (TTCAGGG)$_3$ probe to other repeats, as shown by the lack of hybridization to the cloned DNA lacking TTCAGGG repeats in the negative control lane.

variant also exhibits strain specificity; the (TTCAGGG)$_3$ variant probe does not hybridize to any telomere in the *A. thaliana* strain, Columbia.

7. Implications and future directions

The demonstration that strain- and chromosome-specific telomere repeat variants exist has a few interesting implications. The TTCAGGG variant may have been amplified in the Landsberg strain since its divergence from Columbia, or Columbia telomeres may have lost the TTCAGGG variant arrays. In either case, the rate at which variant repeat domains within telomeres can change must be relatively rapid. The presence of chromosome-specific variants also suggests that different telomeres are genetically isolated from each other, since frequent inter-telomere exchange would homogenize telomere repeat arrays on different chromosomes. Perhaps exchange does occur but is limited to more distal regions of the terminal arrays which would not involve the variant repeats. Nonetheless, the YpAtT7 TAS contains a repetitive element present at several *A. thaliana* telomeres (Richards *et al.*, 1992) and exchange mediated by these sequences might be expected. It will be of interest to use the Landsberg TTCAGGG variants as a chromosome-specific marker to follow the genetic behaviour of telomere repeat arrays and to monitor the stability of the variant repeat array.

In regards to technical matters, it is clear that study of telomere repeat structure is limited by the inability to isolate large telomere repeat arrays intact. The use of recombination-deficient *rad S. cerevisiae* hosts may help stabilize the arrays (Neil *et al.*, 1990). Alternatively, with the improvement in isolation and manipulation of large DNA molecules, it may be possible to directly characterize intact chromosome termini (e.g. by end-labelling and DNA sequencing) from larger chromosomes.

Knowledge of telomere DNA structure has been fundamental to our current understanding of telomere function. While considerable focus in the telomere field is now placed on characterizing telomerase and protein components which interact with the telomeric repeats, there is still more to learn regarding the structure and behaviour of telomeric DNA. As discussed here, telomeres provide a good system for study of the dynamics of genome fluidity and the evolution of an essential repetitive DNA family.

Acknowledgements

We thank the members of the Delbrück plant group for their support and critical input. Special thanks to Carol Greider and Lea Harrington for comments on the manuscript. This work was supported in part by a National Science Foundation grant (DMB-8905589), and a Public Health Service grant from the National Institues of Health (GM43518).

References

Agard DA, Sedat JW. (1983) Three-dimensional architecture of a polytene nucleus. *Nature*, **302**: 676-681.

Allshire RC, Dempster M, Hastie ND. (1989) Human telomeres contain at least three types of G-rich repeats distributed non-randomly. *Nucl. Acids Res.* **17**: 4611-4627.

Biessmann H, Mason JM, Ferry K, d'Hulst M, Valgeirsdottir K, Traverse KL, Pardue M-L. (1990) Addition of telomere associated HeT DNA sequences "heals" broken chromosome ends in *Drosophila. Cell*, **61**: 663-673.

Beissmann H, Valgeirsdottir K, Lofsky A, Chin C, Ginther B, Levis RW, Pardue M-L. (1992) HeT-A, a transposable element specifically involved in "healing" broken chromosome ends in *Drosophila melanogaster. Mol. Cell. Biol.* **12**: 3910-3918.

Blackburn EH. (1984) Telomeres: do the ends justify the means? *Cell* **37**: 7-8.

Blackburn EH, Gall JG. (1978) A tandemly repeated sequence at the termini of the extrachromosomal ribosomal RNA genes in *Tetrahymena. J. Mol. Biol.* **120**: 33-53.

Blackburn EH, Szostak JW. (1984) The molecular structure of centromeres and telomeres. *Annu. Rev. Biochem.* **53**: 163-194.

Brown WRA. (1989) Molecular cloning of human telomeres in yeast. *Nature*, **338**: 774-776.

Brown WRA, MacKinnon PJ, Villasante A, Spurr N, Buckle VJ, Dobson MJ. (1990) Structure and polymorphism of human telomere-associated DNA. *Cell*, **63**: 119-132.

Cheng J-F, Smith CL, Cantor CR. (1991) Structural and transcriptional analysis of a human subtelomeric repeat. *Nucl. Acids Res.* **19**: 149-154.

Cross S, Lindsey J, Fantes J, McKay S, Cooke H. (1990) The structure of a subterminal repeated sequence present on many human chromsomes. *Nucl. Acids Res.* **18**: 6649-6657.

Dawson D, Herrick G. (1984) Telomeric properties of C4A4-homologous sequences in micronuclear DNA of *Oxytricha fallax*. *Cell*, **36**: 171-177.

de-Lange T, Shiue L, Myers R, Cox DR, Naylor SL, Killery AM, Varmus HE. (1990) Structure and variability of human chromosome ends. *Mol. Cell. Biol.* **10**: 518-527.

Forney JD, Blackburn EH. (1988) Developmentally controlled telomere addition in wild type and mutant *Paramecia*. *Mol. Cell. Biol.* **8**: 251-258.

Fussell CP. (1984) Interphase chromosome order: a proposal. *Genetica* **62**: 193-201.

Greider CW, Blackburn EH. (1985) Identification of a specific telomere terminal transferase activity in *Tetrahymena* extracts. *Cell*, **43**: 405-413.

Greider CW, Blackburn EH. (1989) A telomeric sequence in the RNA of *Tetrahymena* telomerase required for telomere repeat synthesis. *Nature*, **337**: 331-337.

Gruenbaum Y, Hochstrasser M, Mathog D, Saumweber H, Agard DA, Sedat JW. (1984) Spatial organization of the *Drosophila* nucleus: a three-dimensional cytogenetic study. *J. Cell Sci.* **Suppl. 1**: 224-234.

Henderson E, Blackburn EH. (1989) An overhang 3' terminus is a conserved feature of telomeres. *Mol. Cell. Biol.* **9**: 345-348.

Henderson E, Hardin C, Wolk S, Tinoco I, Blackburn EH. (1987) Telomeric DNA oligonucleotides form novel intramolecular structures containing guanine–guanine base pairs. *Cell*, **51**: 899-908.

Kang C, Zhang X, Ratliff R, Moyzis R, Rich A. (1992) Crystal structure of four stranded *Oxytricha* telomeric DNA. *Nature*, **356**: 126-131.

Klobutcher LA, Swanton MT, Donini P, Prescott DM. (1981) All gene-sized DNA molecules in four species of Hypotrichs have the same terminal sequence and an unusual 3' terminus. *Proc. Natl Acad. Sci. USA*, **78**: 3015-3019.

McClintock B. (1938) The fusion of broken ends of sister half-chromatids following chromatid breakage at meiotic anaphases. *Missouri Agric. Exp. Stat. Res. Bull.* **290**: 1-48.

McClintock B. (1939) The behavior in successive nuclear divisions of a chromosome broken at meiosis. *Proc. Natl Acad. Sci. USA*, **25**: 405-416.

McClintock B. (1941) The stability of broken ends of chromosomes in *Zea mays*. *Genetics*, **26**: 234-282.

McClintock B. (1942) The fusion of broken ends of chromosomes following nuclear fusion. *Proc. Natl Acad. Sci. USA*, **28**: 458-463.

Morin G. (1989) The human telomere terminal transferase is a ribonucleoprotein that synthesizes TTAGGG repeats. *Cell*, **59**: 521-529.

Morin GB, Cech TR. (1986) The telomeres of the linear mitochondrial DNA of *Tetrahymena thermophila* consists of 53 bp tandem repeats. *Cell*, **46**: 873-883.

Morin GB, Cech TR. (1988) Mitochondrial telomeres: surprising diversity of repeated telomeric DNA sequences among six species of *Tetrahymena*. *Cell*, **52**: 367-374.

Muller F, Wicky C, Spicher A, Tobler H. (1991) New telomere formation after developmentally regulated chromosomal breakage during the process of chromatin dimunition in *Ascaris lumbricoides*. *Cell*, **67**: 815-822.

Muller HJ. (1938) The remaking of chromosomes. *The Collecting Net—Woods Hole*, **13**: 181-198.

Narayanswami SA, Doggett NA, Clark LM, Hildebrand CE, Weier H-U, Hamkalo BA. (1992) Cytological and molecular characterization of centromeres in *Mus domesticus* and *Mus spretus*. *Mammalian Genome*, **2**: 186-194.

Neil DL, Villasante A, Fisher RB, Vetrie D, Cox B, Tyler-Smith C. (1990) Structural instability of human tandemly repeated DNA sequences cloned in yeast artificial chromosome vectors. *Nucl. Acids Res.* **18:** 1421-1428.

Olovnikov AM. (1973) A theory of marginotomy. *J. Theor. Biol.* **41:** 181-190.

Orr-Weaver TL, Szostak JW, Rothstein RJ. (1981) Yeast Transformation: A model system for the study of recombination. *Proc. Natl Acad. Sci. USA*, **78:** 6354-6358.

Petracek M, Lefebvre P, Silflow C, Berman J. (1990) *Chlamydomonas* telomere sequences are A+T-rich but contain three consecutive G-C base pairs. *Proc. Natl Acad. Sci. USA*, **87:** 8222-8226.

van der Ploeg LTH, Liu AYC, Borst P. (1984) Structure of growing telomeres of trypanosomes. *Cell*, **36:** 459-468.

Pluta AF, Kaine BP, Spear BB. (1982) The terminal organization of macronuclear DNA in *Oxytricha fallax. Nucl. Acids Res.* **10:** 8145-8154.

Pluta AF, Zakian VA. (1989) Recombination occurs during telomere formation in yeast. *Nature*, **337:** 429-433.

Richards EJ, Ausubel FM. (1988) Isolation of a higher eukaryotic telomere from *Arabidopsis thaliana. Cell*, **53:** 127-136.

Richards EJ, Goodman HM, Ausubel FM. (1991) The centromere region of *Arabidopsis thaliana* chromosome 1 contains telomere-similar sequences. *Nucl. Acids Res.* **19:** 3351-3357.

Richards EJ, Chao S, Vongs A, Yang J. (1992) Characterization of *Arabidopsis thaliana* telomeres isolated in yeast. *Nucl. Acids Res.* **20:** 4039-4046.

Romero DP, Blackburn EH. (1991) A conserved secondary structure for telomerase RNA. *Cell*, **67:** 343-353.

Sadhu C, McEachern MJ, Rustchenko-Bulgac EP, Schmid J, Soll DR, Hicks JB. (1991) Telomeric and dispersed repeat sequences in *Candida* yeasts and their use in strain identification. *J. Bacteriol.* **173:** 842-850.

Shampay J, Szostak JW, Blackburn EH. (1984) DNA sequences of telomeres maintained in yeast. *Nature*, **310:** 154-157.

Shippen-Lentz D, Blackburn EH. (1989) Telomere terminal transferase activity in the hypotrichous ciliate *Euplotes crassus. Mol. Cell. Biol.* **9:** 2761-2764.

Shippen-Lentz D, Blackburn EH. (1990) Functional evidence for an RNA template in telomerase. *Science*, **247:** 546-552.

Smith GP. (1976) Evolution of repeated DNA sequences by unequal crossover. *Science*, **191:** 528-535.

Sundquist WI, Klug A. (1989) Telomeric DNA dimerizes by formation of guanine tetrads between hairpin loops. *Nature*, **342:** 825-829.

Watson JD. (1972) Origin of concatameric T4 DNA. *Nature New Biol.* **239:** 197-201.

Wilkie AOM, Lamb J, Harris PC, Finney RD, Higgs DR. (1990) A truncated human chromosome 16 associated with a thalassaemia is stabilized by addition of telomeric repeat (TTAGGG)n. *Nature*, **346:** 868-871.

Williamson JR, Raghuraman MK, Cech TR. (1989) Monovalent cation-induced structure of telomeric DNA: the G-quartet model. *Cell*, **59:** 871-880.

Yu G-L, Blackburn EH. (1991) Developmentally programmed healing of chromosomes by telomerase in *Tetrahymena. Cell*, **67:** 823-832.

Yu G-L, Bradley JD, Attardi LD, Blackburn EH. (1990) *In vivo* alteration of telomere sequences and senescence caused by mutated telomerase RNAs. *Nature*, **344:** 126-132.

Zahler AM, Prescott DM. (1988) Telomere terminal transferase activity in the hypotrichous ciliate *Oxytricha nova* and a model for replication of the ends of linear DNA molecules. *Nucl. Acids Res.* **16:** 6953.

Zakian VA. (1989) Structure and function of telomeres. *Annu. Rev. Genet.* **23:** 579-604.

Chapter 8

Telomeres and telomerase: biochemistry and regulation in senescence and immortalization

Carol W Greider, Chantal Autexier, Ariel A Avilion, Kathleen Collins,
Lea A Harrington, Lin L Mantell, Karen R Prowse, Stephanie K Smith,
Rich C Allsopp, Chris M Counter, Homayoun Vaziri, Silvia Bacchetti and
Calvin B Harley

1. Introduction

The natural ends of chromosomes, or telomeres, are stable and rarely undergo re-arrangements or fusions with other chromosome ends. In contrast, chromosomes broken at meiosis or through X-irradiation, endonucleases, or by the movement of transposable elements are unstable. Through frequent re-arrangements, these unnatural chromosome ends wreak havoc on genome organization (reviewed in Blackburn, 1991; Greider, 1991a; Zakian, 1989). If a broken end fuses to another end, a dicentric chromosome will be created which may break at the next mitosis, initiating a cycle of chromosome breakage and fusion (McClintock, 1941) which may lead to cell death or possibly to cancerous transformation. The stability of natural chromosome ends relative to broken ends led Muller and McClintock to define telomeres as special structures required for chromosome stability (McClintock, 1938; Muller, 1938).

Telomeres also play an important role in chromosome replication. Special mechanisms must operate at telomeres to maintain chromosome length. Conventional DNA polymerases may not complete replication of chromosome ends because they travel only in the 5′ to 3′ direction and require a primer (Fig. 1). After primer removal and Okazaki fragment ligation, a region would be left unreplicated; thus one would predict that chromosomes would shorten at each round of division (Olovnikov, 1973; Watson, 1972). Telomerase is a novel DNA polymerase which synthesizes telomeric sequences *de novo* and that may compensate for the deficit of conventional replication (Greider and Blackburn, 1985, 1987, 1989; Yu *et al.*, 1990). Thus chromosome ends are dynamic structures which are maintained through an equilibrium between sequence loss and addition.

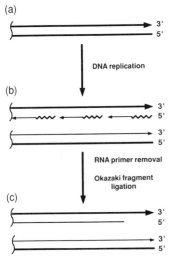

Figure 1. The 'end replication problem'. (a) The molecular end of a DNA molecule is shown. (b) Leading strand replication proceeds to the end of the DNA molecule, while lagging strand replication utilizes RNA primers and Okazaki fragment synthesis. (c) Removal of the RNA primers and Okazaki fragment ligation leaves a region at one end of each daughter molecule unreplicated. If there is no mechanism to fill the gap, the chromosome will get shorter with each round of replication.

2. Telomeric DNA sequences

Telomeric sequences are highly conserved in eukaryotes (Table 1). The first sequences were identified in the ciliates *Tetrahymena* and *Oxytricha*; they consist of repeats of the sequences TTGGGG and TTTTGGGG, respectively. The first telomere cloned from a multicellular organism was from the plant *Arabidopsis thaliana* which has tandem repeats of the sequence TTTAGGG (Richards and Ausubel, 1988). Subsequently, the human telomere sequence was cloned and shown to consist of hundreds of tandem TTAGGG repeats (Brown *et al.*, 1990; Cheng *et al.*, 1989; Cross *et al.*, 1989; de-Lange *et al.*, 1990; Moyzis *et al.*, 1988). In addition to sequence similarities, strand orientation is also highly conserved at telomeres. The G-rich strand is oriented 5' to 3' toward the end of the chromosome. The yeasts *Saccharomyces cerevisae* and *S. pombe* have irregular repeated G-rich sequences as does *Dictyostelium* (Table 1). The differences between the precise repeats in organisms like *Tetrahymena* and the irregular repeats of yeast suggests possible differences in their mode of synthesis. The number of repeats on any given chromosomes varies, giving telomeric restriction fragments a characteristic heterogeneous or 'fuzzy' appearance on Southern blots (for review see Blackburn, 1991; Greider, 1990; Zakian, 1989).

3. *De novo* telomere synthesis by telomerase

The end replication problem described above suggested that chromosomes might shorten at each round of replication. Nature has solved this problem in a number of different ways; some viruses like *Vaccinia* have covalently closed hairpin ends, while others like *Adenovirus* utilize protein primers to initiate replication. The structure of chromosomal telomeres, their heterogeneity and conservation of function, suggested that *de novo* synthesis of telomere repeats is involved in chromosome end maintenance (Shampay *et al.*, 1984). The search for an enzyme activity which might carry out this *de novo* addition led to the identification of telomerase in *Tetrahymena* (Greider and Blackburn, 1985). Telomerase synthesizes TTGGGG telomeric sequences on to single-stranded telomere sequence DNA primers *de novo*. *Tetrahymena* telomerase will add hundreds of TTGGGG repeats on to telomeric sequence primers *in vitro*. The enzyme is a ribonucleoprotein complex and the information for TTGGGG addition comes from an essential RNA component which contains the template sequence CAACCCCAA (Greider and Blackburn, 1989). When mutants in the CAACCCCAA region of the *Tetrahymena* telomerase RNA are expressed *in vivo*, telomeres containing the mutant sequence are generated (Yu *et al.*, 1990). This demonstrates not only that the RNA provides the template for TTGGGG repeat synthesis, but also that telomerase is responsible for the synthesis of telomeres *in vivo*.

Telomerase activities have also been identified in the ciliates *Oxytricha*, *Euplotes* and in immortal human cell lines (Morin, 1989; Shippen-Lentz and Blackburn, 1989; Zahler and Prescott, 1988). To date the RNA components have been cloned only from *Tetrahymena* and *Euplotes*. The *Euplotes* RNA contains the sequence CAAAACCCCAAAACC which is complementary to the TTTTGGGG telomere repeats of this ciliate (Shippen-Lentz and Blackburn, 1990). Although there is no cross-hybridization between the *Tetrahymena* and

Table 1. Eukaryotic telomeric DNA sequences. Telomeres consist of tandem repeats of G-rich sequences. One repeat of the G-rich strand from a variety of organisms is shown here. For references see Forney *et al.* (1987), Ganal *et al.* (1991), Le Blancq *et al.* (1991), Mueller *et al.* (1991), Petracek *et al.* (1990), Schechtman (1987), Zakian (1989)

Ciliates	Tetrahymena	TTGGGG
	Glaucoma	TTGGGG
	Paramecium	TT[T/G]GGG
	Oxytricha	TTTTGGGG
	Euplotes	TTTTGGGG
Slime Molds	Physarum	TTAGGG
	Dictyostelium	AG_{1-8}
	Didymium	TTAGGG
Flagellates	Trypanosomes	TTAGGG
	Giardia	TAGGG
Sporozoans	Plasmodium	TT[T/C]AGGG
Fungi	Neurospora	TTAGGG
	S. cerevisiae	TG_{1-3}
	S. pombe	$TTAC(A)G_{2-5}$
Nematodes	Ascaris	TTAGGC
Algae	Chlamydomonas	TTTTAGGG
Higher Plants	Arabidopsis	TTTAGGG
	Tomato	TTTAGGG
Vertebrates	Human	TTAGGG
	Mouse	TTAGGG

Euplotes RNAs, phylogenetic comparisons between different *Tetrahymena* species indicate a conserved secondary structure for the *Tetrahymena* telomerase RNA. This structure may be generally conserved in telomerase RNAs (Romero and Blackburn, 1991).

4. The telomerase cycle

Primer challenge experiments showed that tandem TTGGGG repeats are added processively without dissociation and re-association of the enzyme from the primer (Greider, 1991b). The fact that there were 1.5 telomere repeats in the template initially suggested a mechanism for primer elongation (Greider and Blackburn, 1989). A model for telomerase primer recognition and elongation is illustrated in Fig. 2. Telomerase initially recognizes a G-rich telomere sequence primer, an event which may or may not involve the telomerase RNA. The primer is then positioned in the catalytic site and the 3' end may base pair with the RNA if it is complementary. The primer is then elongated using the CAACCCCAA sequence as a template. After elongation to, or near to, the end of the template sequence, translocation re-positions the 3' end of the elongated primer relative to the template, allowing another round of synthesis.

How telomerase carries out the predicted translocation step is currently under investigation. There is no requirement for a high energy co-factor such as rATP or rGTP in the reaction. *In vitro* reaction products typically show a very characteristic six base repeat pausing pattern. The most frequent pause occurs after the addition of the first dG residue in the sequence TTGGGG (Greider, 1991b). However, the pausing pattern depends in part on the concentration of nucleotide in the reaction and other factors (Greider and Blackburn, 1987). Because of the lack of energy requirement we suggest that the potential nine positions of base pairing are not maintained during primer elongation. Instead, the base pairing between the template and elongating primer may be only transient.

Tetrahymena telomerase may not be as processive *in vivo* as it is *in vitro*. Telomeres cloned from a *Tetrahymena* strain expressing both mutant and wild-type telomerase RNAs display an interspersed pattern of mutant and wild-type sequence repeats (Yu and Blackburn, 1991; Yu *et al.*, 1990). This suggests that either the telomerase is less processive *in vivo* than *in vitro*, or a single telomerase enzyme unit may contain more than one RNA component (Greider, 1991b).

5. Chromosome healing

In addition to maintaining telomere length, telomerase is also implicated in the phenomenon known as chromosome healing. Examples of chromosome healing were documented as early as McClintock's cytological and genetic work on maize chromosomes in the 1930s (McClintock, 1938, 1939, 1941). Unstable broken ends generated by a number of mechanisms sometimes become spontaneously healed, presumably by the acquisition of a new telomere (reviewed in Greider, 1991a). In addition to spontaneous events, chromosome healing is a developmentally regulated process in organisms such as ciliates and ascarid worms. Ciliates are single cell eukaryotes that contain two different types of nuclei, the germline micronucleus and the somatic macronucleus. After conjugation, a mitotic product of the zygotic nucleus undergoes macronuclear development involving chromosome

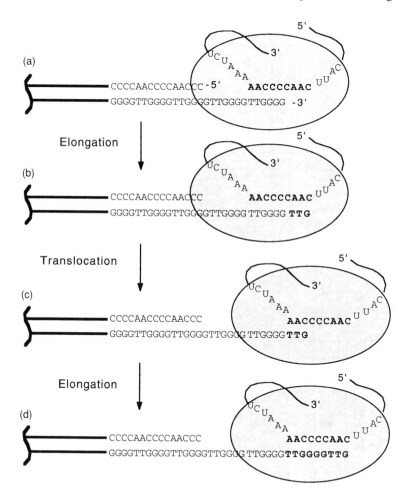

Figure 2. Model for elongation of telomeres by telomerase. The *Tetrahymena* telomere is shown containing an overhang on the d(TTGGGG)$_n$ strand. (a) After recognition of the d(TTGGGG)$_n$ strand by telomerase, most nucleotides are base paired with the CAACCCCAA RNA template. (b) The RNA is copied to the end of the template region. (c) Translocation re-positions the terminal TTGGGGTTG sequence and exposes additional template sequences. (d) Another round of template copying produces additional TTGGGG repeats.

fragmentation, DNA elimination, DNA amplification and telomere addition, (reviewed in Yao, 1989). The ability of telomerase to synthesize telomeric sequence *de novo* make it a prime candidate for the activity that heals chromosomes *in vivo*. However, *in vitro* telomerase shows a high degree of specificity for telomeric sequence primers over random sequence primers. This specificity raises the question of whether telomerase could add telomeric repeats on to non-telomeric sequences.

Experiments using a series of chimaeric oligonucleotides containing non-telomeric sequence at the 3′ end and telomeric d(TTGGGG) repeats at the 5′ end demonstrated that *Tetrahymena* telomerase can add telomere repeats on to non-telomeric 3′ ends. Chimaeric oligonucleotides containing two d(TTGGGG) repeats at the 5′ end and from 6 to 36 bases of pBR sequence at the 3′ end are efficiently elongated by telomerase. Maxam–Gilbert sequencing of end-labelled telomerase products showed that the elongation of chimaeric oligonucleotides was not due to prior removal of non-telomeric sequences (Harrington and Greider, 1991). The requirement for a telomeric sequence near but not at the site of telomere addition is similar to that found during *de novo* healing of yeast chromosomes *in vivo* (Murray *et al.*, 1988). Similar experiments with the human telomerase indicate that this enzyme is also capable of elongating non-telomeric sequences. Human telomerase will elongate oligonucleotides corresponding to a site of chromosome breakage and healing associated with α-thalassaemia (Morin, 1991). These studies provide the first biochemical evidence that telomerase is capable of chromosome healing by *de novo* addition of telomere sequences onto non-telomeric DNA.

The recent work of Yu and Blackburn (1991) shows that telomerase carries out chromosome healing *in vivo*. Mutations in the telomerase RNA were made and transformed into *Tetrahymena* to mark the RNA *in vivo*. Cells containing the mutant telomerase RNAs were then mated and allowed to undergo macronuclear development and associated chromosome fragmentation. Some telomeres cloned from the progeny cells contained the mutant sequence as the first repeat added on to AT-rich macronuclear ends. Thus, telomerase must directly synthesize new telomeres during developmentally programmed chromosome healing in *Tetrahymena* (Yu and Blackburn, 1991).

Because telomere addition is a developmentally regulated event in *Tetrahymena*, we reasoned that telomerase levels may increase during this process. A developmental study of telomerase RNA expression using Northern blots and direct quantitative Phosphor imaging technology (Johnston *et al.*, 1990), showed that telomerase RNA levels increased two- to five-fold during the starvation period prior to the initiation of mating and chromosome fragmentation. A similar increase in the level of telomerase activity was found at this time (Avilion *et al.*, 1992). Thus in ciliates where new telomere addition is an integral part of the life cycle, telomerase RNA expression is developmentally regulated.

6. Human telomeres

The first molecular description of human telomeres came when Howard Cooke and colleagues showed that a sequence from the pseudo-autosomal region of the human X and Y chromosomes, which is located very close to the telomere, hybridized to a heterogeneous-sized terminal band on a Southern blot (Cooke *et al.*, 1985). This provided the first evidence that human telomeres might be structurally related and thus maintained in a manner analogous to that found in unicellular eukaryotes. This work also described the intriguing observation that sperm telomeres are longer than those of other somatic tissues. Subsequently human telomeres have been cloned and shown to contain between 4 and 15 kb of TTAGGG repeats (Brown, 1989; Cheng *et al.*, 1989; Cross *et al.*, 1989; de-Lange *et al.*, 1990; Riethman *et al.*, 1989). The length difference in sperm versus blood DNA is due to a larger number of TTAGGG repeats present at sperm telomeres (Allshire *et al.*, 1989). When telomere length from tumour tissue or established tissue culture cells were compared with normal somatic cell DNA, the telomeres in the tumour tissue were significantly shorter. This indicates that telomere length regulation may differ in normal and tumour tissues (de-Lange *et al.*, 1990; Hastie *et al.*, 1990).

7. Telomere shortening in cell senescence and immortalization

The mechanisms which determine telomere length in normal human tissue and tumour tissue appear to be complex. To understand these length differences, we initiated studies of telomeres in primary human cells in culture. Primary human fibroblasts have a limited life span *in vitro*. Fibroblasts taken from older individuals undergo fewer divisions than those from younger donors (Hayflick, 1965). When primary human fibroblasts are cultured *in vitro*, telomeric sequences are lost with each round of replication (Harley *et al.*, 1990). Telomere shortening occurs *in vivo* as well as *in vitro*; primary fibroblasts from older donors have shorter telomeres than those from younger donors. Furthermore the life span remaining for fibroblasts from different aged donors is proportional to their initial telomere length (Allsopp *et al.*, 1992). In fact, telomere length is a better predictor of *in vitro* replicative capacity than is donor age. This correlation suggests that telomere length could be used as a clock which may signal the cell cycle exit characterized as senescence.

To determine what role telomerase may play in the length regulation of human cells, we have assayed for enzyme activity in both normal and immortalized cultures. Telomerase activity is found in all immortal human cell lines tested. In these cultures telomere length is maintained at a constant, but shorter size than the tissue of origin. We followed both telomere length and telomerase activity in the establishment of an immortal human cell line. Primary human embryonic kidney (HEK) cells undergo only 10–12 doublings *in vitro*. Clones of HEK cells transfected with SV40 T-antigen have an extended life span; after about 100 doublings, the clones undergo 'crisis'. Most cells die at crisis, but the few that survive can give rise to immortalized clones. Telomere length decreased when the primary HEK cells were passaged and continued to decrease in the extended life span clones. However, in the immortal clones, telomere length was short but stable after many rounds of division. Telomerase activity was absent in the primary cells and in the extended life span clones but was present in the immortalized clones which survived crisis. These results suggest that one of the events which occurs during immortalization may be the re-activation of telomerase activity. Cytogenetic analysis showed that the number of ring chromosomes and dicentrics increases dramatically at crisis, indicating that there is a correlation between telomere shortening and the chromosome instability characteristic of cancer cells (Counter *et al.*, 1992).

From the above observations of human telomeres, we have proposed the telomere hypothesis for cellular senescence (Counter *et al.*, 1992; Harley *et al.*, 1992) (Fig. 3). The hypothesis states that human telomerase is active in the

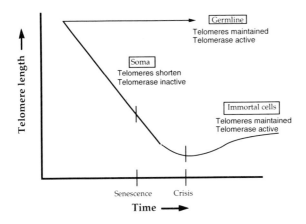

Figure 3. The telomere hypothesis. A schematic illustration of the hypothesis for the role of telomere length in human cell senescence and immortalization. The vertical axis represents 'telomere length' which is measured as the mean length of the terminal restriction fragments on Southern blots. The horizontal axis 'time' represents numbers of cell divisions either *in vivo* or *in vitro*. At the beginning of the time axis, representing embryonic development, stem cells and germ cells, telomerase is active, and telomere length is maintained. In somatic cells telomerase is less active or inactive and telomere length shortens by the loss of TTAGGG repeats with each round of cell division. When telomeres reach some critical length a 'checkpoint' may be signalled; cell division will stop and senescence will occur. If senescence is overcome by SV40 T-antigen or other transforming agents, telomeres will continue to shorten until crisis is reached. Most cells die at crisis, but a few will escape crisis and become immortalized. Telomerase activity is up-regulated or re-activated in the immortalized lines and telomere length is maintained or increased.

germline where telomeres length is maintained. Telomerase is inactive or down-regulated in somatic cells and thus telomeric sequences are lost through incomplete replication at each round of division. Telomere shortening may signal a checkpoint which affects the cell cycle exit characteristic of senescent cells. Cells which bypass the normal checkpoints regulating senescence survive 'crisis' and become immortal. Telomere length is maintained in immortal clones by the activation of telomerase. At present this hypothesis is highly speculative. However, key experiments in the next few years should establish whether telomerase and telomere length play a 'leading role' or a 'bit part' in the drama of cellular senescence and immortalization.

References

Allshire RC, Dempster M, Hastie ND. (1989) Human telomeres contain at least three types of G-rich repeats distributed non-randomly. *Nucl. Acids Res.* **17**: 4611-4627.
Allsopp RC, Vaziri H, Patterson C, Goldstein S, Younglai EV, Futcher AB, Greider CW, Harley CB. (1992) Telomere length predicts replicative capacity of human fibroblasts. *Proc. Natl Acad. Sci. USA*, **89**: 10114-10118.
Avilion AA, Harrington LA, Greider CW. (1992) *Tetrahymena* telomerase RNA levels during macronuclear development. *Dev. Genet.* **13**: 80-86.
Blackburn EH. (1991) Structure and function of telomeres. *Nature*, **350**: 569-573.
Brown WRA. (1989) Molecular cloning of human telomeres in yeast. *Nature*, **338**: 774-776.
Brown W, Dobson M, MacKinnon P. (1990) Telomere cloning and mammalian chromosome analysis. *J. Cell Sci.* **95**: 521-526.
Cheng J-F, Smith CL, Cantor CR. (1989) Isolation and characterization of a human telomere. *Nucl. Acids Res.* **17**: 6109-6127.
Cooke HJ, Brown WRA, Rappold GA. (1985) Hypervariable telomeric sequences from the human sex chromosomes are pseudoautosomal. *Nature*, **317**: 687-692.
Counter CM, Avilion AA, LeFeuvre CE, Stewart NG, Greider CW, Harley CB, Bacchetti S. (1992) Telomere shortening associated with chromosome instability is arrested in immortal cells which express telomerase activity. *EMBO J.* **11**: 1921-1929.
Cross SH, Allshire RC, McKay SJ, McGill NI, Cooke HJ. (1989) Cloning of human telomeres by complementation in yeast. *Nature*, **338**: 771-774.
de-Lange T, Shiue L, Myers R, Cox DR, Naylor SL, Killery AM, Varmus HE. (1990) Structure and variability of human chromosome ends. *Mol. Cell. Biol.* **10**: 518-527.
Forney J, Henderson ER, Blackburn EH. (1987) Identification of the telomeric sequence of the acellular slime molds *Didymium iridis* and *Physarum polycephalum*. *Nucl. Acids Res.* **15**: 9143-9152.
Ganal MW, Lapitan NLV, Tanksley SD. (1991) Macrostructure of tomato telomeres. *Plant Cell* **3**: 87-94.
Greider CW. (1990) Telomeres, telomerase and senescence. *Bioessays*, **12**: 363-369.
Greider CW. (1991a) Chromosome first aid. *Cell*, **67**: 645-647.
Greider CW. (1991b) Telomerase is processive. *Mol. Cell. Biol.* **11**: 4572-4580.
Greider CW, Blackburn EH. (1985) Identification of a specific telomere terminal transferase activity in *Tetrahymena* extracts. *Cell*, **43**: 405-413.
Greider CW, Blackburn EH. (1987) The telomere terminal transferase of *Tetrahymena* is a ribonucleoprotein enzyme with two kinds of primer specificity. *Cell*, **51**: 887-898.
Greider CW, Blackburn EH. (1989) A telomeric sequence in the RNA of *Tetrahymena* telomerase required for telomere repeat synthesis. *Nature*, **337**: 331-337.
Harley CB, Futcher AB, Greider CW. (1990) Telomeres shorten during ageing of human fibroblasts. *Nature*, **345**: 458-460.

Harley CB, Vaziri H, Counter CM, Allsopp RC. (1992) The telomere hypothesis of cellular aging. *Exp. Gerontol.* **27**: in press.

Harrington LA, Greider CW. (1991) Telomerase primer specificity and chromosome healing. *Nature*, **353**: 451-454.

Hastie ND, Dempster M, Dunlop MG, Thompson AM, Green DK, Allshire RC. (1990) Telomere reduction in human colorectal carcinoma and with ageing. *Nature*, **346**: 866-868.

Hayflick L. (1965) The limited *in vitro* lifetime of human diploid strains. *Exp. Cell Res.* **37**: 614-636.

Johnston RF, Pickett SC, Barker DL. (1990) Autoradiography using storage phosphor technology. *Electrophoresis*, **11**: 355-360.

Le Blancq SM, Kase RS, VanderPloeg LHT. (1991) Analysis of a *Giardia lamblia* rRNA encoding telomere with [TAGGG]$_n$ as the telomere repeat. *Nucl. Acids Res.* **19**: 5790.

McClintock B. (1938) The fusion of broken ends of sister half-chromatids following chromatid breakage at meiotic anaphases. *Missouri Agric. Exp. Stat. Res. Bull.* **290**: 1-48.

McClintock B. (1939) The behavior in successive nuclear divisions of a chromosome broken at meiosis. *Proc. Natl Acad. Sci. USA*, **25**: 405-416.

McClintock B. (1941) The stability of broken ends of chromosomes in *Zea mays*. *Genetics*, **26**: 234-282.

Morin G. (1989) The Human telomere terminal transferase is a ribonucleoprotein that synthesizes TTAGGG repeats. *Cell*, **59**: 521-529.

Morin GB. (1991) Recognition of a chromosome truncation site associated with α-thalassaemia by human telomerase. *Nature*, **353**: 454-456.

Moyzis RK, Buckingham JM, Cram LS, Dani M, Deaven LL, Jones MD, Meyne J, Ratliff RL, Wu J-RA. (1988) A highly conserved repetitive DNA sequence, (TTAGGG)$_n$, present at the telomeres of human chromosomes. *Proc. Natl Acad. Sci. USA*, **85**: 6622-6626.

Mueller F, Wicky C, Spicher A, Tobler H. (1991) New telomere formation after developmentally regulated chromosomal breakage during the process of chromatin diminution in *Ascaris lumbricoides*. *Cell*, **67**: 815-822.

Muller HJ. (1938) The remaking of chromosomes. *The Collecting Net—Woods Hole* **13**: 181-198.

Murray AW, Claus TE, Szostak JW. (1988) Characterization of two telomeric DNA processing reactions in *Saccharomyces cerevisiae*. *Mol. Cell. Biol.* **8**: 4642-4650.

Olovnikov AM. (1973) A theory of marginotomy. *J. Theor. Biol.* **41**: 181-190.

Petracek M, Lefebvre P, Silflow C, Berman J. (1990) *Chlamydomonas* telomere sequences are A+T-rich but contain three consecutive G–C base pairs. *Proc. Natl Acad. Sci. USA*, **87**: 8222-8226.

Richards ER, Ausubel FM. (1988) Isolation of a higher eukaryotic telomere from *Arabidopsis thaliana*. *Cell*, **53**: 127-136.

Riethman HC, Moyzis RK, Meyne J, Burke DT, Olson MV. (1989) Cloning human telomeric DNA fragments into *Saccharomyces cerevisiae* using a yeast-artificial-chromosome vector. *Proc. Natl Acad. Sci. USA*, **86**: 6240-6244.

Romero DP, Blackburn EH. (1991) A conserved secondary structure for telomerase RNA. *Cell*, **67**: 343-353.

Schechtman MG. (1987) Isolation of telomeric DNA from *Neurospora crassa*. *Mol. Cell. Biol.* **7**: 3168-3177.

Shampay J, Szostak JW, Blackburn EH. (1984) DNA sequences of telomeres maintained in yeast. *Nature*, **310**: 154-157.

Shippen-Lentz D, Blackburn EH. (1989) Telomere terminal transferase activity in the hypotrichous ciliate *Euplotes crassus*. *Mol. Cell. Biol.* **9**: 2761-2764.

Shippen-Lentz D, Blackburn EH. (1990) Functional evidence for an RNA template in telomerase. *Science*, **247**: 546-552.

Watson JD. (1972) Origin of concatameric T4 DNA. *Nature*, **239**: 197-201.

Yao M-C. (1989) Site specific chromosome breakage and DNA deletion in ciliates. In: *Mobile DNA* (Berg DE, Howe MM, eds). Washington DC: American Society for Microbiology Press. pp 715-743.

Yu G-L, Blackburn EH. (1991) Developmentally programmed healing of chromosomes by telomerase in *Tetrahymena. Cell,* **67**: 823-832.

Yu G-L, Bradley JD, Attardi LD, Blackburn EH. (1990) *In vivo* alteration of telomere sequences and senescence caused by mutated telomerase RNAs. *Nature,* **344**: 126-132.

Zahler AM, Prescott DM. (1988) Telomere terminal transferase activity in the hypotrichous ciliate *Oxytricha nova* and a model for replication of the ends of linear DNA molecules. *Nucl. Acids Res.* **16**: 6953.

Zakian VA. (1989) Structure and function of telomeres. *Annu. Rev. Genet.* **23**: 579-604.

Chapter 9

DNA methylation and CpG islands

Francisco Antequera and Adrian Bird

1. Introduction

Animal DNA is modified by a cytosine DNA methyltransferase which methylates the sequence 5' CpG 3'. The purified methyltransferase appears to methylate CpGs independent of their sequence context. In the genome, however, there is a reproducible pattern of methylated and non-methylated CpGs, implying that the enzyme only has access to a restricted subset of potential sites. The most obvious pattern associated with DNA methylation occurs in invertebrate animals (Bird *et al.*, 1979) as well as some fungi (Antequera *et al.*, 1984). In plants, the only example of this kind reported is *Arabidopsis thaliana* (Simoens *et al.*, 1988). In all these cases, the genome can be divided into heavily methylated domains in which most or all CpGs are in the methylated form, and long non-methylated domains. So far, all the genes in invertebrate organisms of this kind are found within the non-methylated domains. The methylated fraction of the genome, on the other hand, contains transcriptionally inert DNA sequences. In the slime mould *Physarum*, for example, where the same domain pattern of methylation occurs, analysis of the methylated fraction indicates that more than 50% is made up of a single transposable element (Rothnie *et al.*, 1991). This and other evidence suggests that potentially damaging DNA sequences, such as transposable elements, may be held in check by a system involving DNA methylation. An analogous situation has been observed in mammalian embryonic cells, where infecting retroviral proviruses are silenced by a process that involves DNA methylation (Jahner and Jaenisch, 1984). The ancestral role of cytosine methylation may therefore have been at least partly to detect and neutralize invading DNA sequences.

2. CpG islands

For most organisms with DNA methylation, this hypothesis is easy to sustain. For vertebrates, however, it is untenable in its simplest form, as the majority of their genome is methylated at CpG. Approximately 70% of all CpGs in mammalian

DNA are methylated, and many of these are at sequences that are transcribed. Thus DNA methylation has spread through the genome in vertebrates, affecting many genes. Non-methylated domains do persist, however, as closer examination shows that small islands of methylation-free DNA are present in vertebrate genomes. These so-called CpG islands appear to be the descendants of the large tracts of non-methylated DNA found in non-vertebrate animals (see Antequera and Bird, 1993; Bird, 1987 for reviews). They are easily identified as clusters of CpG because this dinucleotide is rare in the rest of the genome, occurring at about 20% of the expected frequency on the basis of the G+C content. The reason for this is the tendency of 5-methylcytosine (5 mC) to deaminate to yield thymine. Despite the existence of a specific repair mechanism (Brown and Jiricny, 1987), many m5CpGs have been transformed into TpG (or CpA in the complementary strand) over evolutionary time (Sved and Bird, 1990). As CpG islands are non-methylated they do not show CpG suppression. Another aspect that enhances the differences between CpG islands and bulk DNA is their elevated G+C content (67% in humans compared to 40% genome average). These two factors mean that the density of CpG is 5–10 times higher in CpG islands than in the remainder of the genome. These characteristics have made them useful landmarks for genome mapping using rare cutter restriction enzymes (Bickmore and Bird, 1992). CpG islands are relatively homogenous in length, ranging in size from about 0.5 kb up to about 3 kb, with an average size of 1 kb.

A very important property of CpG islands is that they are associated with genes. Their location most often coincides with the transcriptional initiation. They tend to be asymmetrically placed about the initiation site, with the majority of the CpG island typically extending downstream into the transcribed region of the gene.

3. CpG islands and genes

Not all genes have CpG islands. Examination of the database indicates that about 56% of human genes are CpG island-associated, including all known housekeeping genes. Other studies have suggested that 40% of the genes with patterns of expression restricted to one or a few cell types are associated with CpG islands also (Larsen et al., 1992). Genes not associated with CpG islands are invariably highly tissue specific in their expression patterns. Figure 1 summarizes the methylation patterns at CpG island genes and non-CpG island genes during development. Three kinds of genes are indicated: (1) a housekeeping gene (HK) that is CpG island-associated and expressed in all tissues; (2) a tissue specific gene (TS) with no CpG island that is expressed in one of the five tissues; (3) a tissue specific gene with a CpG island that is expressed in one tissue. It can be seen that the CpG island is non-methylated in all tissues whether the associated gene is expressed or not. An example of an island-associated tissue specific gene is the human α-globin

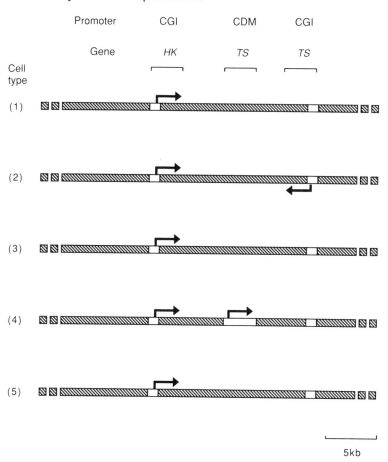

Figure 1. Two sorts of promoter at mammalian genes. The diagram shows DNA sequences that are heavily methylated at CpG (hatched) and sequences that are methylation free (open) at two tissue-specific genes (TS) and one housekeeping gene (HK) in five differentiated cell types of a mammal (1-5). CGI, CpG-island promoter; CDM, CpG deficient, methylated promoter. Arrows indicate transcription in that cell type. CpG-island promoters are not methylated regardless of expression, while CpG-deficient promoters lose methylation when transcribed. The density of CpGs in CpG island promoters is 5-10 fold higher than in CpG-deficient promoters. (Reprinted from Bird (1992) with permission from Cell Press).

gene, the CpG island of which remains in a non-methylated condition in all tissues although the gene is only expressed in erythroid tissues. Unlike CpG island genes, genes without CpG island show changing patterns of methylation with expression. They are methylated in tissues that do not express the gene, but typically lose methylation in tissues where expression is observed. Whether this change in methylation is involved in control of expression or is a passive consequence of it is still a matter for debate.

4. Chromatin structure of CpG islands

Studies on isolated nuclei have shown that CpG island chromatin is different from bulk chromatin. Using methyl-sensitive restriction enzymes, Tazi and Bird (1990) were able to isolate oligonucleosomes derived exclusively from the CpG island fraction. When these were compared with oligonucleosomes from bulk chromatin, several differences were found. On average, each CpG island contained a region of DNA 100–200 base pairs in length that was not wrapped up in nucleosomes. The origin of this nucleosome-free DNA was traced in specific genes to regions that are bound to transcription factors and other regulatory proteins. Such regions are frequently identified as DNase I hypersensitive sites associated with the promoters of many active genes. CpG island chromatin was also different from bulk chromatin with respect to histone composition. Core histones H3 and H4 were in a hyperacetylated form, and histone H1 was very deficient. These experiments show that CpG islands are not only distinct at the level of DNA sequence, but also adopt different structures in the nucleus. Most probably both differences are associated with their position near the 5′ ends of genes.

5. *De novo* methylation of CpG islands

We have stressed the stable lack of methylation at CpG islands during development. Like many biological generalizations this is mostly but not completely true. At present we know of four kinds of situation in which CpG islands do in fact become heavily methylated. In each case a consequence appears to be severe repression of the associated gene. The only example so far that is associated with normal development concerns X chromosome inactivation, which occurs in female mammals. Most CpG islands on the inactive X chromosome in female placental mammals are in a heavily methylated condition whereas those on the active homologue are in the typical non-methylated form (Tribioli *et al.*, 1992).

Methylation is not the primary inactivating signal (Lock *et al.*, 1987), but it clearly ensures stability of the inactive state, as genes can be reactivated under certain circumstances by inhibitors of DNA methylation. Moreover in marsupials CpG islands on the inactive X chromosome are not methylated and the inactivation process is leaky (Kaslow and Migeon, 1987).

The other known examples of CpG island methylation occur in unusual circumstances. One is associated with trinucleotide repeat expansion at *FMR-1*, the gene responsible for fragile X syndrome (Verkerk *et al.*, 1991). The gene is known to be repressed in those suffering from the syndrome, and it may be that methylation of the CpG island itself is responsible for this (Bell *et al.*, 1991; Vincent *et al.*, 1991). The third case concerns tissue culture cells, where CpG island methylation has often occurred on a massive scale. Analysis of randomly selected CpG islands from a number of common permanent cell lines showed that often more than half of all CpG islands, corresponding to thousands of genes, are methylated (Antequera *et al.*, 1990). Since in all known cases CpG island methylation leads to repression of the associated gene, the conclusion is that vast numbers of genes in cultured cells are irrevocably silenced by methylation. The fourth and final case of *de novo* methylation at CpG islands has been observed in cells derived from tumours. In the case of the calcitonin gene, there is as yet no causal link between *de novo* methylation and the process of tumorigenesis (De Bustros *et al.*, 1988). More intriguing is the observation that recessive oncogenes can be *de novo* methylated in cancer cells. In tumours of the retina (retinoblastomas) genetic studies have strongly implicated inactivation of the retinoblastoma gene as a primary cause of tumorigenesis. Analysis of the CpG island at this gene in fresh tumour tissue has shown that in a small but significant fraction of cases *de novo* methylation occurs at the island (Greger *et al.*, 1989; Sakai *et al.*, 1991). Although there is no direct evidence that the gene is silenced by methylation, it seems likely that this will be the case based upon studies of the effects of CpG island methylation on other genes. If so, this would be the first evidence of a role for epigenetic phenomena in tumorigenesis.

6. DNA methylation and transcriptional repression

The mechanism by which methylation leads to transcriptional suppression is becoming clearer. We have identified two proteins that are likely mediators of repression (Lewis *et al.*, 1992; Meehan *et al.*, 1989). One of these, MeCP1, exerts its effect by binding preferentially to DNA that contains a high density of methylated CpGs. Since CpG islands are CpG-rich, they are high affinity substrates for MeCP1 binding when methylated, and repression is correspondingly severe. Studies with sparsely methylated sequences have shown that MeCP1 binding is weaker, and can be displaced if the promoter of the gene is sufficiently strong (Boyes and Bird, 1992). The characteristics of MeCP1 and MeCP2 and their possible roles in repression have been the subject of recent review (Bird, 1992).

7. Origin and maintenance of the CpG islands

Despite intensive work on CpG islands, the biological rationale for their presence remains a mystery. The high frequency of CpGs can be adequately explained by the absence over millions of years of methylation in these regions, but this begs the question of why the regions are methylation-free at all. An early hypothesis was that the binding of the transcription initiation machinery at promoters rendered these sequences inaccessible to the DNA methyltransferase and consequently precluded methylation. Two observations lead us to question this hypothesis. Firstly, we have found that nucleases have very little difficulty in accessing the

DNA within CpG islands; in fact they are hypersensitive to nuclease attack (Antequera *et al.*, 1989; Tazi and Bird, 1990). Secondly a number of genes, such as the human α-globin gene, are expressed in only one tissue, but remain non-methylated in all tissues of the organism. The absence of hypersensitive sites and long range DNase I sensitivity at the unexpressed α-globin gene implies that the initiation complex is not present in most tissues (Vyas *et al.*, 1992). Why then is the CpG island nevertheless methylation free? Another puzzling aspect of CpG islands is their location. The footprint hypothesis must explain why the non-methylated domain extends normally downstream into the transcribed part of the gene, but very little upstream. We do not at present know of macromolecules that would produce such an asymmetrical footprint. A radically different hypothesis is that the lack of methylation of CpG islands is due to a demethylating activity which consistently works faster than *de novo* methylation. Removal of methylation from artificially methylated CpG islands has been observed in early embryonic cells (Frank *et al.*, 1991). The process appears to be rapid, suggesting an active demethylation reaction, although so far no cell extract capable of mimicking this reaction has been isolated. Further work is clearly required to understand the presence of CpG islands in the mammalian genome. The likelihood is that the final explanation will require new knowledge about the principles of genome organization.

References

Antequera F, Tamame M, Villanueva J, Santos T. (1984) DNA methylation in the fungi. *J. Biol. Chem.* **259:** 8033-8036.

Antequera F, MacLeod D, Bird A. (1989) Specific protection of methylated CpG in mammalian nuclei. *Cell,* **58:** 509-517.

Antequera, F, Boyes J, Bird A. (1990) High levels of *de novo* methylation and altered chromatin structure at CpG islands in cell lines. *Cell,* **62:** 503-514.

Antequera F, Bird A. (1993) CpG Islands. In: *DNA Methylation: Molecular Biology and Biological Significance* (Jost JP, Saluz HP, eds). Basel: Birkhauser Verlag. pp 169-185.

Bell M, Hirst M, Nakahori Y, MacKinnon R, Roche A, Flint T, Jacobs P, Tommerup N, Tranebjaerg L, Froster-Iskenius U, Kerr B, Turner G, Lindenbaum R, Winter R, Pembrey M, Thibodeau S, Davies K. (1991) Physical mapping across the fragile X: hypermethylation and clinical expression of the fragile X syndrome. *Cell,* **64:** 861-866.

Bickmore W, Bird A. (1992) The use of restriction endonucleases to detect and isolate genes from mammalian cells. *Methods Enzymol.* **216:** 224-244.

Bird, A. (1987) CpG islands as gene markers in the vertebrate nucleus. *Trends Genet.* **3:** 342-347.

Bird A. (1992) The essentials of DNA methylation. *Cell,* **70:** 508.

Bird A, Taggart M, Smith B. (1979) Methylated and unmethylated DNA compartments in the sea urchin genome. *Cell,* **17:** 889-901.

Boyes J, Bird A. (1992) Repression of genes by DNA methylation depends on CpG density and promoter strength: evidence for involvement of a methyl-CpG binding protein. *EMBO J.* **11:** 327-333.

Brown T, Jiricny, J. (1987) A specific mismatch repair event protects mammalian cells from loss of 5-methylcytosine. *Cell,* **50:** 945-950.

De Bustros A, Nelkin B, Silverman A, Ehrlich G, Poiesz B, Baylin S. (1988) The short arm of chromosome 11 is a "hot spot" for hypermethylation in human neoplasia. *Proc. Natl Acad. Sci. USA*, **85:** 5693-5697.

Frank D, Heshet I, Shani M, Levine A, Razin A, Cedar H. (1991) Demethylation of CpG islands in embryonic cells. *Nature*, **351:** 239-241.

Greger V, Passarge E, Hopping W, Messmer E, Horsthemke B. (1989) Epigenetic changes may contribute to the formation and spontaneous regression of retinoblastoma. *Hum. Genet.* **83:** 155-158.

Jahner D, Jaenisch R. (1984) DNA methylation in early mammalian development. In: *DNA Methylation: Biochemistry and Biological Significance* (Razin A, Cedar H, Riggs A, eds). New York: Springer Verlag. pp 189-219.

Kaslow D, Migeon B. (1987) DNA methylation stabilizes X chromosome inactivation in eutherians but not in marsupials: evidence for multistep maintenance of mammalian X dosage compensation. *Proc. Natl Acad. Sci. USA*, **84:** 6210-6214.

Larsen G, Gundersen G, Lopez R, Prydz H. (1992) CpG islands as gene markers in the human genome. *Genomics*, **13:** 1095-1107.

Lewis J, Meehan R, Henzel W, Maurer-Fogey I, Jeppesen P, Klein F, Bird A. (1992) Purification, sequence and cellular localisation of a novel chromosomal protein that binds to methylated DNA. *Cell*, **69:** 905-914.

Lock L, Takagi N, Martin G. (1987) Methylation of the Hprt gene on the inactive X occurs after chromosome inactivation. *Cell*, **48:** 39-46.

Meehan RR, Lewis JD, McKay S, Kleiner EL, Bird AP. (1989) Identification of a mammalian protein that binds specifically to DNA containing methylated CpGs. *Cell*, **58:** 499-507.

Rothnie H, McCurrach K, Glover L, Hardman N. (1991) Retrotransposon-like nature of Tp1 elements: implications for the organization of highly repetitive, hypermethylated DNA in the genome of *Physarum polycephalum. Nucl. Acids Res.* **19:** 279-286.

Sakai T, Toguchida J, Ohtani N, Yandell D, Rapaport J, Dryja T. (1991) Allele-specific hypermethylation of the retinoblastoma tumor suppressor gene. *Am. J. Hum. Genet.* **48:** 880-888.

Simoens C, Gielen J, Van Montagu M, Inze D. (1988) Characterization of highly repetitive sequences of *Arabidopsis thaliana. Nucl. Acids Res.* **16:** 6753-6766.

Sved J, Bird A. (1990) The expected equilibrium of the CpG dinucleotide in vertebrate genomes under a mutational model. *Proc. Natl Acad. Sci. USA*, **87:** 4692-4696.

Tazi J, Bird A. (1990) Alternative chromatin structure at CpG islands. *Cell*, **60:** 909-920.

Tribioli C, Tamanini F, Patrosso C, Milanesi L, Villa A, Pergolizzi R, Maestrini C, Rivella S, Bione S, Mancini M, Vezzoni P, Toniolo D. (1992) Methylation and sequence analysis around EagI sites: identification of 28 new CpG islands in XQ24-XQ28. *Nucl. Acids Res.* **20:** 727-733.

Verkerk AJMH, Pieretti M, Sutcliffe JS *et al.* (1991) Identification of a gene (FMR-1) containing a CGG repeat coincident with a breakpoint cluster region exhibiting length variation in fragile X syndrome. *Cell*, **65:** 905-914.

Vincent A, Heitz D, Petit C, Kretz C, Oberle I, Mandel J. (1991) Abnormal pattern detected in fragile X patients by pulsed field gel electrophoresis. *Nature*, **349:** 624-626.

Vyas P, Vickers M, Simmons D, Ayyub H, Craddock C, Higgs D. (1992) Cis-acting sequences regulating expression of the human α-globin cluster lie within constitutively open chromatin. *Cell*, **69:** 781-793.

Chapter 10

Chromatin remodelling by nucleoplasmin

Gregory H Leno, Anna Philpott and Ronald A Laskey

1. Introduction

In eukaryotic cells, changes in chromatin structure have a profound impact on many cellular processes. This is particularly evident at fertilization when the sperm chromatin is extensively remodelled in the cytoplasm of the egg. Upon entering egg cytoplasm, the highly condensed sperm chromatin decondenses and the sperm-specific basic proteins, which are synthesized during spermatogenesis to aid compaction of the male genome, are replaced by histones present in the egg. This restructuring of sperm chromatin results in the formation of nucleosomes and is prerequisite for pronuclear formation and normal nuclear function (for review see Poccia, 1986).

The development of *in vitro* systems, which mimic the events at fertilization, has provided a unique opportunity to explore these changes in chromatin structure and to elucidate the mechanisms of chromatin remodelling at the molecular level. Using a cell-free system derived from the eggs of the frog *Xenopus laevis*, it has recently been shown that the acidic protein nucleoplasmin, previously shown to function as a nucleosome assembly factor (for review see Laskey and Leno, 1990), also plays a central role in the decondensation and remodelling of sperm chromatin at fertilization (Ohsumi and Katagiri, 1991a; Philpott *et al.*, 1991; Philpott and Leno, 1992). In immunodepletion/protein addback studies, nucleoplasmin was found to be both necessary and sufficient for the initial phase of chromatin decondensation in egg extract (Philpott *et al.*, 1991). Furthermore, it was shown that this same protein restructures sperm chromatin during decondensation by removing sperm-specific basic proteins and assembling histones H2A and H2B from the egg on to the DNA forming nucleosome cores (Philpott and Leno, 1992). Thus, taken together, these studies suggest that nucleoplasmin functions as both an assembly factor and a disassembly factor remodelling sperm chromatin in the egg extract.

2. Nucleoplasmin as a nucleosome assembly factor

The earliest event in the construction of chromatin is the formation of the nucleosome. Insight into the nature of this process has been greatly facilitated by the development of *in vitro* chromatin assembly systems which have allowed a detailed analysis of the assembly process and, in certain cases, the identification of specific assembly factors (for review see Dilworth and Dingwall, 1988, and Laskey and Leno, 1990). Cell-free systems derived from amphibian eggs have been particularly useful for analysis of nucleosome assembly for at least two reasons. First, they faithfully mimic the events that occur in the fertilized egg and, therefore, are good models for *in vivo* assembly. Secondly, they possess a stockpile of histones, which are required to support chromatin assembly during the very rapid cell cycles that follow fertilization. This histone pool allows for rapid chromatin assembly on newly synthesized DNA during early embryogenesis, and on exogenous DNA added to the extract for experimental analysis.

In cell-free extracts derived from the eggs of *X. laevis*, two proteins have been identified which are essential for nucleosome core assembly *in vitro* (Dilworth *et al.*, 1987). The first of these proteins to be discovered was nucleoplasmin, an acidic, thermostable protein that exists as a pentamer in solution (Dingwall *et al.*, 1982; Earnshaw *et al.*, 1980; Laskey *et al.*, 1978). In the egg extract, nucleoplasmin is complexed with the histones H2A and H2B (Dilworth *et al.*, 1987; Kleinschmidt *et al.*, 1985, 1990) while the second protein, N1 (and possibly a related protein, N2) is found associated with histones H3 and H4 (Dilworth *et al.*, 1987; Kleinschmidt and Franke, 1982). Recent evidence indicates that nucleosome assembly occurs in two steps (Kleinschmidt *et al.*, 1990; Sapp and Worcel, 1990; and for review Laskey and Leno, 1990). First, N1/N2 donates histones H3 and H4 to the DNA and secondly, nucleoplasmin transfers H2A and H2B, resulting in the formation of the nucleosome core (Fig. 1). Nucleoplasmin (and most likely N1) is thought to function as a 'molecular chaperone', preventing non-specific interactions between the histones and DNA and yet allowing the specific subset of

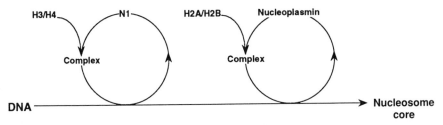

Figure 1. A two-step model for nucleosome core assembly in eggs of *Xenopus laevis*. In egg cytoplasm, N1 is complexed with histones H3 and H4 while nucleoplasmin is bound to H2A and H2B. Assembly of the nucleosome core is thought to occur by the sequential addition of H3/H4 and H2A/H2B to the DNA (see text for references). Figure modified from Laskey and Leno (1990) with permission from Elsevier Trends Journals.

interactions required for the assembly of the nucleosome core (Laskey *et al.*, 1978; and for review Dingwall and Laskey, 1990). Indeed, the concept of 'molecular chaperones' was originally formulated for nucleoplasmin (Laskey *et al.*, 1978). Thus, in *Xenopus* egg extract, nucleoplasmin functions as an assembly factor, donating histones H2A and H2B to the DNA forming the nucleosome core. However, as discussed below, nucleoplasmin's role is not limited to one of assembly alone.

3. Decondensation of sperm chromatin in egg cytoplasm

Following fertilization, the sperm nucleus undergoes three ultrastructural changes in the egg cytoplasm. These are nuclear envelope breakdown, chromatin dispersal and pronuclear envelope assembly (for review see Longo and Kunkle, 1978). Breakdown of the sperm nuclear envelope occurs immediately upon entry into egg cytoplasm and is rapidly followed by chromatin dispersal (also termed decondensation). Decondensation occurs in two stages. The first stage, stage I, is a very rapid dispersal of the chromatin which is accompanied by an increase in volume. Stage II, is a slower, membrane-dependent swelling that occurs following nuclear envelope re-assembly and is associated with protein import into the pronucleus (Longo and Kunkle, 1978).

Evidence for the existence of specific decondensation factors in the egg came from early microinjection studies in which a variety of cell nuclei decondensed when injected into the oocyte nucleus and egg cytoplasm (Graham *et al.*, 1966; Gurdon, 1968, 1976). However, decondensation did not occur when nuclei were injected into enucleated oocytes unless oocyte nuclear material was re-introduced (Lohka and Masui, 1983a). Thus, it appeared that the putative decondensation factor or factors were present in the oocyte nucleus and remained in the egg cytoplasm after breakdown of the germinal vesicle during oocyte maturation.

Major steps forward in the analysis of chromatin decondensation occurred with the development of cell-free systems from amphibian eggs (Barry and Merriam, 1972; Iwao and Katagiri, 1984; Lohka and Masui, 1983a, b, 1984). Using crude cytoplasmic preparations from the eggs and oocytes of *X. laevis*, Barry and Merriam (1972) found that metabolically quiescent chicken erythrocyte nuclei decondensed in egg cytoplasm but not in oocyte cytoplasm, consistent with the results obtained in the earlier microinjection experiments (see above). Subsequently, amphibian egg extract was separated into both soluble supernatant and particulate fractions by high speed centrifugation and demembranated sperm nuclei were incubated in these preparations. From these studies it became clear that stage I decondensation of sperm chromatin which was observed in whole egg extract required only the soluble cytoplasmic phase but that both soluble and particulate fractions were required for pronuclear envelope formation and complete stage II decondensation (Lohka and Masui, 1983a, b, 1984; Newport, 1987).

In spite of the availability of these *in vitro* systems, the identification of a specific decondensation factor remained elusive until very recently when Philpott *et al.* (1991) showed that the nucleosome assembly factor, nucleoplasmin, also functions as the major decondensation factor in *Xenopus* egg extract.

Nucleoplasmin was a logical candidate for a decondensation factor for two reasons. First, it is the most abundant protein in the oocyte nucleus and remains at high concentration in the egg cytoplasm after breakdown of the germinal vesicle. Thus, nucleoplasmin's distribution was coincident with the decondensation activity previously reported. Secondly, nucleoplasmin is highly acidic and assembles histones on to DNA in the egg extract. Thus, considering that extensive chromatin remodelling occurs on sperm chromatin as a prerequisite to pronuclear formation and that this remodelling involves the assembly of maternal histones on the sperm chromatin (see below), nucleoplasmin could facilitate decondensation through its ability to bind and transfer histones to the DNA.

Stage 1 decondensation of *Xenopus* sperm chromatin occurs very rapidly in egg high speed supernatant (HSS) (Fig. 2). Following a 5 minute incubation, nuclear length increases dramatically with more moderate changes observed in nuclear width (c). By 10 minutes (d), decondensation in HSS is nearly complete (Philpott *et al.*, 1991). HSS is devoid of the particulate (membrane) fraction and therefore cannot form nuclear envelopes and carry out stage II decondensation.

To determine whether nucleoplasmin played a role in sperm decondensation,

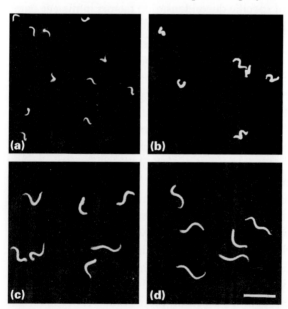

Figure 2. Decondensation of sperm chromatin in *Xenopus* egg extract. Demembranated *Xenopus* sperm nuclei were incubated in buffer (a), or in egg high speed supernatant for 3 (b), 5 (c) or 10 minutes (d) and photographed without fixation. By 5 minutes (c) stage I decondensation is extensive and by 10 minutes, it is nearly complete (d). Bar = 50 μm.

Philpott *et al.* (1991) used an immunodepletion/protein re-addition approach similar to that used by Dilworth *et al.* (1987) to define nucleoplasmin's role in nucleosome assembly. In this procedure, highly specific monoclonal antibodies against nucleoplasmin were incubated in egg extract and subsequently combined with protein A sepharose (PAS). The PAS was removed and the depletion process was repeated, removing virtually all the detectable nucleoplasmin from the egg extract.

Figure 3 shows the results of an experiment where sperm nuclei were incubated in depleted extract (b, '−NPL'), in mock-depleted extract (a, 'MOCK') and in depleted extract following re-addition of purified nucleoplasmin to the physiological concentration (c, '+NPL'). Very little chromatin decondensation is observed in depleted extract (b) compared to the mock-depleted control (a). However, when nucleoplasmin was re-added to a depleted extract, extensive decondensation was observed (c), indicating nucleoplasmin is necessary for sperm decondensation in egg extract. In addition, when sperm nuclei were incubated in purified nucleoplasmin alone, they decondensed to an extent similar to that observed for sperm nuclei incubated in egg extract for the same length of time. Taken together, these results demonstrated that nucleoplasmin is both necessary and sufficient for stage I decondensation of sperm chromatin in *Xenopus* egg extract (Philpott *et al.*, 1991).

Support for nucleoplasmin's role in decondensation has recently come from two other laboratories. Studying nuclear envelope assembly, Newport and Dunphy (1992) found that *Xenopus* sperm chromatin decondensed very rapidly in a heat-treated egg extract, which was highly enriched in nucleoplasmin. In these extracts, decondensation occurred in the absence of ATP hydrolysis, a finding entirely consistent with the results of Philpott *et al.* (1991), who showed that decondensation occurred in a solution of purified nucleoplasmin without added ATP. In separate studies, Ohsumi and Katagiri (1991b) have recently isolated a protein, from extracts of *Bufo japonicus* eggs, that is similar to *X. laevis* nucleoplasmin and appears to be responsible for decondensation of sperm chromatin. Indeed, in light of the results obtained by Philpott *et al.* (1991), it seems likely that a protein similar to nucleoplasmin would also bring about stage I decondensation in *B. japonicus* egg extracts.

In addition to amphibian cell-free systems, extracts from *Drosophila* embryos are also able to decondense completely sperm chromatin from a variety of species and assemble pronuclei *in vitro* (Berrios and Avilion, 1990; Crevel and Cotterill, 1991; Ulitzur and Gruenbaum, 1989). Thus, considering the similarities in both decondensation and nuclear assembly between these extracts and amphibian extracts, it would be interesting to know whether a protein similar to *Xenopus* nucleoplasmin plays a role in decondensation in *Drosophila*. Indeed, the fact that both amphibian and *Drosophila* extracts can decondense sperm chromatin from a wide variety of species, including humans (Brown *et al.*, 1987) suggests that the decondensation process has been highly conserved during evolution and that a

MOCK   -NPL

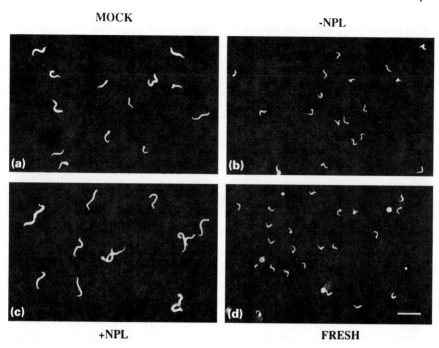

+NPL   FRESH

Figure 3. Decondensation of sperm chromatin in nucleoplasmin-depleted extract and in depleted extract reconstituted with purified nucleoplasmin. Sperm nuclei were photographed without incubation (d, FRESH) or following a 10 minute incubation in a mock-depleted extract (a, MOCK), an extract immunodepleted of nucleoplasmin (b, −NPL) or a depleted extract reconstituted with purified nucleoplasmin to physiological levels (c, +NPL). Virtually no decondensation was observed in depleted extract (b) as compared to the mock depleted control (a). However, re-addition of nucleoplasmin to a depleted extract restores decondensation of sperm chromatin (c). Bar = 50 μm. Figure reproduced from Philpott *et al.*(1991) with permission from Cell Press.

protein, or proteins, similar to nucleoplasmin may play a role in the fertilization process in many other species.

4. Chromatin remodelling during decondensation

During gametogenesis and development of the early embryo, numerous changes in chromatin structure occur. These changes are thought to be the result of post-translational modification of existing histones and/or the replacement of existing histones with specific histone variants or other highly basic, non-histone proteins (for review see Wolffe, 1991). Indeed, some of the most dramatic changes in chromatin composition and structure occur at fertilization when the highly compact sperm chromatin is decondensed and remodelled in the egg cytoplasm.

In many organisms, restructuring of sperm chromatin begins during spermatogenesis with the progressive replacement of certain somatic core histones with core histone variants or other highly basic proteins. This structural transition is thought to compact and stabilize the chromatin of the mature spermatozoon prior to entry into egg cytoplasm (Poccia, 1986). However, these changes must be reversible at fertilization to allow normal nuclear activity during the first cell cycle.

Mature sperm from *X. laevis* possess at least six basic proteins not found on somatic chromatin (Risley and Eckhardt, 1981; Mann *et al.*, 1982). These sperm-specific basic proteins almost entirely replace the somatic histones H2A and H2B on sperm chromatin during spermatogenesis. However, the two core histones, H3 and H4, remain at much higher concentration. In addition, histone H1 appears to be entirely absent from mature sperm chromatin (Risley and Eckhardt, 1981). Based on electrophoretic characteristics (Bols and Kasinsky, 1972) and amino acid analyses (Yokota *et al.*, 1991) most of the sperm-specific basic proteins found in *X. laevis* represent a class intermediate between somatic histones and protamines.

Numerous polyanions including polyglutamic acid (Dean, 1983) and heparin (Jager *et al.*, 1990) are able to decondense sperm chromatin from a variety of species *in vitro*. Furthermore, polyglutamic acid can also mediate the assembly of mouse sperm chromatin into nucleosomes (Dean, 1983), suggesting that highly acidic molecules may mediate decondensation by the removal of sperm-specific proteins and the addition of maternal histones to the DNA. Although Philpott *et al.* (1991) showed that polyglutamic acid was not able to decondense *Xenopus* sperm chromatin at a concentration equivalent to that used for purified nucleo-plasmin, the studies with polyanions have provided strong evidence for the role of acidic molecules in remodelling sperm chromatin at fertilization.

Nucleoplasmin's acidic nature and its roles in decondensation and nucleosome assembly made it a prime candidate for the remodelling of sperm chromatin. Indeed, direct evidence for this has recently been presented (Philpott and Leno, 1992).

Two striking changes in sperm protein composition occur during the rapid decondensation event in *Xenopus* egg extract (Philpott and Leno, 1992). First, there is a dramatic increase in the amounts of histones H2A and H2B on the chromatin resulting in near stoichiometric levels of the four core histones. Secondly, at least two sperm-specific basic proteins (designated 'X' and 'Y') are lost, to varying degrees, in egg extract. In addition, several other proteins from the egg extract are associated with sperm chromatin during decondensation. Although their identities are not yet clear, the proteins HMG A (Kleinschmidt *et al.*, 1983) and the histone H1-like proteins B4 (Smith *et al.*, 1988) or H1X (Ohsumi and Katagiri, 1991b), are possible candidates.

In addition to the changes observed in sperm chromatin composition, dramatic structural changes also occur during decondensation in the extract. This was determined using micrococcal nuclease to assay the subunit structure of sperm chromatin before and after decondensation. Digestion of sperm chromatin incubated in buffer showed no protected DNA fragments at any time point tested, however, within as little as 3 minutes in egg extract, a clear protected fragment of approximately 220 bp was resolved. By 10 minutes, a ladder of fragments was observed (Philpott and Leno, 1992). These results indicated two important points. First, the lack of a discernible protected fragment present on mature sperm chromatin suggested that it is not organized into nucleosomes and, secondly, the very rapid appearance of a protected ladder of fragments on decondensed sperm indicated that nucleosomes are formed during this process. Furthermore, these results suggested that the exchange of sperm basic proteins X and Y for core histones H2A and H2B resulted in the formation of nucleosomes on the DNA (Philpott and Leno, 1992).

To determine whether nucleoplasmin was responsible for the remodelling of sperm chromatin in egg extract, nucleoplasmin was immunoprecipitated from supernatants following incubation of sperm chromatin in egg HSS and in a solution of purified nucleoplasmin alone. In each case, the sperm-specific basic proteins X and Y co-precipitated with nucleoplasmin indicating an association of these proteins during decondensation. Furthermore, when sperm chromatin was incubated in a purified system containing nucleoplasmin and exogenous H2A and H2B, X and Y were still co-precipitated with nucleoplasmin even with the assembly of the histones on to the DNA. Micrococcal nuclease treatment of sperm chromatin, following incubation in this purified system, revealed a protected 146 bp DNA fragment indicating the presence of nucleosome cores. Thus, taken together, these experiments indicate that nucleoplasmin can mimic the egg extract in removing sperm proteins X and Y while assembling histones H2A and H2B onto the chromatin forming nucleosome cores (Philpott and Leno, 1992).

Immunodepletion experiments have confirmed nucleoplasmin's role in remodelling sperm chromatin during decondensation (Philpott and Leno, 1992). Figure 4 shows the proteins present on sperm chromatin following a 3 minute incubation in buffer (a), egg low speed supernatant (LSS) (b), egg depleted of nucleoplasmin (c) and in depleted extract following re-addition of purified nucleoplasmin (d). As previously discussed, decondensation of sperm chromatin in egg extract is accompanied by the loss of at least two sperm-specific proteins, X and Y, and the addition of H2A and H2B forming nucleosome cores (compare panel b with a). (The majority of maternal histone H2A assembled on the chromatin is probably an H2A variant referred to as H2A. X by West and Bonner, 1980; see Dilworth *et al.*, 1987.) In contrast, when sperm chromatin is incubated in nucleoplasmin-depleted extract, decondensation does not occur (Fig. 3) and no changes in the levels of X and Y or H2A and H2B are observed (compare panel c with a). Yet, when nucleoplasmin is added back to a depleted extract, the sperm chromatin decondenses (Fig. 3) and proteins X and Y are removed, mimicking the events occurring in complete extract (compare panels d and b). Thus, nucleoplas-

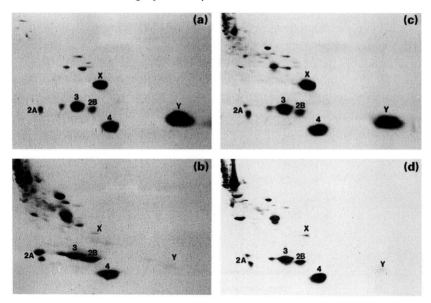

Figure 4. Remodelling of sperm chromatin in nucleoplasmin-depleted and nucleoplasmin-reconstituted extract. Sperm nuclei were incubated for 3 minutes in buffer (a), egg low speed supernatant (LSS) (b), a nucleoplasmin-depleted LSS (c) or a depleted LSS reconstituted with nucleoplasmin (d). Basic proteins were extracted with 0.5 N HCl and analysed by triton-acid-urea (first dimension)/SDS-PAGE (second dimension). Gels were stained with Coomassie blue. Within 3 minutes in LSS, sperm-specific basic proteins 'X' and 'Y' were removed from the chromatin while histones H2A and H2B are assembled on to the DNA (compare a with b). Much of the H2A assembled in extract is probably the variant H2A. X (b). In depleted extract, proteins X and Y are not removed from the chromatin (compare c with a) nor are H2A and H2B assembled on to the DNA (compare c with b). Re-addition of nucleoplasmin to a depleted extract results in removal of X and Y mimicking whole extract (compare d with b). Figure reproduced from Philpott and Leno (1992) with permission from Cell Press.

min functions as both an assembly factor and a disassembly factor, remodelling sperm chromatin during decondensation in egg extract (Philpott and Leno, 1992).

It is interesting to note that *Xenopus* egg LSS was much more efficient than HSS at removing proteins X and Y during decondensation (Philpott and Leno, 1992). Indeed, removal of X and Y in LSS was more efficient than in purified nucleoplasmin alone. Although the reason for this difference has not been determined, possibilities include the requirement for another factor or factors in LSS that work in conjunction with nucleoplasmin to bring about the complete removal of X and Y during decondensation or, alternatively, nucleoplasmin itself could be more efficient at removing X and Y in LSS than in HSS.

Ohsumi and Katagiri (1991a) have recently shown that the nucleoplasmin-like molecule from *Bufo* extracts which appears to be responsible for decondensation of sperm chromatin (see above), also binds *Bufo* sperm protamines during this period. In addition, the fact that these protamines are lost during decondensation in *Bufo* egg extract strongly suggests that this protein, like nucleoplasmin, functions as a disassembly factor during decondensation.

A model for the modulation of *Xenopus* sperm chromatin by nucleoplasmin has been proposed by Philpott and Leno (1992). In this model, nucleoplasmin assembles H2A and H2B on the DNA while removing sperm-specific proteins X and Y. This results in the formation of nucleosome cores (Fig. 5). The possibility

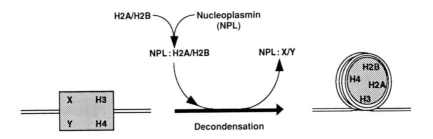

Figure 5. A model for the modulation of sperm chromatin during decondensation in egg extract. Nucleoplasmin is complexed with histones H2A and H2B in egg cytoplasm. During decondensation, nucleoplasmin assembles H2A and H2B on to the sperm chromatin and removes sperm-specific proteins X and Y. This remodelling results in the formation of the nucleosome core. Figure reproduced from Philpott and Leno (1992) with permission from Cell Press.

exists that this remodelling event may be the result of a concerted reaction, with individual molecules of nucleoplasmin both donating H2A and H2B and removing X and Y. This speculation is based largely on the observation that proteins X and Y are co-immunoprecipitated with nucleoplasmin, using a single anti-nucleoplasmin antibody, while histones H2A and H2B are not (Philpott and Leno, 1992). Co-immunoprecipitation of H2A and H2B with nucleoplasmin requires both an anti-nucleoplasmin and an anti-histone antibody, neither of which causes immunoprecipitation on its own (Dilworth *et al.*, 1987). This suggests that the interaction between nucleoplasmin and X and Y is stronger than the interaction between nucleoplasmin and H2A and H2B. Thus, the stronger binding of X and Y to nucleoplasmin would initially favour the release of H2A and H2B from the nucleoplasmin–H2A/H2B complex and subsequently promote the binding of X and Y to nucleoplasmin and their removal from the DNA. In this way, nucleoplasmin would add H2A and H2B to the chromatin and subsequently remove X and Y, completely remodelling sperm chromatin in preparation for pronuclear assembly and early embryogenesis.

5. Concluding remarks

Sperm chromatin from many species, including mammals, is highly compacted and shows no nucleosomal organization. However, following fertilization, sperm chromatin is decondensed and sperm-specific proteins or protamines, and some somatic histones present on the chromatin, are replaced with certain histone variants from the egg cytoplasm. This remodelling results in the formation of nucleosomes and presumably facilitates pronuclear assembly (for reviews see Longo and Kunkle, 1978, and Poccia, 1986). The interspecies similarities observed in the remodelling event suggests that a common mechanism may be responsible for these changes in chromatin structure. The recent identification of nucleoplasmin functioning in this capacity in *Xenopus* egg extract invites the speculation that a similar protein, or proteins, may bring about the chromatin remodelling observed in many species. In addition, it remains to be seen whether the role of nucleoplasmin as a decondensation factor only typifies the events in early embryonic systems or whether similar proteins in somatic cells are responsible for chromosome decondensation following mitosis.

References

Barry JM, Merriam RW. (1972) Swelling of hen erythrocyte nuclei in cytoplasm from *Xenopus* eggs. *Exp. Cell Res.* **71**: 90-96.

Berrios M, Avilion AA. (1990) Nuclear formation in a *Drosophila* cell-free system. *Exp. Cell Res.* **191**: 64-70.

Bols NC, Kasinsky HE. (1972) An electrophoretic comparison of histones in anuran testes. *Can. J. Zool.* **51**: 203-208.

Brown DB, Blake EJ, Wolgemuth DJ, Gordon K, Ruddle FH. (1987) Chromatin decondensation and DNA synthesis in human sperm activated *in vitro* by using *Xenopus laevis* egg extracts. *J. Exp. Zool.* **242**: 215-231.

Crevel G, Cotterill S. (1991) DNA replication in cell-free extracts from *Drosophila melanogaster*. *EMBO J.* **10**: 4361-4369.

Dean J. (1983) Decondensation of mouse sperm chromatin and reassembly into nucleosomes mediated by polyglutamic acid *in vitro*. *Dev. Biol.* **99**: 210-216.

Dilworth SM, Dingwall C. (1988) Chromatin assembly *in vitro* and *in vivo*. *Bioessays*, **9**: 44-49.

Dilworth SM, Black SJ, Laskey RA. (1987) Two complexes that contain histones are required for nucleosome assembly *in vitro*: role of nucleoplasmin and N1 in *Xenopus* egg extracts. *Cell*, **51**: 1009-1018.

Dingwall C, Laskey RA. (1990) Nucleoplasmin: the archetypal molecular chaperone. *Semin. Cell Biol.* **1**: 11-17.

Dingwall C, Sharnick SV, Laskey RA. (1982) A polypeptide domain that specifies migration of nucleoplasmin into the nucleus. *Cell*, **30**: 449-458.

Earnshaw WC, Honda BM, Laskey RA, Thomas JO. (1980) Assembly of nucleosomes: the reaction involving *X. laevis* nucleoplasmin. *Cell*, **21**: 373-383.

Graham CF, Arms D, Gurdon JB. (1966) The induction of DNA synthesis in frog egg cytoplasm. *Dev. Biol.* **14**: 349-381.

Gurdon JB. (1968) Changes in somatic cell nuclei inserted into growing and maturing amphibian oocytes. *J. Embryol. Exp. Morphol.* **20**: 401-414.

Gurdon JB. (1976) Injected nuclei in frog eggs: fate, enlargement and chromatin dispersal. *J. Embryol. Exp. Morphol.* **36**: 523-540.

Iwao Y, Katagiri C. (1984) *In vitro* induction of sperm nucleus decondensation by cytosol form mature toad eggs. *J. Exp. Zool.* **230**: 115-124.

Jager S, Wijchman J, Kremer J. (1990) Studies on the decondensation of human, mouse and bull sperm nuclei heparin and other polyanions. *J. Exp. Zool.* **256**: 315-322.

Kleinschmidt JA, Franke WW. (1982) Soluble acidic complexes contain histones H3 and H4 in nuclei of *Xenopus laevis* oocytes. *Cell,* **29**: 799-809.

Kleinschmidt JA, Scheer U, Dabauvalle MC, Bustin M, Franke WW. (1983) High mobility group proteins of amphibian oocytes: a large storage pool of a soluble high mobility group-1-like protein and involvement in transcriptional events. *J. Cell Biol.* **97**: 838-848.

Kleinschmidt JA, Fortkamp E, Krohne G, Zentgraf H, Franke WW. (1985) Co-existence of two different types of soluble histone complexes in nuclei of *Xenopus laevis* oocytes. *J. Biol. Chem.* **260**: 1166-1176.

Kleinschmidt JA, Seiter A, Zentgraf H. (1990) Nucleosome assembly *in vitro*: separate histone transfer and synergistic interaction of native histone complexes purified from nuclei of *Xenopus laevis* oocytes. *EMBO J.* **9**: 1309-1318.

Laskey RA, Leno GH. (1990) Assembly of the cell nucleus. *Trends Genet.* **6**: 406-410.

Laskey RA, Honda BM, Mills AD, Finch JT. (1978) Nucleosomes are assembled by an acidic protein which binds histones and transfers them to DNA. *Nature* **275**: 416-420.

Lohka MJ, Masui Y. (1983a) The germinal vesicle material required for sperm pronuclear formation is located in the soluble fraction of egg cytoplasm. *Exp. Cell Res.* **148**: 481-491.

Lohka MJ, Masui Y. (1983b) Formation *in vitro* of sperm pronuclei and mitotic chromosomes induced by amphibian ooplasmic components. *Science,* **220**: 719-721.

Lohka MJ, Masui Y. (1984) Roles of cytosol and cytoplasmic particles in nuclear envelope assembly and sperm pronuclear formation in cell-free preparations from amphibian eggs. *J. Cell Biol.* **98**: 1222-1230.

Longo FJ, Kunkle M. (1978) Transformations of sperm nuclei upon insemination. *Curr. Topics Dev. Biol.* **12**: 149-184.

Mann M, Risley MS, Eckhardt RA, Kasinsky HE. (1982) Characterization of spermatid/ sperm basic chromosomal proteins in the genus *Xenopus* (Anura, Pipidae). *J. Exp. Zool.* **222**: 173-186.

Newport J. (1987) Nuclear reconstitution *in vitro*: stages around assembly of protein free-DNA. *Cell,* **48**: 205-217.

Newport J, Dunphy W. (1992) Characterization of the membrane binding and fusion events during nuclear envelope assembly using purified components. *J. Cell Biol.* **116**: 295-306.

Ohsumi K, Katagiri C. (1991a) Characterization of the ooplasmic factor inducing decondensation of and protamine removal from toad sperm nuclei: involvement of nucleoplasmin. *Dev. Biol.* **148**: 295-305.

Oshumi K, Katagiri C. (1991b) Occurrence of H1 subtypes specific to pronuclei and cleavage-stage cell nuclei of anuran amphibians. *Dev. Biol.* **147**: 110-120.

Philpott A, Leno GH. (1992) Nucleoplasmin remodels sperm chromatin in *Xenopus* egg extracts. *Cell,* **69**: 759-767.

Philpott A, Leno GH, Laskey RA. (1991) Sperm decondensation in *Xenopus* egg cytoplasm is mediated by nucleoplasmin. *Cell,* **65**: 569-578.

Poccia D. (1986) Remodeling of nucleoproteins during gametogenesis, fertilization and early development. *Int. Rev. Cytol.* **105**: 1-65.

Risley MS, Eckhardt RA. (1981) H1 histone variants in *Xenopus laevis.* *Dev. Biol.* **84**: 79-87.

Sapp M, Worcel A. (1990) Purification and mechanism of action of a nucleosome assembly factor from *Xenopus* oocytes. *J. Biol. Chem.* **265**: 9357-9365.

Smith RC, Dworkin-Rastl E, Dworkin MB. (1988) Expression of a histone H1-like protein is restricted to early *Xenopus* development. *Genes Dev.* **2**: 1284-1295.

Ulitzer N, Gruenbaum Y. (1989) Nuclear envelope assembly around sperm chromatin in cell-free preparation from *Drosophila* embryos. *FEBS Lett.* **259**: 113-116.

West MHP, Bonner WM. (1980) Histone H2A, a heteromorphous family of eight protein species. *Biochemistry* **19**: 3238-3245.

Wolffe AP. (1989) Transcriptional activation of *Xenopus* class III genes in chromatin isolated from sperm and somatic nuclei. *Nucl. Acids Res.* **17**: 767-780.

Wolffe AP. (1991) Developmental regulation of chromatin structure and function. *Trends Cell Biol.* **1**: 61-66.

Yokota T, Takamune K, Katagiri C. (1991) Nuclear basic proteins of *Xenopus laevis* sperm: Their characterization and synthesis during spermatogenesis. *Dev. Growth Diff.* **33**: 9-17.

Chapter 11

Regulation of the human β-globin expression domain

Niall Dillon, Michael Antoniou, Meera Berry, Ernie DeBoer, Dubravka Drabek,
James Ellis, Peter Fraser, Olivia Hanscombe, Ali Imam, Marie-Ange Koken,
Michael Lindenbaum, Dies Meijer, Sjaak Philipsen, Sara Pruzina, Selina
Raguz, John Strouboulis, Dale Talbot, David Whyatt and Frank Grosveld

1. Introduction

The human β-globin gene cluster spans a region of 70 kb containing five developmentally regulated genes in the order 5′εGγAγ,δ,β 3′ (Fig. 1). In the early stages

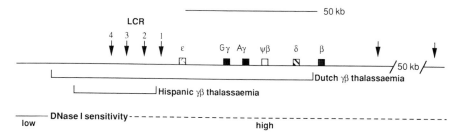

Figure 1. Structure of the human β-globin locus showing the deletions responsible for
Dutch and Hispanic γβ thalassaemia. Regions of high and low DNase I sensitivity in
erythroid cells are also indicated.

of human development, the embryonic yolk sac is the haematopoietic tissue and
expresses the ε-globin gene. This is followed by a switch to the γ-globin genes in
the fetal liver and the δ- and β-globin genes in adult bone marrow (for review, see
Collins and Weissmann, 1984). The globin genes are expressed at exceptionally
high levels giving rise to 90% of the total soluble protein in circulating red blood
cells.

The complete β-like gene locus has been sequenced and a large number of
structural defects, collectively known as the β-thalassaemias, have been documented in and around the β-globin gene (for review, see Collins and Weissman,
1984; Poncz *et al.*, 1989). In the condition known as hereditary persistence of fetal

haemoglobin (HPFH), γ-globin gene expression and hence HbF (fetal haemoglobin) production persist into adult life. These diseases are not only clinically important but they also provide natural models for the study of transcriptional regulation during development (see later). The entire locus is regulated by the locus control region (LCR) which first became apparent from the study of a human thalassaemia (Dutch γδβ). This particular thalassaemia contained an intact β-globin gene but had a deletion of the upstream part of the locus which prevented activation of the β gene (Fig. 1; Kioussis *et al.*, 1983; Wright *et al.*, 1984). The LCR is located upstream of the ε-globin gene (Fig. 1) and is characterized by a set of developmentally stable, hypersensitive sites (HS), 5′ HS 1, 2, 3 and 4 (Forrester *et al.*, 1987; Grosveld *et al.*, 1987; Tuan *et al.*, 1985). The importance of these sites is confirmed by the deletion in Hispanic γδβ-thalassaemia which removes sites 2, 3 and 4, leaving the rest of the locus intact but inactive (Driscoll *et al.*, 1989).

2. Functional analysis of the β-globin LCR

Linkage of the LCR to a cloned β-globin gene resulted in high levels of erythroid-specific expression of the transgene which was dependent on copy number and independent of the site of integration (Grosveld *et al.*, 1987). The original experiment was designed to test the role of the 5′ and 3′ flanking regions of the β-globin gene cluster. Subsequent experiments have shown that the LCR activity resides entirely in the 5′ hypersensitive sites (Blom van Assendelft *et al.*, 1989; Ryan *et al.*, 1989; Talbot and Groveld, 1991) and appears to be a dominant activation effect rather than a locus boundary element (LBE) of the type defined by Kellum and Schedl (1991). It has also been shown that the presence of the LCR creates a region of DNase I sensitivity extending over at least 150 kb (Forrester *et al.*, 1990).

Extensive functional dissection of the LCR has shown that small fragments containing single hypersensitive sites can also confer position-independent, copy-dependent expression on a linked β-globin gene. Functional analysis of HS 2, 3 and 4 has been used to narrow the regions containing the LCR activity to core fragments of 200–300 bp. Footprinting of these fragments has shown that they contain a large number of binding sites for erythroid-specific and ubiquitous DNA binding proteins (Fig. 2; Ney *et al.*, 1990; Philipsen *et al.*, 1990; Pruzina *et al.*, 1991; Talbot *et al.*, 1990). HS 2 contains a double AP1 binding site which has been shown to bind the erythroid-specific factor NFE2 as well as binding sites for the erythroid-specific factors GATA1 and the ubiquitous factors USF and J-BP (Talbot *et al.*, 1990; Talbot and Grosveld, 1991). Deletion of the NFE2 sites has been shown to reduce greatly the level of expression of a linked β-globin gene but the remaining binding sites can still give position-independent, copy-dependent expression at a much lower level (Talbot and Grosveld, 1991). Mutagenesis of the GATA1, USF and J-BP binding sites showed that none of these binding sites was

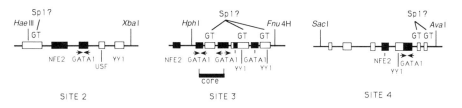

Figure 2. Core fragments from hypersensitive sites 2, 3 and 4 which exhibit LCR activity. Boxes represent binding sites for DNA binding proteins.

essential for LCR activity. However, the LCR effect was only observed in multi-copy integrations of HS2 core fragment constructs. Single copy integrants failed to give the LCR effect, suggesting that interactions between hypersensitive sites are necessary for the LCR function (Ellis *et al.*, 1993).

Footprinting of HS3 showed that it contained binding sites for GATA1 interspersed with several copies of a G-rich motif (Philipsen *et al.*, 1990). There are also multiple GATA1 sites in the enhancer located 3′ to the β-globin gene (Wall *et al.*, 1988). This enhancer does not confer position-insensitive, copy-dependent expression on linked genes indicating that GATA1 binding alone is not responsible for the LCR effect. Functional analysis has indicated that the minimum combination required for LCR activity is a GATA1 site flanked by two of these G-rich motifs (Philipsen *et al.* 1993).

3. Developmental regulation of the β-globin locus

The β-globin gene cluster has long been a prototype system for studying the regulation of multigene loci in general and the finding that transcription of all of the genes in the locus is potentiated by a single discrete regulatory region (the LCR) has raised new and interesting questions about the mechanisms by which the genes are sequentially activated during development. The fact that all of the sequences required for full expression of the genes in chromatin have been identified, offers a unique opportunity for investigating the mechanisms by which switching in a multigene locus actually occurs *in vivo*.

Linkage of any of the genes from the locus to the LCR results in expression in transgenic mice at levels which are far higher than are ever observed in the absence of the LCR (Behringer *et al.*, 1990; Dillon and Grosveld, 1991; Grosveld *et al.*, 1987; Shih *et al.*, 1990). This implies that, in addition to its chromatin opening function, the LCR also acts directly on the genes to enhance transcription. It is difficult to imagine how this could occur over distances as great as 50 kb except through direct physical interaction by looping out. There is now considerable evidence for this type of mechanism in different systems (Amouyal *et al.*, 1989; Bickel and Pirotta, 1990; Gellert and Nash, 1987; Mueller *et al.*, 1989).

How would the formation of such interactions be controlled so that only the genes which are specific for a given stage of development are activated while the others remain silent? One possibility is that stage-specific *trans*-acting factors binding to sequences in the promoters actively prevent activation of genes at particular stages of development. A similar effect could also result from the absence of stage-specific positive factors. A second possibility is that the genes compete for interaction with the same regulatory element as first suggested for the chicken globin genes (Choi and Engel, 1988). According to this scenario, each gene would retain an intrinsic capability for interaction with, and activation by, the LCR at all stages of development but interaction could only occur with one gene at a time. Stage-specific factors binding to the promoter would result in one gene interacting more effectively with the LCR and preventing the expression of the other genes through competition. Evidence for some type of cross-talk between the genes of the locus comes from the observation that genetic conditions which affect the expression of one gene can modify the expression of another (see later). However, competition models which envisage two genes competing for interaction with the LCR introduce a new set of parameters resulting from the kinetics of loop formation. These will be considered later in this review.

Functional analysis of human globin gene switching has depended to a large degree on the use of transgenic mice. Although mice switch directly from embryonic to adult globin gene expression at 12.5 days gestation and do not have separate fetal globin genes, a human γ-globin gene without the LCR is developmentally regulated in transgenic mice. The γ-globin transgene is expressed as an embryonic gene in mice and is switched off at the same time as the mouse embryonic genes. A human β-globin transgene is switched on at the same time as the mouse adult β genes (Chada *et al.*, 1985; Kollias *et al.*, 1986). Both genes were found to express at levels which were only a few percent of their normal level. A human ε-gene without the LCR is not expressed at all in transgenic mice (Shih *et al.*, 1990).

When the ε, γ and β genes were linked individually to the LCR, they were now found to express at levels which were similar to those of the endogenous mouse genes (Behringer *et al.*, 1990; Dillon and Grosveld, 1991; Enver *et al.*, 1990; Shih *et al.*, 1990). Developmental analysis of transgenic mice carrying the ε-gene showed that while high levels of expression were observed in the embryonic yolk sac, the gene was completely silenced in the fetal liver and adult bone marrow (Raich *et al.*, 1990; Shih *et al.*, 1990). A similar analysis of low-copy animals carrying a γ gene linked to the LCR again found high-level embryonic expression but in this case, the expression was found to extend into the early fetal liver before being completely switched off in the later fetal and adult stages (Dillon and Grosveld, 1991). This was a surprising result since the mouse does not have separate fetal genes. The conclusion must be that there are developmental changes in factor patterns in the mouse erythroid compartment which are not immediately apparent from the pattern of mouse globin expression.

We have already mentioned the existence of genetic conditions which affect the expression of different genes in the locus. One particularly interesting group of conditions are known by the term hereditary persistence of fetal haemoglobin (HPFH) and result in a continuation of γ expression into adult life at varying levels. While some of these conditions are associated with large deletions in the locus, a subset of patients have intact loci but have mutations in one of the γ promoters. We have tested whether one of these mutations (Greek HPFH; −117) is the direct cause of the failure of γ silencing by introducing a γ gene carrying the mutation into mice. The mice were found to express the mutated transgene at high levels in the adult stage in comparison to a wild-type gene which was completely silenced in adults (Berry *et al.*, 1992). Gel mobility analysis of the mutated sequence showed that, while a number of proteins bound to both wild-type and mutated sequences, GATA1 bound only to the wild-type sequence, suggesting that in this context it may be acting as a supressor.

A striking feature of the mutational HPFH syndromes is the fact that in heterozygotes the expression of the γ genes in adult life is associated with a specific down-regulation of the β gene on the same chromosome. This down-regulation of β is roughly equivalent to the rise in γ expression. For example, in southern Italian HPFH (−202), an increase in γ expression to 40% of the normal level results in a decrease in β expression to around 60% (Giglioni *et al.*, 1984). An even more dramatic example comes from experiments where a chromosome carrying the Greek Aγ HPFH mutation at −117 was introduced into murine erythroleukaemia cells. The HPFH γ gene was expressed at close to full level in these cells, reducing the level of β expression to less than 10% of that observed for a normal human chromosome (Papayannopoulou *et al.*, 1988). This equivalence between up-regulation of the γ gene resulting from mutations in different parts of the γ promoter and down-regulation of the β gene is in marked contrast to the situation in β-thalassaemias resulting in mutations or small deletions in the β promoter. In these conditions, in spite of a drastic reduction or even abolition of β-expression, γ expression in adult life generally remains at less than 5% (Poncz *et al.*, 1989).

It is clear from the genetic data that expression of a γ gene can reduce β-expression by more than 90%. It is also known that β has at least some capacity to be expressed during fetal life; low levels of β are observed quite early in fetal development and there is a gradual increase in β-expression during fetal life as γ expression decreases. This suggests that competition between the genes for activation by the LCR may play a role in modulating switching in the locus and that this competition is polar, with the γ genes competing with the β genes but not *vice versa*. Further evidence supporting a polar competition model comes from a recent functional study of globin switching in transgenic mice in which the order of the γ and β genes relative to the LCR was varied (Hanscombe *et al.*, 1991). When the genes were placed in their normal order (i.e. γ placed 5′ to β) expression of β was suppressed in the embryonic stage. When the order was reversed placing β in the 5′ position, both genes were now expressed in the embryonic stage (Fig. 3). Interpretation of this experiment was complicated by the arrangement of the

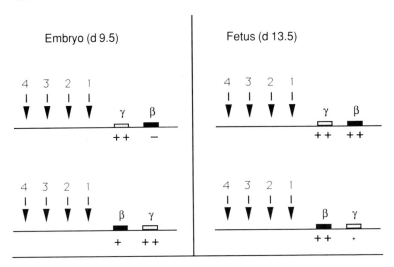

Figure 3. Effect of gene order on the expression of the γ- and β-globin genes at the embryonic yolk sac and fetal liver stages in transgenic mice.

genes in tandem repeats but testing of different constructs suggested that in a situation where there are two genes for every LCR, relative distance of the genes from the nearest LCR was an important parameter. A more proximally located gene appeared to have a strong advantage in competing for activation by the LCR while a more distally located gene appeared to be unable to compete effectively.

The data summarized earlier lead us to propose the model illustrated in Fig. 4. According to this model, competition occurs in one direction only with the early

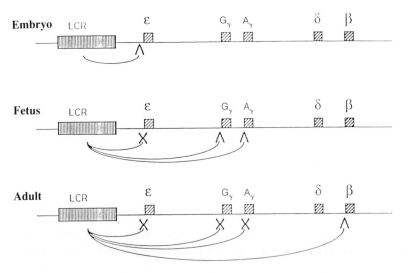

Figure 4. Model for developmental switching in the human β-globin locus. Arrows represent activation of genes by the LCR. An X indicates a blocked interaction.

genes modulating the expression of the later expressing genes but not *vice versa*. One question arising from such a model is that with stage-specific factors clearly sufficient for silencing of the ε and γ genes, why should it be necessary to invoke competition at all? Firstly the data we have discussed strongly favour a role for competition *in vivo*. In addition, the evidence to date suggests that eukaryotic genes in general are not regulated by simple single mechanisms but rather by a number of different elements working in combination and that such mechanisms are developed through an *ad hoc* recruitment of available factors (reviewed in He and Rosenfeld, 1991). Our model suggests that competition from a transcribing γ gene is one component which has been recruited into a multicomponent mechanism for silencing the β gene. Competition would also have the specific advantage of ensuring a smooth change-over between the genes, preventing the damage to the red cells which would result from over-production of β-like chains.

It is interesting to speculate on the type of mechanism which might give rise to polar competition. Loop models have already been discussed and are generally considered to be the most likely mechanism by which genetic elements act over long distances, although it must be stressed that there is, as yet, little direct evidence for looping as a mechanism for enhancer activity. A loop exclusion model in which only one gene at a time could interact with the LCR could explain the experimentally observed competition (Fig. 5). Such a model would also require that the LCR function as a single entity, presumably as a result of interaction between the individual hypersensitive sites. How would a loop exclusion model give rise to a polar competition effect? Tracking of a loop along the

Fetus, γ high, β low

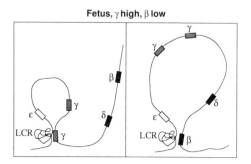

Adult, γ off, β high

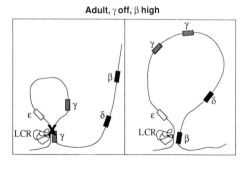

Figure 5. Schematic representation of how differential activation of the genes might occur during development through the formation of loops between the LCR and the promoters of individual genes.

chromosome until the first gene is encountered is one mechanism which might give rise to a polar effect. Such a mechanism would probably be energy requiring and require complex structures. Alternatively, polarity could be due to simple differences in the frequency of interaction of the LCR with the promoters as a result of differences in distance between promoters and LCR. A difference in frequency of interaction would always work in favour of a proximal gene, enhancing any competitive advantage provided by a stronger promoter at a specific developmental stage. The same effect would work against a distal gene reducing any advantage of a stronger promoter. The result would be to create a strong polar effect.

Nothing is known at present about how chromatin moves in the dense mixture of soluble proteins and nuclear matrix which makes up the nucleoplasm. It is known that the β-globin locus is part of a large DNase I-sensitive region which extends at least 100 kb downstream from the β gene (Forrester *et al.*, 1990). This suggests that the locus is part of a single large chromatin loop. Attachment to the nuclear matrix would obviously constrain the free movement of a DNA loop. Although several scaffold attachment sites have been mapped in the locus by LIS extraction (Jarman and Higgs, 1988), these do not appear to have the functional characteristics of locus boundary elements (D Greaves, unpublished) suggesting that they may not be attached to the nuclear scaffold in the activated locus. Directly relating chromatin structure to transcriptional function presents major technical problems. Functional analysis of competition between genes of equal promoter strength placed at varying distances from the LCR is an alternative approach to the study of chromatin dynamics which could provide new insights into this problem.

4. Developmental role of the LCR

The data reviewed earlier have concentrated on the role of sequences located in the immediate vicinity of the genes in the developmental regulation of the locus with competition playing a modulating role during switching. So far we have considered the LCR only as an element which is responsible for high-level expression of all of the genes. However, the complex organization of the hypersensitive sites also suggests the possibility that some developmental specificity may reside in the LCR itself. To test this possibility, an experiment was carried out in which each hypersensitive site was linked individually to a large fragment containing the two γ genes and the δ and β genes in their normal spacing. When these constructs were tested in transgenic mice, significant differences were observed between the developmental patterns obtained with the different hypersensitive sites (Fraser *et al.*, 1993). Particularly striking was the difference observed between HS3 and 4. HS3 was found to be the only site which was capable of giving rise to high-level expression of the γ genes in the fetal liver. In contrast, HS4 gives only low levels of γ expression but is able to potentiate adult β-expression much more efficiently than HS3. This result adds a new level of complexity to the regulation of the locus and raises the possibility that highly specific interactions occur between the individual hypersensitive sites and the promoters of the different genes in the locus. It is possible that the point mutations which give rise to the HPFH syndromes alter the configuration of the γ promoters so that they can now interact efficiently with HS4.

5. Regulation of the intact β-globin locus in transgenic mice

In order to extend the analysis of the β-globin locus to allow the functional dissection of the entire domain, we have recently developed a technique which allows us to introduce the entire locus into mice on a 70 kb fragment. This was achieved by generating single-stranded homopolymer tails at a unique restriction site at a point of overlap between two cosmids which together cover the entire locus (Strouboulis *et al.*, 1992). When this fragment was injected into mouse oocytes, transgenic mice carrying intact copies of the complete locus were generated at high frequency. Developmental analysis of human globin expression in these mice revealed essentially the same pattern as was obtained with smaller constructs, with the exception that the γ genes were expressed earlier than ε in the embryonic stage, and also at a much higher level. This indicates that the evolution of the present expression pattern of the ε and γ genes in humans was the result of multiple changes in both the sequences of the promoters and the factor balance in developing erythroid tissues. The ability to introduce the intact locus into mice should facilitate further analysis of the developmental regulation of the genes and the role of the LCR in regulating the whole locus.

References

Amouyal M, Mortensen L, Buc H, Hammer, K. (1989) Single and double loop formation when deoR repressor binds to its natural operator sites. *Cell*, **58**: 545-551.

Behringer RR, Ryan TM, Palmiter RD, Brinster RL, Townes TM. (1990) Human γ to β globin gene switching in transgenic mice. *Genes Dev.* **4**: 380-389.

Berry M, Grosveld F, Dillon N. (1992) A single point mutation is the cause of the Greek form of hereditary persistence of fetal haemoglobin. *Nature*, **358**: 499-502.

Blom van Assendelft G, Hanscombe O, Grosveld F, Greaves DR. (1989) The β-globin domain control region activates homologous and heterologous promoters in a tissue-specific manner. *Cell*, **56**: 969-977.

Bickel S, Pirotta V. (1990) Self association of the *Drosophila* zeste protein is responsible for transvection effects. *EMBO J.* **9**: 2959-2967.

Chada K, Magram J, Costantini F. (1985) An embryonic pattern of expression of a human fetal globin gene in transgenic mice. *Nature*, **319**: 685-689.

Choi O-R, Engel JD. (1988) Developmental regulation of human β globin gene switching. *Cell*, **55**: 17-26.

Collins FS, Weissman SM. (1984) The molecular genetics of human hemoglobin. *Prog. Acid Res. Mol. Biol.* **31**: 315-462.

Dillon N, Grosveld F. (1991) Human γ-globin genes silenced independently of other genes in the β-globin locus. *Nature,* **350**: 252-254.

Driscoll C, Dobkin C, Alter B. (1989) γδβ-Thalassemia due to a *de novo* mutation deleting the 5′ β-globin gene activation-region hypersensitive sites. *Proc. Natl Acad. Sci. USA,* **86**: 7470-7474.

Ellis J, Talbot D, Dillon N, Grosveld F. (1993) Synthetic human β-globin 5′HS2 constructs function as partially active locus control regions. *EMBO J.* **12**: 127-134.

Enver T, Raich N, Ebens AJ, Papayannopoulou T, Costantini F, Stamatoyannopoulos G. (1990) Developmental regulation of human fetal-to-adult globin gene switching in transgenic mice. *Nature,* **344**: 309-313.

Forrester W, Takegawa S, Papayannopoulou T, Stamatoyannopoulos G, Groudine M. (1987) Evidence for a locus activator region. *Nucl. Acids Res.* **15**: 10159-10177.

Forrester W, Epner E, Driscoll C, Enver T, Brice M, Papayannopoulou T, Groudine M. (1990) A deletion of the human β globin locus activation region causes a major alteration in chromatin structure and replication across the entire β globin locus. *Genes Dev.* **4**: 1637-1649.

Fraser P, Pruzina S, Antoniou M, Grosveld F. (1993) Each hypersensitive site of the human β-globin locus control region confers a different developmental pattern of expression on the globin genes. *Genes Dev.* **7**: 106-113.

Gellert M, Nash H. (1987) Communication between segments of DNA during site-specific recombination. *Nature,* **325**: 401-404.

Giglioni B, Casini C, Mantovani R, Merli S, Comp P, Ottolenghi S, Saglio G, Camaschella C, Mazza U. (1984) A molecular study of a family with Greek hereditary persistence of fetal hemoglobin and β-thalassaemia. *EMBO J.* **11**: 2641-2645.

Grosveld F, Blom van Assendelft G, Greaves D, Kollias G. (1987) Position-independent high level expression of the human β-globin gene in transgenic mice. *Cell,* **51**: 975-985.

Hanscombe O, Whyatt D, Fraser P, Yannoutsos N, Greaves D, Dillon N, Grosveld F. (1991) Globin gene order is important for correct developmental expression. *Genes Dev.* **8**: 1387-1394.

He X, Rosenfeld MG. (1991) Mechanisms of complex transcriptional regulation: implications for brain development. *Neuron,* **7**: 183-196.

Jarman A, Higgs D. (1988) Nuclear scaffold attachment sites in the human globin gene complexes. *EMBO J.* **7**: 3337-3344.

Kellum R, Schedl P. (1991) A position-effect assay for boundaries of higher order chromosomal domains. *Cell,* **64**: 941-950.

Kioussis D, Vanin E, deLange T, Flavell RA, Grosveld F. (1983) β-globin gene inactivation by DNA translocation in γ-thalassaemia. *Nature,* **306**: 662-666.

Kollias G, Wrighton N, Hurst J, Grosveld F. (1986) Regulated expression of human ^Aγ-, β-, and hybrid γβ-globin genes in transgenic mice: manipulation of the developmental expression patterns. *Cell,* **46**: 89-94.

Mueller H, Sogo J, Schaffner W. (1989) An enhancer stimulates transcription in trans when attached to the promoter via a protein bridge. *Cell,* **58**: 767-777.

Ney PA, Sorrentino BP, Lowrey CH, Nienhuis AW. (1990) Inducibility of the HS II enhancer depends on binding of an erythroid specific nuclear protein. *Nucl. Acids Res.* **18**: 6011-6017.

Papayannopoulou T, Enver T, Takegawa S, Anagou NP, Stamatoyannopoulos G. (1988) Activation of developmentally mutated human globin genes by cell fusion. *Science,* **238**: 1056-1058.

Philipsen S, Talbot D, Fraser P, Grosveld F. (1990) The β-globin dominant control region: hypersensitive site 2. *EMBO J.* **9**: 2159-2167.

Philipsen S, Pruzina S and Grosveld F. (1993) A specific combination of a G rich motif and GATA1 binding sites is sufficient to provide position independent expression of a β globin gene in transgenic mice. *EMBO J.* in press.

Poncz M, Henthorn P, Stoeckert C, Surrey S. (1989) Globin gene expression in hereditary persistence of fetal hemoglobin and β thalassaemia. In: *Oxford Surveys on Eukaryotic Genes,* vol. V (McClean N, ed). Oxford: University Press.

Pruzina S, Hanscombe O, Whyatt D, Grosveld F, Philipsen S. (1991) Hypersensitive site 4 of the human β-globin locus control region. *Nucl. Acids Res.* **19**: 1413-1419.

Raich N, Enver T, Nakamoto B, Josephson B, Papayannopoulou T, Stamatoyannopoulos G. (1990) Autonomous developmental control of human embryonic globin switching in transgenic mice. *Science*, **250**: 1147-1149.

Ryan TM, Behringer RR, Martin NC, Townes TM, Palmiter RD, Brinster RL. (1989) A single erythroid specific DNase I super-hypersensitive site activates high levels of human β-globin gene expression in transgenic mice. *Genes Dev.* **3**: 314-323.

Shih D, Wall R, Shapiro S. (1990) Developmentally regulated and erythroid-specific expression of the human embryonic β globin gene in transgenic mice. *Nucl. Acids Res.* **18**: 5465-5472.

Strouboulis J, Dillon N, Grosveld F. (1992) Developmental regulation of a complete 70-kb human β-globin locus in transgenic mice. *Genes Dev.* **6**: 1857-1864.

Talbot D, Philipsen S, Fraser P, Grosveld F. (1990) Detailed analysis of the site 3 region of the human β-globin dominant control region. *EMBO J.* **9**: 2169-2178.

Talbot D, Grosveld F. (1991) The 5'HS2 of the β-globin locus control region enhances transcription through the interaction of a multimeric complex binding at two functionally distinct NF-E2 binding sites. *EMBO J.* **10**: 1391-1398.

Tuan D, Solomon W, Li Q, London I. (1985) The "β-like globin" gene domain in human erythroid cells. *Proc. Natl Acad. Sci. USA*, **82**: 6384-6388.

Wall L, deBoer E, Grosveld F. (1988) The human β-globin gene 3' enhancer contains multiple binding sites for an erythroid-specific protein. *Genes Dev.* **2**: 1089-1100.

Wright S, deBoer E, Rosenthal A, Flavell RA, Grosveld F G. (1984) DNA sequences required for regulated expression of the β-globin genes in murine erythroleukaemia cells. *Phil. Trans. Roy. Soc. Lond. B* **307**: 271-282.

Chapter 12

Nuclear and nucleolar structure in plants

Peter Shaw, David Rawlins and Martin Highett

1. Introduction

Epifluorescence microscopy provides a highly sensitive detection method which allows the localization of structures within biological specimens to a resolution only limited by the resolution of the optical system employed. (In fact in many cases it is possible to *detect* objects far below the optical resolution limit.) When fluorescence detection is combined with three-dimensional (3-D) microscopy, either by computer processing of conventional (wide-field) focal section image data, or by the use of confocal microscopy, the result is a very powerful tool for the analysis of subcellular structure and activity. We have used fluorescence *in situ* hybridization on intact tissue sections to localize both DNA and RNA transcripts within the nuclei of plant cells, and have imaged these specimens by both wide-field and confocal microscopy. We have followed this with computer image processing, both to improve the images by deconvolution, and to visualize and interpret the 3-D structures obtained.

2. Telomere arrangement in plant root cells

The telomeres of all eukaryotic chromosomes consist of multiple tandem repeats of short (7–8) nucleotide sequences, which are extended by specific telomerases (see Richards *et al.*, Chapter 7; Greider *et al.*, Chapter 8, this volume, and references therein). We have used the telomere clone described by Richards and Ausubel (1988), isolated from *Arabidopsis thaliana*, to produce DNA probes labelled with biotin by random primer extension (see Rawlins and Shaw, 1991, for full details). We have shown by *in situ* hybridization to mitotic chromosome squash preparations that the *Arabidopsis* probe cross-hybridizes with the species *Pisum sativum* L. and *Vicia faba* L., labelling solely the telomeres at the end of each chromosome (Rawlins and Shaw, 1991). To investigate the location of the telomeres in interphase nuclei, we used tissue slices, approximately 50 μm in thickness, cut using a vibratome from root tips, fixed with freshly prepared formaldehyde. After hybridization with biotin-labelled probe, the target DNA was visualized with a fluorescent antibody. An optical section from *P. sativum*, collected using a Biorad MRC500 confocal laser scanning microscope, is shown in Fig. 1a. The most striking conclusion is that nearly all the telomeres are located at

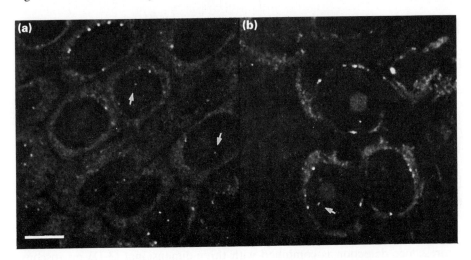

Figure 1. Confocal optical sections of root vibratome tissue sections fluorescently labelled with telomere probe. (a) *Pisum sativum* root tissue. (b) *Vicia faba* root tissue. In both species most of the telomeres are located at the nuclear periphery, often with up to four telomeres around the nucleolus (examples are arrowed). Bar = 5 μm.

the nuclear periphery. In addition we generally observe four telomeres internally near the nucleolus. These may correspond to the telomeres of the nucleolar organizer chromosomes; there are two pairs of such chromosomes in *Pisum* with the nucleolar organizer regions (NORs) occurring close to the ends of their respective chromosomes. Apart from these, very few telomeres are found within the nuclear interior. A very similar peripheral location is seen in *V. faba* interphase cells (Fig. 1b and Fig. 2), again with a few telomeres near the nucleolus.

The close association of telomeres with the nuclear periphery suggests that it is mediated by a specific interaction. This might be a direct interaction of telomere sequences with proteins of the nuclear lamina or other parts of the peripheral nuclear matrix. Alternatively, it may be mediated by other telomere-associated proteins, possibly even by telomerase. De Lange (1992) has recently identified a telomere-binding protein in human nuclear matrices. However there is so far no clear evidence of a specific peripheral location of telomeres in somatic human cells.

The stack of optical sections constitutes a 3-D representation of the specimen. This is best displayed by calculating a series of projections along directions about a common axis of rotation. When rapidly displayed sequentially on a computer screen, this gives a realistic illusion of rotation, which provides very good depth

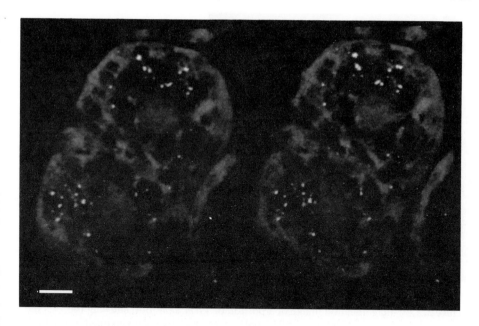

Figure 2. Stereo projection of an entire confocal 3-D data set showing two *Vicia faba* sister cells. Most of the telomeres are clustered on opposite sides of the peripheries of the two nuclei. Bar = 10 μm.

cues and greatly helps interpretation. Stereo projections, which are all that can be displayed in a publication, are not nearly so effective at conveying 3-D information.

When viewed in three dimensions in this way, the arrangement of telomeres around the nuclear periphery appears random in *Pisum*. However, in *Vicia* there is generally a clear clustering of the telomeres. The clusters are often related by an approximate mirror symmetry in sister cells, as in the example shown in Fig. 3, almost always occurring on opposite sides of the nuclei. This polarity of clusters may extend through many cells in a file, suggesting an underlying ordering of the interphase chromosomes which is maintained or re-established through several cycles of cell division.

It is not clear why these two species should exhibit differences in the detailed arrangement of telomeres, even though in both species the telomeres are located predominantly at the nuclear periphery. It may be that the larger *Vicia* chromosomes are more highly constrained, and that reorganization of the gross chromosome location after telophase is more difficult. This would leave the chromosomes in the 'Rabl' configuration (Rabl, 1885), with centromeres at one side of the nucleus and telomeres at the other side. However, in this case, the polarity we have seen, with the telomeres on *opposite* sides of the nuclei, would imply a rotation of the nuclei after telophase.

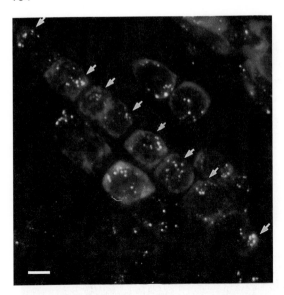

Figure 3. Low-magnification image of a portion of *Vicia faba* root, showing the telomere clustering in a file of cells. The polarity of clustering is maintained through at least eight cells, corresponding to three cycles of cell division. Bar = 20 μm.

3. Image deconvolution in 3-D optical microscopy

The major advantage of confocal microscopy in imaging thick biological structures is the good rejection of out-of-focus information, which is achieved by the use of two apertures at conjugate planes to the focal plane. Thus, accurate focal sections are produced with much reduced blurred contributions from bright structures away from the plane of focus. An alternative to confocal microscopy is to measure focal sections using conventional, wide-field optics, measure the point spread function – the blurred image of a single point – and use computation to reverse the effect of the blurring by the point spread function. This is generally termed deconvolution, and is a familiar problem in several different fields. A number of different methods have been developed, but for the most part only the simpler and less time-consuming ones have been applied to optical imaging, mainly because the vast amount of data in a 3-D optical image makes many of the methods prohibitively expensive in computation (for reviews see Agard *et al.*, 1989; Shaw and Rawlins, 1991a). We routinely use the iterative, constrained algorithm originally devised by Jannson (1984) and developed for optical image processing by Agard *et al.* (1989).

In order to compare confocal imaging with deconvolution of conventional fluorescence images we measured 3-D data from the same specimen by the two methods. A low light level cooled CCD (charge coupled device) camera was used for the measurement of the conventional, wide-field data. The results of one such experiment are shown in Fig. 4. Figure 4a shows a single section from the data set collected by CCD camera, and Fig. 4b shows the equivalent section after deconvolution. The increase in clarity of the fine image detail and the reduction of background out-of-focus flare are clearly seen. In Fig. 4c the equivalent section from the data set collected by confocal microscopy is shown. The agreement with the deconvoluted CCD data is quite remarkable, even down to the finest image details. The fact that two such different imaging techniques agree so well gives us great confidence in the accuracy of these images. Furthermore, we have previously shown that deconvolution can also be applied to confocal data, using an appropriate point spread function (Shaw and Rawlins, 1991b). Fig. 4d shows the equivalent section from the deconvoluted confocal data set. A further substantial improvement in image clarity is evident. In many of the subsequent images we have combined confocal microscopy with deconvolution in this way, to improve their interpretability.

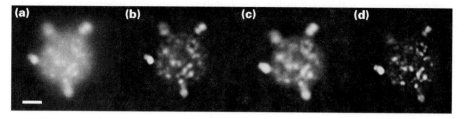

Figure 4. Comparison of wide-field (non-confocal) and confocal imaging. The same nucleolus has been imaged twice, once using wide-field optics, once using confocal optics. An equivalent single section is shown in each case. (a) Wide-field data. (b) Wide-field data after deconvolution. (c) Confocal data. (d) Confocal data after deconvolution. Note the excellent agreement between (b) and (c), and the increased clarity of (d) over either (b) or (c). Bar = 2 μm.

4. Location of rDNA and 5S genes

The nucleolus is well known to be the site of the actively transcribed rRNA genes, and of assembly of the four ribosomal RNAs and the ribosomal proteins into preribosomal small and large subunits. Thus, it provides a spatially well-defined region of the cell for study in which many copies of the rRNA gene are transcribed, where the 45S primary transcript is processed and combined with the 5S transcripts, which are imported into the nucleolus after transcription from the 5S genes organized as tandem arrays elsewhere in the nucleus. The nucleolus should therefore be an ideal system for studying transcription, transcript processing and transport.

The ultrastructure of the nucleolus has been studied extensively over a number of years using electron microscopy (for reviews see Hadjiolov, 1985; Schwarzacher and Wachtler, 1991; Warner, 1989). There is general agreement that, when stained with standard reagents – osmium, uranyl acetate and lead citrate –

several characteristic structures are seen (Goessens, 1984; Jordan, 1984). Firstly there is a number of lightly stained regions, termed fibrillar centres. These are typically a fraction of a micrometre (μm) in diameter but may be much larger in some cells. Often they are surrounded by densely staining material called the dense fibrillar component. The rest of the nucleolus is filled with particles approximately the size of ribosomes – termed the granular component. Condensed chromatin (nucleolar-associated chromatin) is often seen outside the body of the nucleolus, and this is generally agreed to be condensed, inactive copies of the rRNA genes.

In spite of the agreement about the ultrastructure, there is still controversy over its significance (Jordan, 1991). Early studies of the incorporation of ^{3}H into nascent transcripts, localized the label to the dense fibrillar component (Goessens, 1976; Geuskens and Bernhard, 1966). Subsequently, immunogold electron microscopy using antibodies to RNA polymerase I showed labelling only on the fibrillar centres (Scheer and Rose, 1984). Some investigators have localized rDNA to the fibrillar centres (Thiry et al., 1988), others to the dense fibrillar component (Wachtler et al., 1989). Thus, there is disagreement about the location of the rRNA genes and about the likely sites of transcription.

We have used in situ hybridization and 3-D confocal microscopy to address this problem. In our initial studies (Rawlins and Shaw, 1990) we used double-stranded DNA probes to localize the rDNA in P. sativum root tips. This showed that each nucleolus has four perinucleolar knobs, corresponding to the inactive, condensed nucleolar-associated chromatin, and internal nucleolar structure usually consisting of small bright foci of labelling. We have subsequently used single-stranded RNA probes, produced by in vitro transcription of a single repeat of the pea rRNA or 5S genes cloned into the Bluescript vector. In all the clones, the orientation of the insert was such that T7 RNA polymerase produces sense probe, which is complementary to the transcribed strand of the DNA, and T3 RNA polymerase produces antisense probe, which is complementary to the respective transcripts. Non-isotopic labelling was carried out by incorporation of either digoxigenin- or fluorescein-labelled UTP. The probes were imaged after hybridization either using a fluorescently labelled secondary antibody or by direct visualization of the labelled probe, respectively. The single-stranded RNA probes have proved more sensitive with lower background labelling and give the possibility of distinguishing between the labelling of gene and transcript.

Figure 5 (see p. 175) shows a conventional fluorescence micrograph of a field of a typical specimen; the yellow/green fluorescence is from in situ labelling with sense rDNA probe, showing the nucleolus and perinucleolar labelling, the orange fluorescence is produced by the DNA-specific dye 7-aminoactinomycin D to show the nucleus. The whole of the tissue slice is available for in situ labelling, and thus all the cell types present in the root tip can be easily imaged. For all the probes we have used, we have carried out nuclease digestion control experiments, to confirm that either DNA or RNA targets are labelled as expected.

All nucleoli show the four perinucleolar knobs of condensed nucleolar-associated rDNA, and this is a good diagnostic test for the success of the hybridization experiment (e.g. probe penetration and tissue preservation). The deconvoluted images often show substructure of the knobs consistent with tightly packed chromatin fibres. Two of the knobs are larger, two smaller, which is consistent with the known relative sizes of the clusters (Ellis *et al.*, 1984). The internal nucleolar rDNA labelling is greatly dependent on cell type. In the cells of the quiescent centre, often only the four condensed masses of rDNA chromatin are seen, with no identifiable nucleolar body. In the active meristematic region several types of labelling are seen. Most nucleoli contain many bright internal spots or foci of labelling. These are interconnected with a network of extended fainter labelling, which together with the foci permeates the whole body of the nucleolus (see Fig. 6a). The intensity of labelling of the foci is generally less than that of the perinucleolar chromatin. In a proportion of nucleoli there is a central mass of condensed chromatin in which closely appressed strands can be seen (Fig. 6b). Again, the intensity of labelling is less than the perinucleolar chromatin, suggesting a lower degree of condensation. In some nucleoli there are very few foci, or none at all. In these nucleoli the internal labelling is much more diffuse and often shows distinct regions of high and low labelling intensity, giving the impression of 'cavities' in a cage-like structure, as shown in Fig. 6c. In more highly differentiated, and presumably less transcriptionally active cells, such as those of the root cap, the epidermis and the vascular tissue, the nucleoli are smaller and the internal foci are larger and account for a greater proportion of the internal labelling.

In order to quantify the relationship between rDNA distribution and transcriptional activity we have made measurements of the total integrated fluorescence intensity from the 3-D confocal data (using the 'raw', non-deconvoluted data). In each case we measured the total fluorescence intensity for the nucleolus, and that of the perinucleolar knobs. The proportion of fluorescence present in the knobs is plotted against nucleolar volume in Fig. 7. This shows a strong negative correlation between nucleolar volume and proportion of fluorescence in the knobs. Since nucleolar volume is known to be correlated with transcriptional activity (see, for example, Thompson and Flavell, 1988), this graph shows that as the nucleolus becomes more active, more of the condensed perinucleolar rDNA decondenses into the nucleolus enlarging it. The proportion of perinucleolar rDNA varies from nearly 100% (in the quiescent centre) to a very small percentage in the largest meristematic cells. This suggests that virtually all the rDNA is capable of decondensing into the nucleolus, at least in these cells.

The sense probe to the 5S genes shows the nuclear location of the three pairs of 5S gene clusters present in *P. sativum* (Ellis *et al.*, 1988). An interesting observation is that there is generally one 5S gene cluster located at the nucleolar periphery; it is possible that this is the most actively transcribed cluster. Figure 8 (see p. 175) shows a nucleolus doubly labelled with 5S sense probe (green) and rDNA

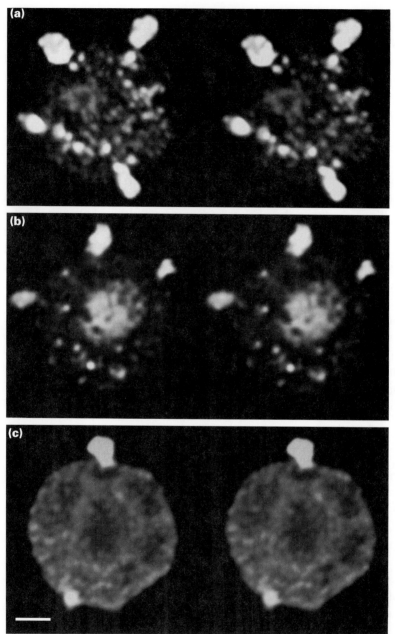

Figure 6. Stereo projections of the central 1 µm slice through nucleoli labelled with sense probe to rDNA (i.e. localizing the rDNA genes). (a) Most nucleoli in the active meristematic region show many bright foci of labelling, together with fainter, more extended labelling. In this nucleolus the four perinucleolar knobs of condensed nucleolar-associated chromatin are all visible in this central slice. (b) Some meristematic nucleoli show a central mass of condensed chromatin fibres. (c) A few nucleoli have a much more uniform, fainter internal labelling, with internal 'cavities' showing little label. Bar = 2 µm.

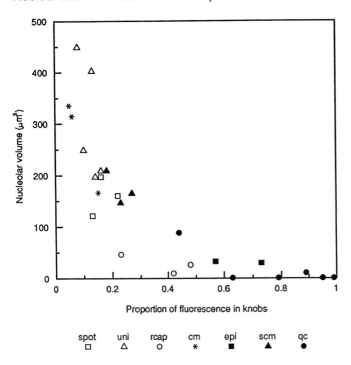

Figure 7. A graph showing the proportion of the total integrated rDNA fluorescence contained in the perinucleolar knobs as a function of nucleolar volume. Each point represents the data from an entire 3-D image of a nucleolus. There is a strong negative correlation, showing that as the nucleolus becomes larger and more active, the proportion of rDNA decondensed into the body of the nucleolus increases.

sense label (red). The antisense probe shows location of both the genes and the transcripts within the nucleolus. A characteristic pattern of bright labelling surrounding dark cavities, very reminiscent of the diffuse rDNA structure, is seen for 5S antisense labelling. This pattern is virtually identical to that seen with antisense rDNA probe, which localizes the 45S transcripts, as shown by the double labelling experiment shown in Fig. 9 (see p. 175).

In summary, *in situ* labelling with sense probe to the rDNA has shown three levels of chromatin condensation: the most condensed chromatin is seen in the perinucleolar knobs of nucleolar-associated chromatin, which must be transcriptionally inactive; an intermediate condensation of internal rDNA is seen in foci and condensed masses within the nucleolus; the most decondensed rDNA is seen as a fainter extended internal network. Either or both of the last rDNA components may be transcriptionally active. The internal arrangement depends on transcriptional activity; as the nucleolus becomes more active, rDNA from the perinucleolar knobs decondenses to increase the nucleolar volume. Labelling with antisense probe to 45S and 5S transcripts shows, in virtually all nucleoli, a characteristic cage-like arrangement. This is also shown in some cases of rDNA (sense) labelling, usually in the largest and transcriptionally most active nucleoli. The cage structure is very reminiscent of immunofluorescence labelling obtained by Corben *et al.* (1989) using a monoclonal antibody to a plant nucleolar matrix protein, and it is therefore possible that this structure represents an underlying matrix on which the transcription and/or subsequent processing is organized. It should be remembered that the antisense probes we have employed will in all probability label RNA up to and including pre-ribosomes in the nucleolus. In order to clarify the different stages of processing we have recently made probes to the external transcribed spacer and non-transcribed spacer regions of *P. sativum* rDNA and to a conserved region of plant U3 snRNA. This should enable us to identify the sites of transcription, or at least the sites of the earliest transcript processing events, and to determine their relationship to the structures we have observed.

Acknowledgements

This work was funded by the Agricultural and Food Research Council of the UK, via a grant in aid to the John Innes Institute, and was also supported by funds from the AFRC plant molecular biology initiative.

References

Agard DA, Hiraoka Y, Shaw PJ, Sedat JW. (1989) Fluorescence microscopy in 3-dimensions. *Methods Cell Biol.* **30:** 353-377.

Corben E, Butcher G, Hutchings A, Wells B, Roberts K. (1989) A nucleolar matrix protein from carrot cells identified by a monoclonal antibody. *Eur. J. Cell Biol.* **50:** 353-359.

Ellis THN, Davies DR, Castleton JA, Bedford ID. (1984) The organization and genetics of rDNA length variants in peas. *Chromosoma*, **91:** 74-81.

Ellis THN, Thomas CM, Simpson PR, Cleary WG, Newman M-A, Burcham KWG. (1988) 5S rRNA genes in Pisum: sequence, long range and chromosomal organization. *Mol. Gen. Genet.* **214:** 333-342.

Geuskens M, Bernhard W. (1966) Cytochemie ultrastructure nucleole. *Exp. Cell Res.* **44:** 579-598.

Goessens G. (1976) High resolution autoradiographic studies of of Ehrlich tumor cell nucleoli. E*xp. Cell Res.* **100:** 88-94.

Goessens G. (1984) Nucleolar structure. *Int. Rev. Cytol.* **87:** 107-158.

Hadjiolov AA. (1985) The nucleolus and ribosome biogenesis. *Cell Biol. Monogr. 12.* Vienna: Springer Verlag.

Jannson PA. (1984) *Deconvolution with Applications in Spectroscopy.* London: Academic Press. (see pp 69-135).

Jordan EG. (1984) Nucleolar nomenclature. *J. Cell Sci.* **67:** 217-220.

Jordan EG. (1991) Interpreting nucleolar structure: where are the transcribing genes? *J. Cell Sci.* **98:** 437-442.

de Lange T. (1992) Human telomeres are attached to the nuclear matrix. *EMBO J.* **11:** 717-724.

Rabl C. (1885) Uber Zelltheilung. *Morphologisches Jahrbuch*, **10:** 214-330.

Rawlins DJ, Shaw PJ. (1990) Three-dimensional organization of ribosomal DNA in interphase nuclei of *Pisum sativum* by *in situ* hybridization and optical tomography. *Chromosoma*, **99:** 143-151.

Rawlins DJ, Shaw PJ. (1991) Localization of telomeres in plant interphase nuclei by *in situ* hybridization and 3D confocal microscopy. *Chromosoma*, **100:** 424-431.

Richards EJ, Ausubel FM. (1988) Isolation of a higher eukaryotic telomere from *Arabidopsis thaliana. Cell*, **53:** 127-136.

Scheer U, Rose K. (1984) Localization of RNA polymerase I in interphase cells and mitotic chromosomes by light and electron microscopic immuno-cytochemistry. *Proc. Natl Acad. Sci. USA*, **81:** 1431-1435.

Schwarzacher HG, Wachtler F. (1991) The functional significance of nucleolar structures. *Annu. Genet.* **34:** 151-160.

Shaw PJ, Rawlins DJ. (1991a). Three-dimensional fluorescence microscopy. *Prog. Biophys. Mol. Biol.* **56:** 187-213.

Shaw PJ, Rawlins DJ. (1991b). The point spread function of a confocal microscope: its measurement and use in deconvolution of 3-D data. *J. Microsc.* **163:** 151-165.

Thiry M, Scheer U, Goessens G. (1988). Localization of DNA within Ehrlich tumor cell nucleoli by immuno-electron microscopy. *Biol. Cell*, **63:** 27-34.

Thompson WF, Flavell RB. (1988) DNase I sensitivity of ribosomal RNA genes in chromatin and nucleolar dominance in wheat. *J. Mol. Biol.* **204:** 535-548.

Wachtler F, Hartung M, Devictor M, Wiegant J, Stahl A, Schwarzacher HG. (1989) Ribosomal DNA is located and transcribed in the dense fibrillar component of human Sertoli cell nucleoli. *Exp. Cell Res.* **184:** 61-71.

Warner JR. (1989) The nucleolus and ribosome formation. *Curr. Opinion Cell Biol.* **2:** 521-527.

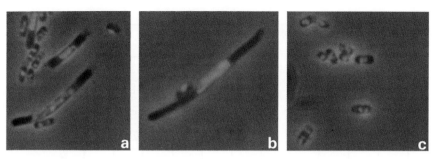

Chapter 3, Figure 1. Fluorescence photomicrographs of mutants deficient in Xer-mediated recombination: (a) XerD⁻, (b) *dif*, (c) wild-type. Figure cited on pp. 27 and 32.

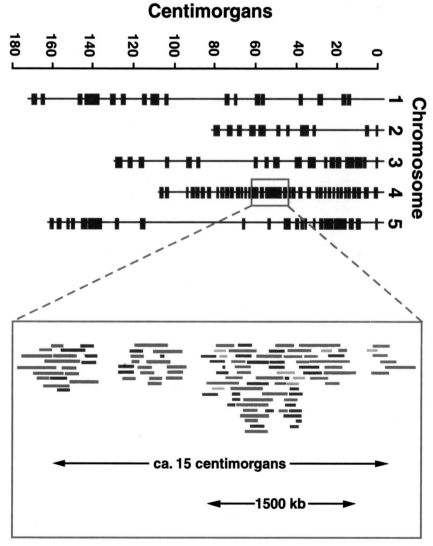

Chapter 18, Figure 2 (refer to p. 181 for caption).

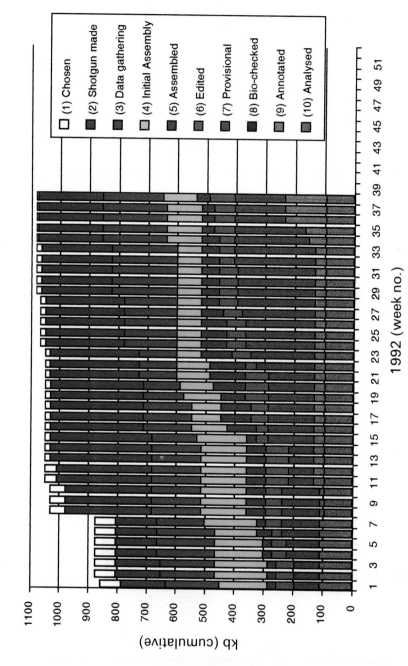

Chapter 4, Figure 11. A summary of the status of sequencing the first megabase of the *E. coli* genome. Figure cited on p. 58.

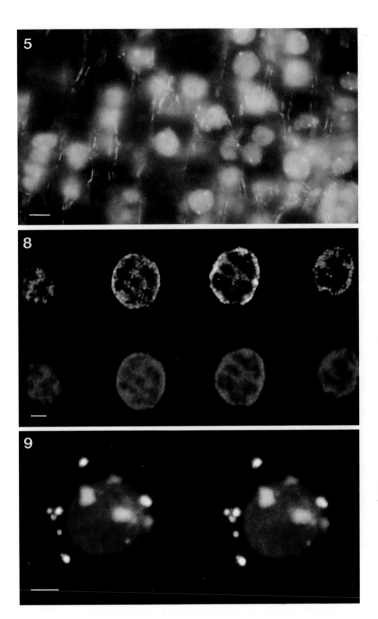

Chapter 12, Figure 5. Conventional epifluorescence micrograph of a portion of *Pisum sativum* root tissue, labelled with sense rDNA probe (green), counterstained with the DNA-specific dye 7-aminoactinomycin D. Bar = 10 μm. Figure cited on p. 166.

Chapter 12, Figure 8. Confocal sections from a specimen labelled with antisense 5S probe (green) and antisense 45S probe (red). The two probes co-localize to a large extent. Sheets of labelling surrounding dark cavities making a cage-like structure are evident. Bar = 2 μm. Figure cited on p. 167.

Chapter 12, Figure 9. A stereo projection of a the 3-D structure of a nucleolus labelled with sense 5S probe (green) and sense rDNA probe (red). The location of the 5S gene clusters in in the nucleoplasm is clear, although one cluster is at the nucleolar periphery. The four perinucleolar rDNA knobs are visible, but much of the internal rDNA substructure is lost in this projection. Bar = 2 μm. Figure cited on p. 169.

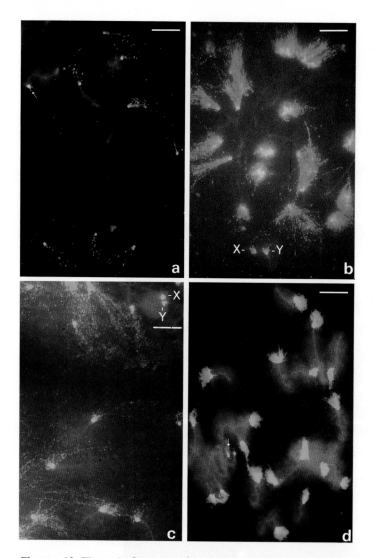

Chapter 13, Figure 3. Organization of minor satellite chromatin in *M. domesticus, M. spretus* and the F1 hybrid. (a) In most pachytene cells, the minor satellite chromatin is confined to the centromere of the SC (shown by arrow). In some cells, such as this cell, the minor satellite also is found in a few chromatin loops. (b) In *M. spretus* the minor satellite is the predominant satellite sequence and occurs in a large number of centromeric chromatin loops. In this species, the X and the Y centromeric chromatin is labelled, X and Y. (c) In the F1 male hybrid, the *M. spretus* Y chromosome and the *M. domesticus* X chromosome centromeres are labelled, X and Y. In the paired pachytene chromosomes, the origin of each centromere should be evident by the abundance of minor satellite but in practice the two centromeres are not distinguishable with this technique. (d) In an alternative method the minor satellite is labelled by an '*in situ* primed' reaction to the denatured pachytene chromosome spreads of *M. spretus* by the addition of polymerase, nucleotides, and a 17 bp oligonucleotide that is specific for the minor satellite. The chromatin is red, the SC is bright red and the minor satellite is yellow-green. The X and Y centromeres are shown by the arrow. Scale bar = 10 μm. Figure cited on pp. 186, 188, and 190.

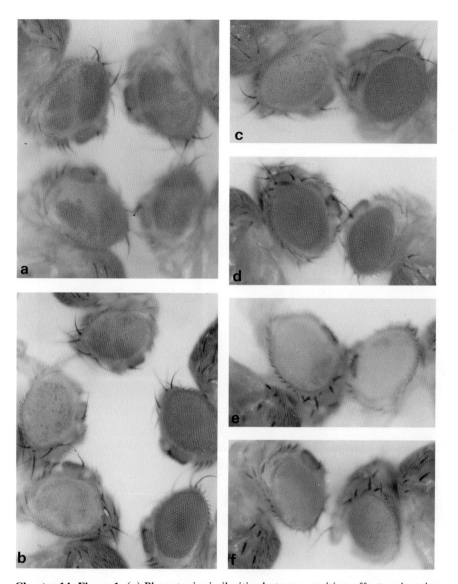

Chapter 14, Figure 1. (a) Phenotypic similarities between position-effect variegation (PEV) and somatic gene loss. A re-arrangement selected for PEV of the *brown* gene present on a transposon, V21, shows random patches and spots (flies below). This phenotype resembles random loss of the transposon when P-transposase is expressed in somatic cells (flies above). Genotypes are T(2;3) V21 P[*bw*+](92C), *bw*^D;*st* V21/*st* (below) and P[*bw*+](8D)/+;*bw*^D;*st* P[*ry*+Δ2,3](99B) (above). (b) Examples of dominant PEV. Genotypes counterclockwise from bottom right: *st* (wild-type for *brown*), *bw*^DX7/+;*st* (a weak allele from Tom Kaufman), *bw*^VI/SMI;*st* (*Plum*/*bw*+), *bw*^D/+;*st* (the strongest *bw*^V allele), *w*^m4 (an example of recessive PEV on the *white* gene. (c) Suppression by *Suppressor-of-Plum*. Genotypes are: *bw*^D/+ (left) and *bw*^D/*px bw* *Su(Pm) sp* (right). (d–f) Strong dominant PEV on a transposed copy of *brown*. Genotypes are: (d) P[*bw*+](92C)/+;*st* (92C hemizygote); (e) T(2;3) V21^extreme P[*bw*+] (92C), *bw*^D;*st* Sb V21^extreme/*st* (V21^extreme hemizygote); (f) T(2;3) V21^extreme P[*bw*+](92C), *bw*^D;*st* Sb V21^extreme/ P[*bw*+](92C), *st* (V21^extreme/92C heterozygote). Except for *w*^m4, all flies carry the *scarlet* mutation leading to eyes that are red for *bw*+ and white for *bw*−. In (a) and (d–f), a female is on the left and a male is on the right. Figure cited on pp. 194, 195, 198 and 201.

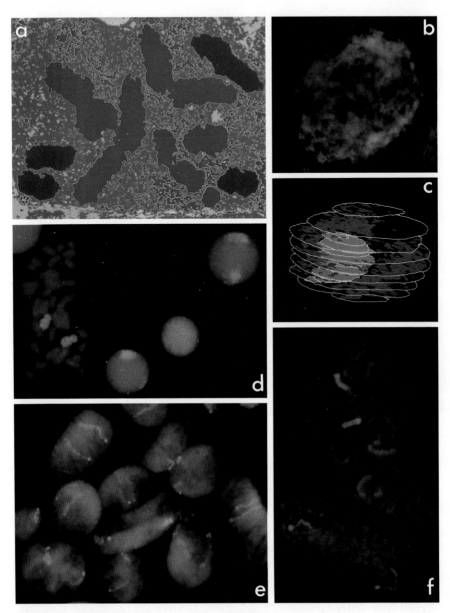

Chapter 16, Figure 2 (refer to p. 181 for caption).

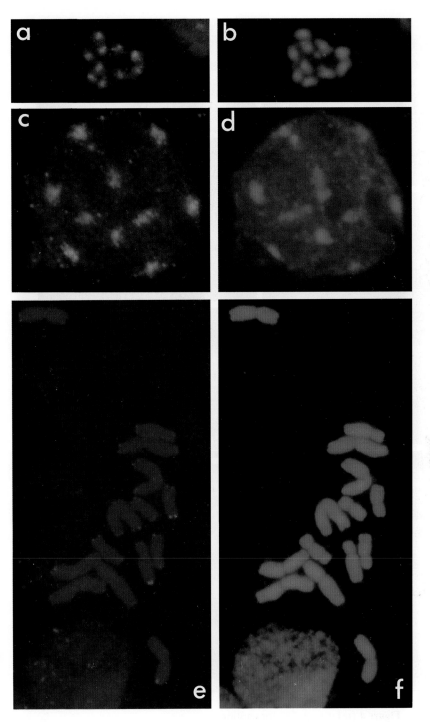

Chapter 16, Figure 3 (refer to p. 181 for caption).

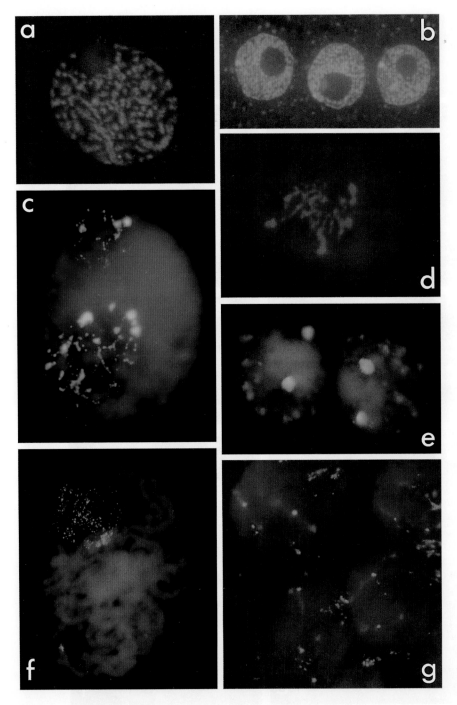

Chapter 16, Figure 4 (refer to p. 182 for caption).

Chapter 18, Figure 2. YAC clone coverage of the *Arabidopsis thaliana* genome. RFLP markers mapping to the five Arabidopsis chromosomes were hybridized to YAC libraries (Hauge *et al.*, 1991; Schmidt and Dean, 1993). The YAC contigs identified in this way are shown as blue blocks along the chromosomes (top portion of the figure). (To convert the size of the YAC contigs into genetic distance, the value of 1 cM for 140 kb was used (Chang *et al.*, 1988)).

Hybridization of newly available RFLP markers mapping to chromosome 4 together with chromosome walking experiments have resulted in greater coverage on this chromosome (R Schmidt, J West, G Cnops and C Dean, unpublished observations). An enlargement of one region of chromosome 4 is shown in the bottom portion of the figure. The coloured blocks represent individual YAC clones from different libraries. Figure cited on pp. 254 and 256.

Chapter 16, Figure 2. (a, b, c) Nuclei of the hybrid barley (*Hordeum vulgare*) × *S. africanum* (a wild rye) show parental genome separation. (a) Processed electron micrograph of a section where chromosomes have been identified by their volumes and morphology (see also Finch *et al.*, 1981). Those of wild rye origin, coloured blue, are peripheral to those of barley origin, coloured red (× 6400). (b) Interphase nucleus of a similar hybrid with the rye-origin chromosomes stained yellow-green by *in situ* hybridization of labelled rye genomic DNA. Barley origin chromosomes are more central and show the orange propidium iodide counterstain (× 1200). (c) A computer reconstruction of probed sections of a hybrid interphase nucleus (nuclear envelope displayed in white on every fourth section). DNA of wild-rye origin (red) is more peripheral than barley-origin chromatin (not displayed); nucleolus (grey; × 4200).

(d, e, f) Individual chromosomes occupy domains at interphase. (d) Human chromosome 8 delineated with pooled probes which label the entire length yellow at metaphase (made using probes and techniques described by Lichter *et al.*, 1988). At interphase, the two chromosome 8 homologues are seen as individual domains that are usually spatially separated (× 900). (e) Root-tip nuclei of a wheat line which carries a rye translocation, 1B/1R, labelled with total genomic rye DNA. The individual rye chromosome arms (yellow) tend to occupy discrete, elongated domains at interphase. Pairs of arms may show similar patterns of coiling but no association of homologues (× 600). (f) Double-target genomic *in situ* hybridization allows identification of two alien chromosomes from a wild couch grass, *Thinopyron bessarabicum* (red-orange), and rye chromosome arms (green) in early and late prophase nuclei of a 1B/1R wheat-addition line (× 900). Figure cited on pp. 222, 223, 224 and 225.

Chapter 16, Figure 3. (a, b, c, d) Heterochromatin in metaphase chromosomes (a, b) and an interphase nucleus (c, d) of Arabidopsis after *in situ* hybridization with a cloned sequence associated with the centromere, pAL1 (a, c) and stained with the DNA fluorochrome DAPI (b, d). There are ten hybridization sites (one per chromosome) that correspond to bright sites in the DAPI-stained nucleus and peripheral, condensed, heterochromatin in the interphase is Fig. 1a. The interphase nucleus is likely to be endopolyploid. (a, b; × 3200) from Maluszynska and Heslop-Harrison (1993) with permission from the Annals of Botany company, and (c, d; × 2700) from Maluszynska and Heslop-Harrison (1991) with permission from Blackwell Scientific Publications. (e, f) Telomeric and sub-telomeric sequences in barley. *In situ* hybridization of synthetic oligomers to the consensus telomere sequence shows that the sequence is located at all ends of the 14 barley chromosomes (*in situ* hybridization signal in red in (e); DAPI DNA stain in (f)). Different strengths of signal indicate the variability in copy number (from Schwarzacher and Heslop-Harrison, 1991, with permission from NRCC). In the

interphase nucleus, the hybridization sites are mostly located in one half of the nucleus (e), that with the lower amount of DNA (f; × 1300).
Figure cited on pp. 222, 225, 226 and 228.

Chapter 16, Figure 4. (a, b) Locations of nucleoli (N) in DAPI (DNA) stained sections of nuclei of cereals. In somatic, root-tip cells of the hybrid *H. chilense* × *S. africanum*, the nucleoli tend to be central in the cell (b), while those in pre-meiotic nuclei from the anthers of a 5AL/5RS wheat line are more peripheral (a) (× 1800).

(c, d) *In situ* hybridization of the rDNA sequence pTa71 to interphase nuclei of (c) wheat, detected by FITC (yellow; × 1500) and (d) barley, detected by rhodamine (red; × 1800). Active genes are decondensed and hardly visible in these preparations. Inactive genes are visible as punctate sites of condensed chromatin within nucleoli, and as larger sites in perinucleolar or other nuclear locations. The orange (propidium iodide; c) and blue (DAPI; d) counterstaining show the sites of the nucleoli as weakly stained areas.

(e) Double target *in situ* hybridization to interphase nuclei of rye showing the locations of the pSc119 sub-telomeric sequence (red), the rDNA (yellow-green) and DNA (DAPI staining, blue) under a triple-wavelength filter system which excites fluorescence with UV, blue and green light. rDNA is located in large condensed sites adjacent to the nucleoli; only fine decondensed fibres run into the nucleoli. The sub-telomeric heterochromatic sites are peripheral and clustered in the region of the nucleus filled with a lower proportion of DNA (× 1800).

(f) The single nucleolus arising from late pachytene bivalents in a 5AL/5RS wheat line. The two major rDNA loci are involved in nucleolus formation, while a minor site is not associated with the nucleolus (lower left; × 1100).

(g) Differential decondensation of a rye chromosome arm in wheat interphase nuclei. The 1R chromosome arm (red) carries an unexpressed rDNA site (yellow). In the interphase nuclei, the average length of the distance from centromere to rDNA site is similar to the distance from rDNA site to sub-telomeric heterochromatin. At metaphase, the latter distance is 20% of the former. Thus interphase decondensation is uneven along the chromosome, and the recombination and gene-rich, distal, non-heterochromatic, region decondenses most. The extra decondensation is in the region less filled with chromatin and including the telomeres and expressed genes, while the more filled region includes little genetic length of chromosomes, most of their DNA (and physical metaphase length) and the centromeres (× 1100).
Figure cited on pp. 222, 226, 227 and 228.

Chapter 13

Probing pachytene chromosomes

Peter B Moens and Ronald E Pearlman

1. Introduction

During prophase of meiosis, each chromosome acquires a proteinaceous axial core to which the two sister chromatids are attached in a series of loops (see Moens and Pearlman, 1988). The cores of homologous chromosomes become aligned in parallel during synapsis and so form the synaptonemal complex, SC (Fig. 1). The SCs are remarkably similar in the meiocytes of most sexually reproducing organisms, but are not found in some fungi and protists and are missing in cases such as male Drosophilids that lack chromosome pairing and recombination at meiosis. The structure of the SC as seen by light and electron microscopy has been reported

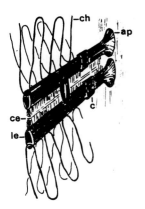

Figure 1. A diagrammatic representation of a set of paired homologous chromosomes in a meiotic prophase nucleus at the pachytene stage of meiosis. The sister chromatids, ch, are attached to a core that becomes the lateral element, le, of the pairing structure, the synaptonemal complex (SC). The parallel aligned lateral elements are connected by transverse filaments and there is a medial structure, the central element, ce. The lateral elements abut the inner nuclear membrane by the attachment plaque, ap. The dense ovoid structure bridging the central element is the type of SC-nodule found in mouse spermatocytes that corresponds to the recombination nodules found in other organisms. Centromere, c. The space between the lateral elements is approximately 80 nm.

183

in prodigious detail since its discovery in 1957 (reviewed by Gillies, 1985; and von Wettstein *et al.*, 1984).

We are interested in the mechanisms that regulate the arrangement of the DNA loops with respect to the SC. In that context we have investigated the nature of DNA sequences associated with the SC, and the positions of given sequences in the loops relative to the SC. In this chapter we summarize and comment on published and recent observations on the chromatin organization of pachytene chromosomes.

2. Mapping with pachytene chromosomes?

The great length of pachytene chromosomes, 2–10 times as long as mitotic chromosomes, should favour their use in high resolution physical gene mapping by *in situ* hybridization (Moens and Pearlman, 1990a). The length advantage of the pachytene chromosomes is negated to an extent by the size of the chromatin loops which are not condensed in meiotic prophase chromosomes. Thus, a sequence at the top of a 3 μm loop could appear a maximum of 3 μm to either side from where the loop is attached to the SC, depending on how the loop settles during chromosome preparation. This uncertainty does not affect the mapping of cosmid sequences close together on the same loop (see later).

We have tested the pachytene mapping system by marking the centromeric end of the unpaired male mouse X chromosome with three probes. The centromere was marked with minor satellite probe, a gift of Dr JB Rattner (University of Calgary, Alberta, Canada). A nearby region, DXWas70 (approximately 50 copies) lies in band A1 and was marked with probe 70-38. DXWas68 (approximately 20 copies) lies about 5.5 map units from DXWas70 in band A3 and was marked with probe 68-36 (Disteche *et al.*, 1989; Moens and Pearlman, 1990a). In mitotic metaphase chromosomes, the signals from the two probes 70 and 68 are close together and frequently overlap. In the meiotic pachytene preparation, all three signals are clearly separated by a distance of several micrometers, even although each probe produces a cluster of signals rather than a single signal (Fig. 2).

Preliminary results with probes to unique sequences show that frequently there are four signals, one from each chromatid, clustered in the appropriate region of the surface spread chromosome (H Heng, Department of Genetics, Hospital for Sick Children, Toronto, Ontario, Canada, personal communication). Such a unique sequence is not necessarily attached to the SC since the signals usually occur some distance from the structure. With this technique, it is possible to detect the positions of given sequences on a chromatin loop relative to a chromosome core. Because the chromatin at meiotic prophase is less condensed than at mitotic metaphase, the sequential order of cosmid probes within a genetic region can be detected (H Heng, personal communication). These results indicate that the

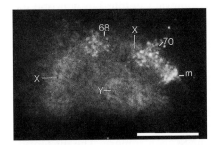

Figure 2. Surface-spread pachytene X-Y chromosomes of the laboratory mouse, *Mus domesticus*. The DNA was denatured and hybridized with three biotinylated probes simultaneously. The minor satellite probe locates to the centromeric end of the X chromosome, m, but not to the centromere of the Y chromosome. The probe, 70-38, identifies the approximately 50 copies of DXWas70 which reside in band A1 of the mitotic metaphase chromosome, and the probe, 68-36, recognizes the approximately 20 copies of DXW68 in band A3. The cores of the X chromosome and the Y chromosome are marked X and Y. The chromatin surrounding the cores is red fluorescent with the DNA binding dye, propidium iodide, and the probes are detected with green fluorescent FITC conjugated to avidin. Scale bar = 10 μm.

probing of meiotic prophase chromosomes has a limited use in mapping of gene sequences along the length of the chromosome, but that it is useful for the fine mapping of cosmid sequences and for the analysis of gene positions on the chromatin loops at meiosis.

3. *In situ* hybridization with mouse centromeric DNA

Although meiotic prophase chromosomes have no kinetochore functions or microtubule attachments, they have well-developed centromeric structures associated with the SCs (Fig. 1). The centromeres are visible in electron microscope preparations of sectioned nuclei or surface-spread pachytene chromosomes. Mammalian somatic and pachytene centromeres can be identified by a positive immune reaction with CREST serum in the light and electron microscope (Moens *et al.*, 1987). CREST serum is obtained from humans suffering from the auto-immune CREST syndrome and the serum contains several anti-centromere antibodies (Earnshaw and Rothfield, 1985).

The mouse (*Mus domesticus*) major satellite DNA, about 5–10% of the mouse genome, is the main component of the constitutive heterochromatin at the centromeric region and occurs abundantly in the chromatin loops near the centromeric region of pachytene SCs (Moens and Pearlman, 1990b). The minor satellite is located at a discrete part of the centromeric regions and has been thought to be a centromere-specific component of the *M. domesticus* centromere (Wong and Rattner, 1988).

In situ hybridization with a minor, or centromeric satellite DNA probe has revealed several interesting associations between pachytene centromeres and centromeric DNA (Moens and Pearlman, 1990b). The mouse minor satellite probe is restricted to the SC centromere region (marked by immunostaining with CREST serum) in most pachytene nuclei of the mouse, *M. domesticus* (Fig. 3a arrow, see p. 176). In other nuclei, however, there are fluorescent signals arising from the centromere as well as one or more loops of chromatin at the centromeric region (Fig. 3a, see p. 176). Apparently the minor satellite sequences are preferentially, but not necessarily, associated with the SC centromere. The nucleotide sequence of the minor satellite shares sequences with the major satellite, which occurs abundantly in the chromatin loops near the centromeric region of the SC. To suppress cross hybridization, we performed *in situ* hybridizations with minor satellite in the presence of excess unlabelled major satellite.

Our explanation for the variations in minor satellite hybridization patterns is that the aggregation of minor satellite in the SC centromere is stage dependent and that as meiotic prophase progresses, more of the minor satellite DNA becomes confined to the SC centromere. The possibility thus exists that centromeric proteins first bind to the minor satellite (a CENP-B binding 17 bp motive of the minor satellite; Haaf *et al.*, 1992; Matsumoto *et al.*, 1989) and are then brought to the centromere by the condensation of the minor satellite chromatin. Such a possibility, however, is not supported by the absence of minor satellite in the centromere of the Y chromosome (Fig. 2), and by the fact that in *M. spretus* the minor satellite is the most abundant satellite (Wong *et al.*, 1990), not at all confined to the SC centromere, but resident in the chromatin loops surrounding all centromere regions, including the Y-chromosome centromere (Fig. 3b, see p. 176).

4. Satellite organization in the *M. domesticus* × *M. spretus* hybrid

The differences in minor satellite distribution among the mouse species offer the opportunity to investigate the organization of chromatin loops in the species hybrid. In captivity, male *M. spretus* can be crossed to female *M. domesticus*. The male offspring is infertile but has sufficient testicular material with spermatocytes for use in pachytene chromosome analysis. The hybrid mice used in these experiments were a gift from Dr Vern Chapman (Roswell Park Cancer Institute, Buffalo NY, USA). If the regulation of the association between minor satellite chromatin and the SC is a function of the chromosome itself, the prediction is that for a set of paired homologous chromosomes in the F1 hybrid, the *M. domesticus* SC centromere should give a compact signal and the *M. spretus* centromere should have the usual bundle of chromatin loops. At pachytene, the results are ambiguous because of the close proximity of the two centromeres and the preparation protocol where the use of SDS causes the loops to disperse (Fig. 3c, see p. 176). At diplotene or at metaphase I, the centromeres are pulled apart and the two types of centromeric satellite are clearly distinguishable (Fig. 4). The *M. spretus* minor satellite has not been compacted by the presence of the *M. domesticus* genome, and vice versa: the *M. domesticus* minor satellite has not been dispersed by the *M. spretus* genome. Hence it would appear that the arrangement of the minor satellite chromatin is a function of the chromosome itself.

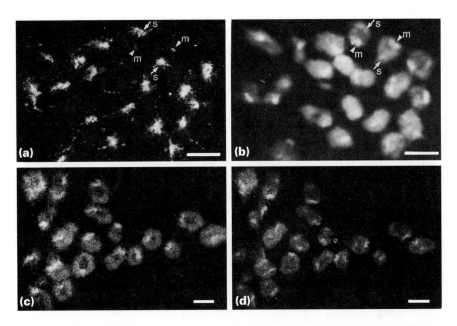

Figure 4. The *M. domesticus* and *M. spretus* centromeres can be distinguished in the bivalents of the F1 hybrid at diplotene and metaphase I when the centromeres have separated. (a) *In situ* hybridization with the minor satellite probe demonstrates the small *M. domesticus* signal, m, at one pole of the bivalent and the large *M. spretus* signal, s, at the other pole. (b) DAPI staining of the same bivalents as in (a) shows the abundant heterochromatin (major satellite) in the *M. domesticus* centromeric region, m, and the paucity of heterochromatin in the centromeric region of *M. spretus*. (c) With the PRINS procedure, the *M. spretus* minor satellite also has a strong signal and, in addition, the chromatin loops have pronounced fluorescence, so much so that the minor satellite of the *M. domesticus* centromere is obscured. (d) The same metaphase bivalents as in (c) also demonstrates the difference in amounts of heterochromatin at the two centromeres. Scale bar = 10 μm.

With primed *in situ* labelling, PRINS (Koch *et al.*, 1989), a short probe is hybridized to the denatured DNA where it initiates the incorporation of biotiny-lated nucleotides with the Klenow fragment of DNA polymerase I. Because the 17 bp motive is unique to the minor satellite, it can be used as a primer for the specific PRINS labelling of the minor satellite. However, with this technique any naturally occurring or artificially induced single strand interruption or RNA–DNA hybrids can also initiate the incorporation of label. As a result, background fluorescence can be pronounced. In *M. spretus*, the minor satellite *in situ* labelling produces a strong signal in the centromeric heterochromatin (Fig. 3d, see p. 176, and Fig. 5a); in *M. domesticus*, the centromeres are barely visible within the high background (Fig. 5b); and in the F1 hybrid only the *M. spretus* centromeric chromatin is well defined (Fig. 5c). Here, too, the *M. domesticus* genome has not caused condensation of the *M. spretus* minor satellite.

The chromatin loops are strongly individualized in the hybrid (Figs 4c and 5c), but not or less so in the pure species (Fig. 5a, b) for as yet unexplained reasons. It is possible that the poor synaptic fit between the homologous chromosomes of the hybrid may interfere with the homology-searching mechanism, leaving single strand breaks unrepaired, or transcription may be impaired, producing initiation sites for the PRINS procedure.

5. Telomeres

The organization of the telomeric DNA in the mouse pachytene chromosomes presents another example of a tight association between specific sequences and

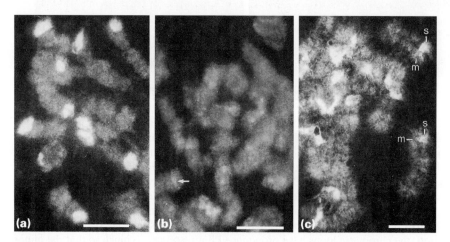

Figure 5. Minor satellite primed *in situ* labelling patterns of surface spread pachytene chromosomes of *M. spretus*, *M. domesticus*, and their F1 hybrid. (a) In *M. spretus*, the abundant minor satellite is brightly fluorescent (also see Fig. 3d, p. 176) while the chromatin loops apparently also incorporated sufficient label to give a fluorescent image. (b) The minor satellite of *M. domesticus* centromeres is barely detectable (arrow) among the fluorescent chromatin. (c) In the F1 hybrid, the *M. spretus* centromere is recognizable by its abundant minor satellite, s, whereas the *M. domesticus* centromere lacks a bright signal, m. A surprising observation is the individualization of chromatin loops in the hybrid bivalents with the PRINS procedure (Figs 4c and 5c). Since the procedure can cause the incorporation of label at any site of a single strand break, it is possible that in the hybrid pairing difficulties may have induced more breaks than are normally present in the pure species. Scale bar = 10 μm.

the SC. With fluorescent *in situ* hybridization, the signal from a telomere probe does not come from the loops, as seen with some other tandemly repeated sequences or with unique sequences, but is localized at the ends of the SCs (Moens and Pearlman, 1990b; Fig. 6). The telomeres of the X and Y pseudoautosomal regions are exceptions as they produce signals in the end of the SC as well as in the chromatin loops (Fig. 6, XY). The concentration of telomeric DNA at the ends of the SCs could support the concept that they are involved in the pairing of chromosomes.

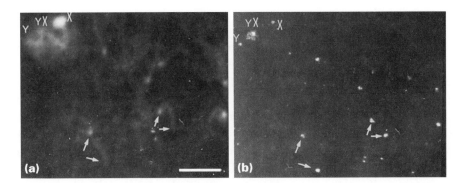

Figure 6. *In situ* hybridization of pachytene chromosomes with a telomere probe. (a) DAPI image of surface-spread mouse sex chromosomes and autosomal chromosomes. Brightest are the SCs of the autosomes and the cores of the X and Y chromosomes, X, Y and XY. The arrows mark the centromeric and the distal ends of two of the SCs.(b) The telomere probe hybridizes to the ends of the SCs and the sex chromosome cores. An exception is the pseudoautosomal end of the X–Y pair where the telomere sequences are present at the end of the SC as well as in the chromatin surrounding the SC. Scale bar = 10 μm.

Normally, several signals are present at the ends of the SC, indicating that the telomeres of the chromatids are not fused together. Furthermore, since telomere sequences are similar for non-homologous chromosomes (Greider *et al.*, Chapter 8, this volume), illegitimate synapsis would occur randomly between chromosome ends. Moreover, numerous reports on chromosomes in the process of pairing show that synapsis is initiated near, but not at, the chromosome ends (Hasty, 1988), perhaps in the telomere associated sequences (see Richards *et al.*, Chapter 7, this volume). It is nevertheless possible that occasional illegitimate synapsis and recombination do occur (V Chapman, personal communication).

6. Heterochromatin in centromere-associated sequences

The presence of tandemly repeated satellite sequences in clusters of chromatin loops around the SC centromere was demonstrated with rat satellite DNA I probe

p93-50 (Moens and Pearlman, 1989). The pericentric regions of most rat chromosomes have more or less of these satellite sequences, while some do not have the
sequences contained in the probe. As a result, some centromeric chromatin
regions are heavily labelled, some lightly, and some not at all. The fluorescence
along the individual loops is not even. Instead, there are numerous small beads,
suggesting that the satellite sequences are interspersed by non-satellite sequences
(Fig. 3b, see p. 176). The combined signals are the brightest where the loops
converge at the SC attachment site but we are not able to tell whether the satellite
sequences were actually associated with, or within, the SC.

7. Sequences contained within the SCs

We observed that following purification of SCs and DNase I treatment to remove
the chromatin loops, the remaining SC structures were still positive when stained
with the DNA binding dyes, DAPI or propidium iodide, and immunoreactive
with anti-DNA antibodies. We assume that the remaining DNA fraction is
protected from DNA digestion by a tight association with SC components. To
analyse that DNA fraction, the purified DNase I-treated rat SCs were digested
with proteinase K and the released DNA was purified and cloned (Pearlman *et al.*,
1991). A control sample of 13 cloned genomic fragments contained two LINE
fragments, a satellite I fragment, and ten fragments of no significant similarity to
known sequences. In contrast, the 21 clones from the SCs contained a dis-
proportionately large number of microsatellites (four GT repeats, one GTGC
repeat, one GGGA repeat), as well as unusually short LINE and SINE fragments
(seven) flanked by non-LINE or non-SINE sequences, and eight unidentified
sequences. The conclusion is that the DNA associated with the SC is not representative of the genomic DNA. The type of sequences recovered from the SCs have
been implicated in recombination processes and may function as such in the
context of the SC.

8. Conclusions

Our curiosity centres around the basic meiotic functions of homologous chromosome pairing, recombination and segregation. Of the numerous experimental
approaches to these phenomena, we have addressed some of the structural
aspects by hybridization of several DNA probes to surface-spread meiotic prophase chromosomes. These first results indicate that the approach is viable but
that improvements in chromosome preservation, labelling techniques and micros-

copy are necessary to fully explore structural analysis of chromosome structure at meiotic prophase.

We have also isolated the DNA that appears to be associated with the SCs. The next experiments will examine the DNA binding properties of the various SC proteins in regard to the genomic DNA and in regard to the types of DNA isolated from the SCs. Specifically, the interaction between SC pairing proteins and various classes of DNA can be tested. If recombination is associated with the SC recombination nodules (RNs) it is expected that the various enzymes will be detected in association with RNs. Conversely, if anti-RN antibodies are found they may identify the proteins of the recombination processes. The application of molecular and cell biology techniques to the meiosis specific chromosome function promise to provide new insights into one of the basic biological processes, chromosome reduction in diploid organisms.

Acknowledgement

Financial support for the research was provided by NSERC of Canada.

References

Disteche CM, McConnell GK, Grant SG, Stephenson DA, Chapman VM, Gandy S, Adler DA. (1989) Comparison of the physical and recombination maps of the mouse X chromosome. *Genomics*, **5:** 177-184.

Earnshaw WC, Rothfield N. (1985) Identification of a family of human centromere proteins using autoimmune sera from patients with scleroderma. *Chromosoma*, **91:** 313-321.

Gillies CB. (1985) An electron microscopic study of synaptonemal complex formation at zygotene in rye. *Chromosoma*, **92:** 165-175.

Haaf T, Warburton PE, Willard HF. (1992) Integration of human α-satellite DNA into simian chromosomes: centromere protein binding and disruption of normal chromosome segregation. *Cell*, **70:** 681-696.

Hasty N. (1988) Highly repeated DNA families in the genome of *Mus musculus*. In: *Genetic Variability and Strains of the Laboratory Mouse*, 2nd edition (Lyons M, Searle A, eds). Oxford: Oxford University Press.

Koch JE, Kolvraa S, Petersen KB, Gregersen N, Bolund L. (1989) Oligonucleotide-priming methods for the chromosome-specific labelling of alpha satellite DNA in situ. *Chromosoma*, **98:** 259-265.

Matsumoto H, Masukata H, Muro Y, Nozaki N, Okazaki T. (1989) A human centromere antigen (CENP-B) interacts with a short specific sequence in alphoid DNA, a human centromeric satellite. *J. Cell Biol.* **109:** 1963-1973.

Moens PB, Heyting C, Dietrich AJJ, van Raamsdonk W, Chen Q. (1987) Synaptonemal complex antigen location and conservation. *J. Cell Biol.* **105:** 93-103.

Moens PB, Pearlman RE. (1988) Chromatin organization at meiosis. *BioEssays*, **9:** 151-153.

Moens PB, Pearlman RE. (1989) Satellite DNA I in chromatin loops of rat pachytene chromosomes and in spermatids. *Chromosoma*, **98:** 287-294.

Moens PB, Pearlman RE. (1990a) *In situ* DNA sequence mapping with surface-spread mouse pachytene chromosomes. *Cytogenet. Cell Genet.* **53:** 219-220.

Moens PB, Pearlman RE. (1990b) Telomere and centromere DNA are associated with the cores of meiotic prophase chromosomes. *Chromosoma*, **100:** 8-14.

Pearlman RE, Tsao N, Moens PB. (1991) Synaptonemal complexes from DNase-treated rat pachytene chromosomes contain (GT)n and LINE sequences, but no MARs/SARs. *Genetics*, **130:** 865-872.

von Wettstein D, Rasmussen SW, Holm PB. (1984) The synaptonemal complex in genetic segregation. *Ann. Rev. Genet.* **18:** 331-431.

Wong AKC, Rattner JB. (1988) Sequence organization and cytological localization of the minor satellite of mouse. *Nucl. Acids Res.* **16:** 11645-11661.

Wong AKC, Biddle FG, Rattner JB. (1990) The chromosomal distribution of the major and minor satellite is not conserved in the genus *Mus*. *Chromosoma*, **99:** 190-195.

Chapter 14

The enigma of dominant position-effect variegation in *Drosophila*

Steven Henikoff, Kate Loughney and Thomas D Dreesen

1. Introduction

In *Drosophila*, chromosome re-arrangements that juxtapose euchromatin and heterochromatin cause mosaic inactivation of neighbouring genes, a phenomenon known as position-effect variegation (PEV). As expected for gene inactivation, PEV alleles are generally recessive. However, in the case of the *brown* gene, PEV is dominant, a problem that has intrigued and frustrated geneticists since 1930. A combination of classical and modern genetic methods have demonstrated that dominant PEV depends upon somatic pairing of homologues. The dominant effect also requires a sequence-specific component which maps to a small DNA segment that includes the *brown* gene. Now it appears that this enigmatic phenomenon can be understood in terms of current ideas about protein components of heterochromatin and transcription factors that regulate genes as heteromultimers.

Among the most important advances in molecular biology have been the elucidation of many enigmatic phenomena discovered by previous generations of geneticists. A famous example is the discovery of movable genetic elements in maize (McClintock, 1951), later explained in terms of transposable DNA segments and now known to be universal features of genomes (Berg and Howe, 1989). In *Drosophila*, there are other examples of puzzling genetic phenomena that have been elucidated by application of modern methods. For example, an allele-specific suppressor locus, suppressor-of-Hairy-wing, is now known to act by encoding a protein that binds to transposable elements found in regulatory regions of affected loci (Parkhurst *et al.*, 1988). However, some other mysterious phenomena appear to be less tractable, and explanations remain speculative. One such phenomenon in *Drosophila* is PEV, discovered by Muller more than 60 years ago. PEV is the mosaic expression of a gene which results from chromosome re-arrangements that create new junctions between euchromatin and heterochro-

193

matin. While certain features of PEV remain mysterious, there has been progress in understanding one aspect of the phenomenon that was especially disturbing to Muller: PEV alleles causing mosaic expression of the *brown* gene are dominant over wild-type *brown*. This chapter describes how a combination of classical genetic analysis and modern technology has contributed to a provisional understanding of these dominant brown alleles.

2. The enigma

A large class of mutations discovered by Muller when he began X-irradiating flies were what he termed 'eversporting displacements' (Muller, 1930), stable chromosome re-arrangements causing genes to appear mutant in some cells and wild-type in others (Fig. 1a, bottom, see p. 177). Shortly thereafter it became clear that re-arrangement breakpoints responsible for this somatic mosaicism are not within the affected genes, but are nearby, hence the term 'position-effect'. This combination of germline stability and somatic instability is still not understood at the mechanistic level, even though the phenomenon has been described in great detail.

Two basic ideas have been proposed to explain PEV. One is that the gene is missing from cells that show the mutant phenotype, first proposed by Schultz (1936). Such somatic loss could explain frequent patches that are very similar in appearance to clonal patterns generated by random somatic excision of a gene (Fig. 1a, top, see p. 177). The other idea is that the gene is still present in cells that show the mutant phenotype, but that an alteration in chromatin structure has spread from the breakpoint between euchromatin and heterochromatin with consequent disruption of gene activity. To distinguish these possibilities it is necessary to determine whether an affected gene is present in cells that show the mutant phenotype. This has been possible in polytene tissues, leading to the conclusion that PEV can occur without gene loss (Henikoff, 1981; Rushlow *et al.*, 1984; Umbetova *et al.*, 1991). However, a controversy has emerged as to whether many PEV phenotypes are explained by under-representation of sequences in polytene tissues (Karpen and Spradling, 1990). While it remains an intriguing possibility that gene loss mechanisms are responsible for aspects of the phenomenon, there can be no doubt that some kind of chromatin spreading leading to gene inactivation is responsible for classical examples of PEV. The molecular basis for such spreading is still obscure, in spite of considerable progress in identifying the genes responsible and in finding parallels to other inactivation phenomena. These aspects of PEV have been discussed in several recent reviews (Eissenberg, 1989; Grigliatti, 1991; Henikoff, 1990, 1992; Spradling and Karpen, 1990).

As one would expect for gene inactivation, PEV is generally recessive to wild-type. For example, dozens of PEV alleles of the *white* gene have been described, and all are recessive to wild-type (Fig. 2). However, in the case of the *brown* gene, PEV is invariably dominant over wild-type to some detectable extent (Fig. 1b, see p. 177). Muller discovered this in his early experiments: because *brown-Variegated (bw^V)* alleles are eye-colour dominants and difficult to miss in a screen, new *Plum* mutations (as Muller first called them) have been found very frequently among progeny of X-irradiated flies. The original *Plum* mutation provided Muller with a clear illustration of what he called an antimorph, a mutation that appears to antagonize the activity of wild-type (Muller, 1932). The basis for this conclusion was the allelism of dominant *Plum* to recessive *bw*, with less red pigment in *Plum/bw* eyes, which appear almost fully mutant, than in *Plum/bw$^+$* eyes which are variegated (Fig. 2). His interpretation of these observations was that *Plum* "has an actively negative value"; this presages the term 'negative dominant', a type of antimorphic effect in which an altered protein product forms heteromultimers with normal proteins, leading to complexes with reduced activity (Herskowitz, 1987). But how could gene inactivation provide "an actively negative value"? This is a puzzle that intrigued a number of *Drosophila* geneticists for another 60 years.

3. The work of Slatis

Of the various early proposals, perhaps that of Ephrussi and Sutton (1944) came closest to what we now understand to be the basis for the dominance of *bw^V* mutations. Their 'structural' hypothesis implicated somatic pairing of

Position-effect variegation (PEV)

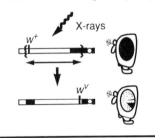

Recessive PEV of *white*:

w^V/w^- w^V/w^+ w^-/w^+

Dominant PEV of *brown*:

bw^V/bw^- bw^V/bw^+ bw^-/bw^+

Figure 2. An example of PEV on the *white* gene. Ionizing radiation causes chromosomal breaks (wavy lines) resulting in aberrations, such as the inversion depicted. The recessive w^V allele resulted from juxtaposition of the gene to heterochromatin (filled boxes). Variegation is seen as numerous spots or patches of wild-type cells on a mutant background. In contrast to *white*, PEV alleles of *brown* are dominant; bw^V/bw^+ is still variegated, though somewhat suppressed relative to bw^V/bw^-. Circles indicate centromeres.

homologous chromosomes as the basis for gene dysfunction in phenomena such as PEV. They proposed that pairing is necessary for normal activity of genes, so that abnormal pairing caused by re-arrangement is the basis for PEV. By this model, the *brown* gene was seen to be especially sensitive to pairing disruptions. In order to test the structural hypothesis, Slatis carried out a series of mutant screens, generating bw^V mutations by X-irradiating either wild-type or *bw* mutant flies, and measuring the degree of variegation in different heteroallelic combinations (Slatis, 1955a). His ability to obtain *brown-Variegateds* starting with the null *bw* allele argued against the notion that Muller's 'actively negative value' was simply an altered product of the *brown* gene. Slatis's results also did not support predictions of the Ephrussi-Sutton model, probably because homologous pairing is not a requirement for normal *brown* activity. Nevertheless, his extensive analysis showed that dominance is a consistent feature of re-arrangements that place the *brown* locus next to heterochromatin, but not for other types of mutant *brown* alleles.

Slatis also carried out a study of the unusual *brown-Dominant (bw^D)* allele, so called because it is so strongly dominant that $bw^D/+$ is indistinguishable from bw^D/bw^D (a 'true' dominant). Besides being unusually strong, this allele is unique in that it is the only PEV mutation to have arisen spontaneously. In polytene nuclei, the bw^D chromosome appears to be nearly normal, with only an extra band seen in the vicinity of the gene at *59E1-2* on the salivary gland map (Lindsley and Grell, 1968). Slatis's genetic and cytological studies (1955a, b), combined with a more recent cytological analysis in diploid cells (Lindsley and Zimm, 1992), confirm that the event was an insertion of a block of heterochromatin very close to or within the gene, causing it to be null. Our molecular analysis of bw^D showed that the brown coding region is broken in two; the adjacent DNA lacks restriction sites and is presumably simple sequence DNA characteristic of centromeric heterochromatin (Dreesen *et al.*, 1991 and unpublished results). This confirmed Slatis's observation that bw^D/bw^D behaves like a null mutation. However, it appears that there has been a change since the original insertion event: until 1950 it was possible to obtain X-ray-induced revertants (Hinton and Goodsmith, 1950), a result not consistent with an insertion of heterochromatin into, rather than adjacent to a gene. No X-ray-induced revertants were found in a similar screen 40 years later (K Loughney, unpublished results). Therefore, we suggest that subsequent to insertion of heterochromatin close to the *brown* gene in 1940, a small inversion occurred with one breakpoint within the *brown* gene and another within the heterochromatic block, probably prior to Slatis's analysis in the early 1950s.

This unusual allele continues to be useful. Because the strong mutant phenotype of $bw^D/+$ is especially consistent for a PEV allele, flies of this genotype are useful in screens for PEV suppressor loci. We have obtained large numbers of suppressors of $bw^D/+$ that also suppress (recessive) *white* PEV, as well as a few suppressors that might be bw^V-specific (P Talbert, C LeCiel and S Henikoff, unpublished results). Features of this allele also have inspired a recent comparison to a human disorder, Huntington's disease. Laird has pointed out the striking similarities between bw^D and Huntington's disease which is also a true dominant that appears to have arisen as an extremely rare spontaneous event; he has proposed that the two mutations have a common genetic basis (Laird, 1990).

4. Molecular analysis of the *brown* gene

The enigma of dominant PEV was nearly forgotten following the decline in *Drosophila* research that began in the 1950s as geneticists moved to model organisms more suitable for molecular studies. Thirty-four years passed between Slatis's work and the next published investigation of bw^V mutations. In the interim, the *karmoisin* gene was shown to be subject to dominant PEV (Henikoff, 1979), making it the third clear case (along with *Punch*; Lindsley and Grell, 1968). We began investigation of this phenomenon by first cloning of the *brown* gene. The cDNA sequence revealed that the encoded protein is closely related to proteins encoded by the *white* and *scarlet* genes, all three belonging to a large family of active transport proteins (Dreesen *et al.*, 1988), now termed the 'ABC' family. This finding provided strong support for the 'permease hypothesis' proposed by Sullivan and Sullivan many years earlier that *white*, *scarlet* and *brown* encode proteins necessary for transport of pigment precursors (Sullivan and Sullivan, 1975). Features of the ABC family also were consistent with an earlier speculation by Farmer, based on his observation of an allele-specific genetic interaction, that proteins encoded by the *brown* and *white* genes might be subunits of a complex (Farmer and Fairbanks, 1986). So the *brown* gene and the *white* gene are very similar; they are expressed in the same cells at the same time during development and encode similar proteins that might interact with one another. However, PEV on the *white* gene is always recessive but on the *brown* gene PEV is always dominant. Therefore, it seemed especially unlikely that dominant variegation is an example of negative dominance at the protein level.

Analysis of *brown* gene transcripts in wild-type heads showed that the gene is relatively simple. Two mRNAs just under 3 kb in length differ in the extent of the 3' untranslated tail (Dreesen *et al.*, 1988), and both have seven very small introns (K Loughney, C LeCiel, L Martin-Morris and S Henikoff, unpublished results). In $bw^V/+$ flies, mRNA accumulation was sharply reduced, both for the copy in *cis* to the re-arrangement breakpoint and for the copy in *trans* (Henikoff and Dreesen, 1989). Therefore the basis for dominance of bw^V alleles is reduction in the number of transcripts from the normal copy of the gene on the unre-arranged chromosome, referred to as '*trans*-inactivation'. These studies also argued against a speculation that antisense RNA transcribed from promoters located in heterochromatic regions could interfere with normal brown expression (Frankham, 1988), since no transcripts from either strand could be detected in bw^D/bw^D homozygotes. Another hypothesis, that dominance is caused by a putative closely-linked dosage-sensitive locus that regulates *brown* (Spofford, 1976), also was found to be inconsistent with our results: *trans*-inactivation could be detected in flies with as many as three normal copies of the *brown* chromosomal region. While this work was helpful in excluding whatever hypotheses remained, it did not provide us with a clear candidate for Muller's 'actively negative value'.

5. *Suppressor-of-Plum*

The most important clue leading to an explanation of this phenomenon came from unpublished work carried out about 25 years earlier. A spontaneous dominant suppressor of Muller's original *Plum* mutation was discovered by a student carrying out a genetics project in a Baltimore high school. *Suppressor-of-Plum* (*Su(Pm)*) was found to be closely linked to the *brown* gene. The student's father, a genetics teacher named Byron Kadel, was intrigued by *Su(Pm)* and carried out an extensive genetic analysis in the laboratory of TRF Wright. Kadel discovered that the suppressor was separable from the *brown* gene and behaved in crosses as if it were associated with a tandem duplication. The brief report of this work in the Redbook (Lindsley and Grell, 1968) led us to contact Wright, who provided the stock and Kadel's detailed reports. There we found his evidence that the duplication, nearly invisible cytologically, consists of two copies of the functional *brown* gene and flanking regions. This suggested that suppression was simply a consequence of having twice as much wild-type *brown* gene product. But suppression wasn't simply the result of having twice the dose. With a crossover, Kadel had exchanged one of the duplicated copies of bw^+ with bw^-. When we looked at $bw^V/Su(-+)$ heterozygotes, we noticed that there was still partial suppression relative to $bw^V/+$, even though the dose of wild-type *brown* is the same (Fig. 1c, see p. 177). The residual suppression mapped to the duplication itself, suggesting that some feature of the duplicated region suppresses *trans*-inactivation (Henikoff and Dreesen, 1989). This revived the idea that somatic pairing was somehow responsible, since *in situ* hybridization analysis showed that in $bw^D/Su(Pm)$ heterozygotes the small duplication caused disruption of pairing in the region.

These studies of *Su(Pm)* led to a model for *trans*-inactivation whereby the altered chromatin structure characteristic of PEV can be transmitted directly from the *cis* copy of the *brown* gene region to the region in *trans* (Fig. 3a). We thought that 'heterochromatinization' might be able to pass from one homologue to the other in the vicinity of the *brown* gene, but not in the vicinity of *white* or most other genes. This defined a genetic element near *brown* that we called a 'transceiver' which could mediate this transmission when present on both homologues. In $bw^V/Su(-+)$ heterozygotes, the pairing of two copies of the region on the suppressor chromosome with one copy on bw^V would cause competitive pairing and interfere with contact between transceiver elements. This model could explain why the bw^D insertion was the strongest *trans*-inactivating allele. Since it

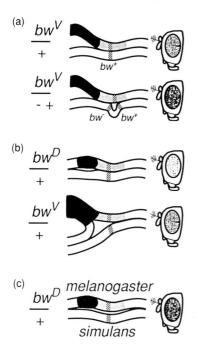

Figure 3. Explanation of phenotypic observations in terms of the transceiver hypothesis, in which paired copies of a genetic element (hatched) near *brown* allow spreading of inactivation (shaded) from the *cis* to the *trans* copy of the gene by side-by-side contact. Heterochromatin is indicated in black. In each case, disruption of pairing in the vicinity of the *brown* gene and separation of the paired elements correlates with suppression. (a) Suppression by *Su(Pm)*. A small tandem duplication of *brown* and flanking regions causes disruption of pairing with partial suppression. (b) The highest degree of *trans*-inactivation is seen for the bw^D allele, a heterochromatic insertion. Unlike gross re-arrangements associated with other bw^v alleles, the bw^D insertion does not cause paired homologues to unzip in salivary gland squashes. (c) Suppression in *Drosophila melanogaster/simulans* hybrids. Paired homologues do not stick together well in salivary gland squashes of hybrids, suggesting that unpairing causes suppression.

was not a gross re-arrangement, there would be less of a tendency for the paired homologues to unzip from the breakpoint and therefore more intimate contact would be possible between the *cis* and *trans* copies of the putative transceiver (Fig. 3b). Another observation that could be explained by the transceiver hypothesis was that dominance was known to be partially suppressed in interspecific hybrids between *Drosophila melanogaster* and *D. simulans* (Schultz, 1937). Suppression therefore correlated with reduced pairing of polytene chromosomes in these hybrids, resulting in reduced contact between paired transceiver elements and more frequent escape from *trans*-inactivation in the eye (Fig. 3c).

6. Some predictions of the transceiver hypothesis are fulfilled

The transceiver hypothesis made predictions as to the behaviour of copies of the *brown* region inserted into the genome at ectopic sites. The 3 kb *brown* transcriptional unit is functional at these sites on an 8.4 kb fragment. These copies lack extensive flanking regions required for homologous pairing with the endogenous gene. If such pairing is necessary for *trans*-inactivation to occur, these copies should escape inactivation (Fig. 4a). This might not be the case for any insertion

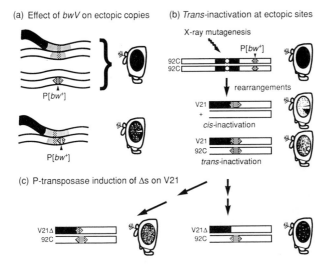

(a) Effect of *bwV* on ectopic copies   (b) *Trans*-inactivation at ectopic sites

(c) P-transposase induction of Δs on V21

Figure 4. Behaviour of transposed copies of *brown* inserted into ectopic sites. (a) New flanking sequences prevent pairing with the chromosomal copy of *brown*, unless the insertion is extremely close to the chromosomal locus. (b) The phenomenon of *trans*-inactivation can be reproduced at sites of P[*bw*⁺] insertion (e.g. 92C) in a classical X-ray screen for PEV. (c) Deletions and excisions of the transposed copy on the V21 chromosome partially suppress *trans*-inactivation as a function of deletion size. Pairs of open boxes represent homologous chromosomes. Triangles represent 5′ and 3′ P-element ends flanking the copy of *bw*⁺ inserted at 92C.

that is extremely close to the *brown* chromosomal locus. These predictions were fulfilled (Dreesen *et al.*, 1991). Of 36 single insert lines with active copies of *brown*, 35 were no closer than about 0.5% of the genome away (Fig. 5); these were completely unaffected by the presence of *bw*ᵛ re-arrangements that strongly *trans*-inactivated the natural chromosomal locus at 59E. The single exception was an insertion into 59E which was indeed subject to *trans*-inactivation. Therefore, an ectopic copy of *brown* must be extremely close to the chromosomal locus in order to be *trans*-inactivated. While it might be argued that the sequences responsible for *trans*-inactivation lie outside of the 8.4 kb fragment, this argument is harder to make for cosmid-based transposons tested which have much more extensive flanking sequences.

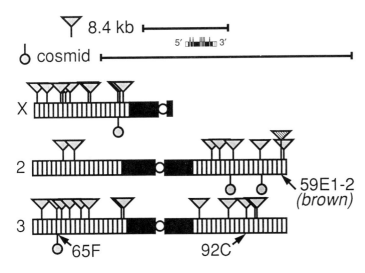

Figure 5. P[*bw*⁺] single insert lines. The *brown* transcription unit is shown schematically along with its approximate map position within the plasmid clones used for transformation. The 3.2 kb transcription unit has seven small introns. The position of each insertion site is shown on the schematic map of divisions 1–100. A single insertion (hatched) is *trans*-inactivated by *brown* PEV alleles (Fig. 4a); it is located at the position of the *brown* locus, *59E1-2*. Two insertion lines, 65F and 92C, were X-irradiated to obtain PEV alleles of P[*bw*⁺] (Fig. 4b).

A second prediction is that it should be possible to reproduce the phenomenon of *trans*-inactivation at sites of transposons (Fig. 4b). This prediction also was fulfilled (Dreesen *et al.*, 1991). Lines having inserts at 65F and 92C were X-irradiated. Selection for *cis*-inactivation yielded four re-arrangements, examples of classical PEV on the inserted copies. In each case, heterozygotes between the re-arranged copy and the original insertion chromosome showed dominant variegation. Control crosses showed that *trans*-inactivation only occurred on the parent insertion line from which the PEV mutation was derived. Other insertion lines were not affected by *trans*-inactivating alleles. Therefore, at all three sites of the *brown* gene, the natural site at 59E and the two ectopic sites at 65F and 92C, a copy of *brown* at the identical position is *trans*-inactivated, but a copy at another position is not. This chromosome-local feature of the phenomenon is further evidence that somatic pairing is involved, since there seems to be no other way that an insertion at the identical position on the normal homologue could be acted upon differently from one elsewhere.

A PEV re-arrangement that placed the 92C insertion next to heterochromatin (called 'V21'; see Fig. 1a, bottom, p. 177), was subjected to further study, including a P-transposase induced deletion analysis described below. In that study, we isolated a strongly enhanced example of *cis*-inactivation (V21ᵉˣᵗʳᵉᵐᵉ, see Fig. 1d–f, p. 177) that left the transposon itself unaltered; it proved to be rescuable from the influence of heterochromatin by mobilization to other sites. Interestingly, *trans*-inactivation of the 92C insertion by V21[extreme] also was strongly enhanced. The high degree of dominance in this case argues that the sequences present on the transposon insertion at 92C are sufficient for full-blown *trans*-inactivation to occur. Excluding vector sequence, this transposon consists of the *brown* gene on an 8.4 kb fragment. Therefore, this experiment maps the putative transceiver to the *brown* gene and immediate flanking regions.

7. An unfulfilled prediction leads to a modified somatic pairing model

Reproduction of the *trans*-inactivation phenomenon at the site of a transposon made possible a detailed *in vivo* deletion analysis using P-transposase as a site-specific mutagen. A frequent result of exposure to P-transposase in the germline is deletion of DNA between transposon ends. For the P[*bw*[+]] transposon next to a heterochromatic breakpoint in V21, deletions that lead to a fully mutant *bw*[-] phenotype are readily selected. This allowed us to test a third prediction of the transceiver model (Fig. 4c), that deletion of the *cis* copy of the *brown* gene segment would prevent transmission of heterochromatinization between paired homologues and thus eliminate *trans*-inactivation (Dreesen *et al.*, 1991). Furthermore, partial deletion of the gene could be used to map the putative transceiver. The basis of the prediction is our expectation that the transceiver would be a site of special somatic pairing present on both the *cis* and *trans* copy of the gene. However, our results were inconsistent with this idea in that they showed that no specific site could be mapped to the *cis* copy. First, removal of the entire transposon from the re-arranged chromosome (V21) only partially suppressed, but never eliminated *trans*-inactivation of the 92C copy of *brown* on the unre-arranged chromosome. Second, partial deletions of *brown* within the V21 transposon also suppressed, with the extent of suppression proportional to the length of sequence deleted. Since no site could be mapped to V21-linked copy of the gene, we concluded that the only specific sequences that are required for *trans*-inactivation of 92C by V21 are on the transposon carried on the 92C homologue.

Similar deletion mapping of the 92C copy of *brown* is hampered by the fact that *brown* is the reporter gene used to assay *trans*-inactivation, so that deletions into the gene are uninformative. We have partially circumvented this problem by generating deletions of flanking sequences *in vivo*, using genotypic selection, that is, by detection of DNA alterations in phenotypically normal flies. This has led to more precise localization of the sequence-specific component for *trans*-inactivation of brown at 92C, apparently within 1 kb of the 5′ end of the gene (LE Martin-Morris, K Loughney, E Kershisnik, G Poortinga, S Henikoff, unpublished results).

The length-dependent suppression of *trans*-inactivation by deletions within the V21-linked copy of *brown* has an interesting parallel to suppression by *Su(Pm)*. In both cases, suppression is accompanied by pairing-dependent looping out of extra material in the vicinity of the *trans*-inactivated copy of the gene (compare Fig. 3a with Fig. 6). For $bw^V/Su(Pm)$, the duplication of a few polytene

Active   Inactive

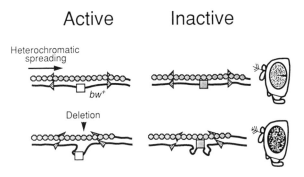

Figure 6. A model for *trans*-inactivation. The transceiver (square) is thought to be a positive regulator of bw^+ (open square) that can be prevented from normal functioning (stippled square) whenever intimate contact is made with heterochromatic proteins (stippled octagons) that spread from the V21 breakpoint. Excisions or deletions within $P[bw^+]$ on the *cis*-inactivated homologue cause the *trans* copy to loop out. As a result, the bw^+ gene is more frequently active, leading to fewer mutant spots in the eye.

bands including *brown* causes looping out of perhaps a few hundred kb of DNA from the suppressor chromosome. Similarly for V21/P[bw^+(92C)] heterozygotes, the few kb of DNA in 92C that has no pairing partner in deleted V21 derivatives is expected to loop out in the same way, although on a much smaller scale. This suggests that suppression has the same basis in both cases: pairing disruption reduces contact between homologues in a critical region in the vicinity of the *brown* gene. If so, then this critical region of contact can be no more than a few kb from the gene itself to explain suppression of *trans*-inactivation at 92C. This leads to the notion that contact is between a protein that binds to the sequence-specific component mapping to the 92C copy of *brown* and a non-sequence-specific component on the V21 chromosome (Fig. 6). Fortunately, current ideas about the nature of heterochromatic spreading provide candidates for this non-sequence-specific component: proteins in heterochromatic-specific complexes encoded by genetic modifiers of PEV. Since these proteins should be present at the *cis*-inactivated copy of *brown*, they might be in position to make contact with the proposed sequence-specific protein bound to the copy of the gene that gets *trans*-inactivated.

To summarize, three components appear to be required for *trans*-inactivation: (1) a heterochromatic breakpoint near the *brown* gene capable of causing *cis*-inactivation (Slatis, 1955b); (2) somatic pairing between homologous copies of the gene and immediately flanking sequences; and (3) a sequence-specific component also in the immediate vicinity of the gene but only on the *trans*-inactivated copy.

These three components can be accounted for by a model in which contact occurs between a sequence-specific DNA-binding protein on the *trans*-inactivated copy of *brown* and a general heterochromatic protein that is present in the complex responsible for *cis*-inactivation. Somatic pairing brings these two putative proteins sufficiently close to one another that they make frequent contact, and this contact interferes with expression of the *trans* copy. Suppression occurs when pairing is locally disrupted, decreasing the frequency of contact (Fig. 6). Therefore, the proposed basis for *trans*-inactivation is heterodimerization of DNA-binding proteins, a type of model that has been shown to be responsible for gene regulation in a growing number of systems (Lamb and McKnight, 1991). However, rather than an interaction between different proteins capable of binding to nearby sites on the same chromosome, *trans*-inactivation results when these different proteins are bound to paired homologues.

What makes *brown* different from genes such as *white* that never show *trans*-inactivation? We suppose that the sequence-specific protein bound to *brown* does not bind to these other genes. It is perhaps significant that both *Punch* and *karmoisin*, the other genes subject to dominant PEV, also affect pigment deposition in the eye. This suggests that the proposed *brown*-specific DNA-binding protein is an eye-specific transcription factor binding to a *brown* gene enhancer, a possibility that we hope will lead to identification of the protein (P Talbert and LE Martin-Morris, unpublished results).

8. Other *trans*-sensing effects

Trans-inactivation is just one of a growing number of '*trans*-sensing effects', whereby a gene senses the state of its homologous copy via somatic pairing (Tartof and Henikoff, 1991). Unlike PEV of euchromatic genes in which the same set of rules apply to nearly all genes, *trans*-sensing effects are amazingly diverse. For example, in *Ubx* transvection, disruption of pairing increases gene dysfunction when breaks are between the affected gene and the centromere, and breaks are often very distant from the gene itself (Smolik-Utlaut and Gelbart, 1987). In contrast, in the *zeste–white* interaction, disruption of pairing *de*creases gene dysfunction only when breaks are very close to the affected gene (Gubb *et al.*, 1990). Unlike *trans*-inactivation, neither phenomenon requires a heterochromatic breakpoint nearby. To confound matters further, some but not all of these phenomena are affected by mutations at the *zeste* locus; whether this is a profound distinction between the phenomena is not known.

Perhaps what this diversity reveals is that gene expression in *Drosophila* has evolved in the context of somatic pairing of homologous sequences; this context can be altered in different ways. For example, in the case of *trans*-inactivation, the context might be altered in a chromosome re-arrangement by bringing together a heterochromatic protein and a normal transcription factor via somatic pairing. In the case of the *zeste–white* interaction, the context is altered without chromosome re-arrangement; rather a transcription factor mutation is thought to lead to aggregates so large that effects can be felt across paired homologues (Bickel and Pirrotta, 1990). By this view, *trans*-sensing effects do not necessarily reflect a somatic pairing requirement for normal gene expression as previously suggested (Judd, 1988). Instead, somatic pairing might be required primarily for chromosome packaging at interphase. Mutations that alter the normal context of gene expression by causing interference between paired homologues are then detected as *trans*-sensing effects. Evidence for frequent somatic pairing of mammalian chromosomes supports the suspicion that similar interference between homologues might be the basis for certain human disorders (Laird, 1990; Tartof and Henikoff, 1991).

References

Berg DE, Howe MM (eds). (1989) *Mobile DNA*. Washington DC: American Society of Microbiology Press.

Bickel S, Pirrotta V. (1990) Self-association of the *Drosophila* zeste protein is responsible for transvection effects. *EMBO J.* **9**: 2959-2967.

Dreesen TD, Johnson DH, Henikoff S. (1988) The brown protein of *Drosophila melanogaster* is similar to the white protein and to components of active transport complexes. *Mol. Cell. Biol.* **8**: 5206-5215.

Dreesen TD, Henikoff S, Loughney K. (1991) A pairing-sensitive element that mediates *trans*-inactivation is associated with the *Drosophila brown* gene. *Genes Dev.* **5**: 331-340.

Eissenberg JC. (1989) Position effect variegation in *Drosophila*: towards a genetics of chromatin assembly. *Bioessays*, **11**: 14-17.

Ephrussi B, Sutton E. (1944) A reconsideration of the mechanism of position effect. *Proc. Natl Acad. Sci. USA*, **30**: 183-197.

Farmer JL, Fairbanks DJ. (1986) Interaction of the *bw* and *w* loci in *D. melanogaster*. *Drosophila Inf. Svc.* **63**: 50-51.

Frankham R. (1988) Molecular hypotheses for position-effect variegation: anti-sense transcription and promoter occlusion. *J. Theoret. Biol.* **135**: 85-107.

Grigliatti T. (1991) Position-effect variegation – an assay for nonhistone chromosomal proteins and chromatin assembly and modifying factors. In: *Functional Organization of the Nucleus: A Laboratory Guide* (Hamkalo BA, Elgin SCR, eds). San Diego: Academic Press. pp 587-627.

Gubb D, Ashburner M, Roote J, Davis T. (1990) A novel transvection phenomenon affecting the *white* gene of *Drosophila melanogaster*. *Genetics*, **126**: 167-176.

Henikoff S. (1979) Position effects and variegation enhancers in an autosomal region of *Drosophila melanogaster*. *Genetics*, **93**: 105-115.

Henikoff S. (1981) Position-effect variegation and chromosome structure of a heat shock puff in *Drosophila*. *Chromosoma*, **83**: 381-393.

Henikoff S, Dreesen TD. (1989) Trans-inactivation of the *Drosophila brown* gene: Evidence for transcriptional repression and somatic pairing dependence. *Proc. Natl Acad. Sci. USA*, **86**: 6704-6708.

Henikoff S. (1990) Position-effect variegation after 60 years. *Trends Genet.* **6**: 422-426.

Henikoff S. (1992) Position effect and related phenomena. *Curr. Opin. Genet. Dev.* **2**: 907-912.

Herskowitz I. (1987) Functional inactivation of genes by dominant negative mutations. *Nature*, **329**: 17-23.

Hinton T, Goodsmith W. (1950) An analysis of phenotypic reversions at the *brown* locus in *Drosophila. J. Exp. Zool.* **114**: 103-114.

Judd BH. (1988) Transvection: allelic cross talk. *Cell*, **53**: 841-843.

Karpen G, Spradling AC. (1990) Reduced DNA polytenization of a minichromosome region undergoing position-effect variegation in *Drosophila. Cell*, **63**: 97-107.

Laird CD. (1990) Proposed genetic basis of Huntington's disease. *Trends Genet.* **6**: 242-247.

Lamb P, McKnight SL. (1991) Diversity and specificity in transcriptional regulation: The benefits of heterotypic dimerization. *Trends Biochem. Sci.* **16**: 417-422.

Lindsley DL, Grell EH. (1968) *Genetic Variations of* Drosophila melanogaster. Washington DC: Carnegie Institute.

Lindsley DL, Zimm GG. (1992) *The Genome of* Drosophila melanogaster. San Diego: Academic Press.

McClintock B. (1951) Chromosome organization and genic expression. *Cold Spring Harbor Symp. Quant. Biol.* **16**: 13-47.

Muller HJ. (1930) Types of visible variations induced by X-rays in *Drosophila. J. Genet.* **22**: 299-334.

Muller HJ. (1932) Further studies on the nature and causes of gene mutations. *Proc. Intl Congr. Genet.* **1**: 213-225.

Parkhurst SM, Harrison DA, Remington MP, Spana C, Kelley RL, Coyne RS, Corces VG. (1988) The *Drosophila su(Hw)* gene which controls the phenotypic effect of the gypsy transposable element, encodes a putative DNA-binding protein. *Genes Dev.* **2**: 1205-1215.

Rushlow CA, Bender W, Chovnick A. (1984) Studies on the mechanism of heterochromatic position effect at the *rosy* locus of *Drosophila melanogaster. Genetics*, **108**: 603-615.

Schultz J. (1936) Variegation in *Drosophila* and the inert heterochromatic regions. *Proc. Natl Acad. Sci. USA*, **22**: 27-33.

Schultz J. (1937) *Carnegie Institute Washington Yearbook* **36**: 298-305.

Slatis HM. (1955a) A reconsideration of the brown-dominant position effect. *Genetics*, **40**: 246-251.

Slatis HM. (1955b) Position effects at the brown locus in *Drosophila melanogaster. Genetics*, **40**: 5-23.

Smolik-Utlaut SM, Gelbart WM. (1987) The effects of chromosomal rearrangements on the *zeste–white* interaction in *Drosophila melanogaster. Genetics*, **116**: 285-298.

Spofford JB. (1976) Position-effect variegation in *Drosophila.* In: *Genetics and Biology of* Drosophila, vol. 2a (Ashburner M, Novitski E, eds). London: Academic Press. pp 955-1019.

Spradling AC, Karpen GH. (1990) Sixty years of mystery. *Genetics*, **126**: 779-784.

Sullivan DT, Sullivan MC. (1975) Transport defects as the physiological basis for eye color mutants of *Drosophila melanogaster. Biochem. Genet.* **13**: 603-613.

Tartof KD, Henikoff S. (1991) Trans-sensing effects from *Drosophila* to humans. *Cell*, **65**: 201-203.

Umbetova GH, Belyaeva ES, Baricheva EM, Zhimulev IF. (1991) Cytogenetic and molecular aspects of position effect variegation in *Drosophila melanogaster.* IV. Underreplication of chromosomal material as a result of gene inactivation. *Chromosoma*, **101**: 55-61.

Chapter 15

Parental imprinting in mouse development

Anne C Ferguson-Smith and M Azim Surani

1. Introduction

In recent years, genetic and embryological studies have proven that the long assumed principle of the functional equivalence of parental genomes is not correct. If it were, then a balanced diploid mammalian organism should develop normally regardless of the parental origin of its two haploid sets of chromosomes. However, in order for mammalian development to proceed normally, both a maternally and paternally inherited set of chromosomes is essential. This is due to the presence of genes whose expression is dependent on their parental origin resulting in differential activity from the homologous parental chromosomes. Parental imprinting is the process which renders the two parental genomes functionally inequivalent and results in the expression of an imprinted gene being dependent on the germline through which it has passed. Neither the question of why such a process should evolve nor the precise mechanisms involved in the regulation of imprinted genes is fully understood. However, it follows that the dosage of an imprinted gene can be doubled or lost completely if there is a uniparental duplication or deficiency of the gene or chromosomal region in which it resides. This can have profound effects on embryonic development.

2. Parental imprinting in early embryogenesis

The effects of duplications/deficiencies of whole parental genomes have been studied in the mouse using pronuclear transfer experiments resulting in the formation of bimaternal (gynogenetic) and bipaternal (androgenetic—AG) conceptuses. Bimaternal embryos (parthenogenones—PG) can be generated more easily by ethanol activation of unfertilized eggs with subsequent diploidization. Parthenogenetic and gynogenetic embryos exhibit identical phenotypes. Neither PG nor AG conceptuses develop to term (Barton *et al.*, 1984; Solter, 1988). A proportion of PG embryos can survive to day 10.5 (d10.5) of gestation (gestation = 21 days) and reach up to the 25 somite stage, but they are growth retarded. In addition, the extraembryonic components are poorly developed with failure of the trophoblast to proliferate properly. Blastocyst reconstitution experiments in which PG inner cell masses (which give rise to the embryo proper) are incorporated into extraembryonic components derived from normal embryos also result in embryonic lethality (Barton *et al.*, 1985; Gardner *et al.*, 1990). Thus a paternal genome is required for development of the extraembryonic lineages and also for the growth and development of the embryo proper especially during the second half of gestation. The absence of a maternal genome seems to have a more severe effect on the development of the AG conceptus. Indeed, the AG embryo is severely retarded resembling at best a d8.5 embryo at d10 of gestation (Solter, 1988; Surani *et al.*, 1986). In contrast to PG embryos, however, the extraembryonic trophoblast appears well developed though histological examination shows some subtle abnormalities (S Barton, MA Surani, unpublished observation). Thus a maternal genome is important for the early embryonic development at this stage. These studies therefore prove a requirement for both parental genomes in normal development and implicate reciprocal roles for the parental genomes in the development of embryonic and extraembryonic tissues.

3. Parental imprinting in later development

The developmental potential of these embryonic and extraembryonic cells can be analysed later in development by generating chimaeras between AG, PG and normal embryos (AG→N and PG→N) for example by injecting an inner cell mass from a manipulated embryo into the blastocoel cavity of a normal host. The resulting chimaeras again show reciprocal phenotypic differences: in growth and also with respect to the lineages in which AG and PG cells are found. PG→N chimaeras will survive to term but are smaller than controls by up to 50%. Analysis of the tissue distribution of PG cells shows a marked selection against them in some tissues of mesodermal origin, notably skeletal muscle, where commencing at d13–d15 there is progressive loss of PG cells such that at birth they are usually absent. This is in contrast to brain, epidermis and germline which retain a relatively high proportion of PG cells at birth (Fundele *et al.*, 1989, 1990). These effects are independent of genetic background.

In contrast, AG→N chimaeras will survive to term only if the AG contribution is less than 20%. An AG contribution of 50% or greater results in lethality prior to d15. These chimaeras, however, exhibit a marked growth enhancement, elongation of the anteroposterior axis, a small head and often an enlarged heart. In addition, the tissue distribution of AG cells seems reciprocal to that described for PG→N chimaeras with preferential retention of AG cells in mesodermal derivatives. Indeed, one of the AG→N chimaeras surviving to term exhibited severe skeletal anomalies associated with axial elongation and an abnormal proliferation of chondrocytes. AG cells rarely contributed to the brain and other neuroectodermal components (Barton *et al.*, 1991).

The reciprocal distributions of AG and PG cells in chimaeras suggest roles for imprinted genes in the proliferation and development of these particular lineages. Moreover, they implicate gene dosage as important in the successful development of tissues such as skeletal muscle, cartilage and brain. More detailed histological analyses of these chimaeras at various stages of gestation should not only provide information on the functions of imprinted genes but also contribute to our understanding of cell–cell interactions during organogenesis.

4. Genetic studies of parental imprinting

Genetic complementation tests have identified at least eight subchromosomal regions on six mouse chromosomes which harbour imprinted genes (Cattanach and Beechey, 1990). These experiments make use of reciprocal and Robertsonian chromosomal translocations which undergo meiotic non-disjunction resulting in unbalanced gametes containing duplications/deficiencies of particular chromosomal regions. Fusion of an unbalanced egg with an unbalanced sperm may result in a balanced zygote which carries a uniparental duplication/deficiency (disomy) for particular chromosomal regions involved in the translocation. Mutant phenotypes associated with the presence of a uniparental disomy may arise if there is a requirement for both a maternally and paternally inherited copy of the region. In this way mouse chromosomes 2, 6, 7, 11, 12, and 17 have been shown to undergo parental non-complementation and thus contain imprinted genes. However, imprinted genes may exist elsewhere. Altered gene dosage may produce more subtle phenotypes which might be missed using the disomy approach. In addition, targeted mutagenesis of several genes, thought crucial for particular cellular processes, has often not resulted in a mutant phenotype, perhaps indicating redundancy within the mouse genome (Rudnicki *et al.*, 1992).

One region, identified from the above studies, which has been the focus of much attention, is the distal portion of chromosome 7. Distal chromosome 7 in the mouse shares syntenic homology with human chromosomes 11 and 15q regions associated with human genetic diseases exhibiting parental origin effects in their pattern of inheritance. These disorders include Beckwith–Weidemann syndrome, Wilm's tumour, Glomus tumour, Prader–Willi syndrome and Angelman syndrome (Ferguson-Smith *et al.*, 1990; Nicholls *et al.*, 1992).

Using animals heterozygous for the reciprocal translocation T(15:7)9H in intercrosses, embryos carrying maternal duplication/paternal deficiency for distal chromosome 7 (MatDi7) can be generated at a low frequency. Such embryos die at d16 of gestation and exhibit 50% growth retardation compared to normal littermates (see later). The reciprocal paternal duplication/maternal deficiency (PatDi7) die at an unknown stage prior to d11 and have never been identified

(Searle and Beechey, 1990). PatDi7 cells have been rescued in the context of normal cells in chimaeras (PatDi7→N). Three PatDi7→N embryos recovered at d15 and d17 of gestation showed 35–55% growth enhancement compared to non-disomic chimaeric and non-chimaeric littermates (Ferguson-Smith *et al.*, 1991). Unlike the experimental cells in PG and AG→N chimaeras, PatDi7 cells can contribute to all lineages. It is not known whether these chimaeras survive to term.

5. Identification of imprinted genes

The regions involved in the disomy studies, discussed above, span many megabases of DNA. In order to identify the imprinted genes themselves and conduct molecular analyses on their mode of regulation, higher resolution molecular studies must be undertaken. An important requirement for the identification of imprinted genes, as well as for subsequent investigations on the mechanism of their mode of parental origin-specific regulation, is an experimental system in which gene activity from maternally and paternally inherited chromosomes can be easily distinguished. The disomy system is most useful in this regard as comparisons between normal and disomic embryos allow any differences to be attributed to the deficient parental chromosome. We have shown that the insulin-like growth factor II (*Igf2*) gene located on distal chromosome 7 is repressed in MatDi7 embryos (Ferguson-Smith *et al.*, 1991). This is because the absent paternally inherited allele is normally expressed. The growth retardation seen in MatDi7 embryos is most likely due to the repression of the *Igf2* gene. However, the lethality in MatDi7 embryos at d16 of gestation is not due to the reduced *Igf2* activity, as mice which have inherited a mutated paternal allele of *Igf2* are viable and fertile though growth retarded (DeChiara *et al.*, 1990). Thus the lethality of MatDi7 embryos is due to imprinted factors other than (or in addition to) *Igf2*. Targeted mutagenesis by homologous recombination, such as that cited for the *Igf2* gene, is useful in the identification of imprinted genes as mutant phenotypes showing a parental origin effect in the pattern of inheritance would indicate that the mutated gene was imprinted (DeChiara *et al.*, 1991). However, to date, the *Igf2* gene is the only gene showing this effect after targeted mutagenesis.

The first imprinted gene to be identified was found using an approach combining genetic and molecular methods. The only endogenous mutation in the mouse known to display clear parental origin effects in its pattern of inheritance is *Tme* (T maternal effect) which maps to the *T* locus on chromosome 17. *Tme* is associated with deletions in the imprinted portion of this chromosome and the deletion is lethal around d17 of gestation but only when maternally inherited. Inheritance of the mutation via the male germline has no major effect (Johnson, 1974, 1975; Winking and Silver, 1984). A reverse genetic approach was taken to physically map a portion of the deletion and analyse the transcription of four genes in the region including two, *Igf2r* (the insulin-like growth factor two receptor) and *Sod-2* (superoxidase dismutase), located within the deletion. In this way Barlow and co-workers were able to show absence of *Igf2r* expression when the deletion was maternally inherited; transcription was normal when the deletion was inherited from the male. Therefore, in normal embryos, the paternal allele of the *Igf2r* is repressed and the maternally inherited allele is expressed (Barlow *et al.*, 1991). Subsequent studies, however, have shown that the lethality can be rescued on some genetic backgrounds with the imprinting of *Igf2r* being unaffected (Forejt and Gregorova, 1992). This suggests that, at least under some circumstances, the gene is not required for fetal viability and also that there may be another imprinted gene in the *Tme* region that may play a role in the parental origin-specific lethality. The function of the *Igf2r* is not well understood. The gene is developmentally regulated (Senior *et al.*, 1990). The Igf2r protein acts as a mannose 6-phosphate receptor and is believed to target proteins with such residues to lysosomes for degradation (Morgan *et al.*, 1987). It also has good affinity for the ligand Igf2 but is not believed to mediate the mitogenic role of the growth factor which actually has higher affinity for the Igf1r. The role of the Igf2/Igf2r interaction is not clear at present.

A further approach has proved fruitful in the identification of an imprinted gene. Intercrosses between two genetically different outbred strains of mice generate progeny which may carry parental origin-specific sequence length polymorphisms in coding sequences. These RNA variants allow the maternally and/or paternally inherited transcripts to be identified. This approach was used to show that only the maternal allele of the *H19* gene was expressed with the paternally inherited gene being repressed (Bartolomei *et al.*, 1991). The *H19* gene is a developmentally regulated gene located, like *Igf2*, on the distal portion of chromosome 7. The function of this gene is unknown; however, two of the three groups that independently cloned it, identified it as being activated in embryonal carcinoma cells and 10T1/2 cells after differentiation (Davis *et al.*, 1987; Poirier *et al.*, 1991). The only site of expression in the adult is in skeletal muscle where the imprint is maintained (Bartolomei *et al.*, 1991). Transgenic animals for *H19*, in which the exogenous gene is expressed, die during embryogenesis (Brunkow and Tilghman, 1991). It is likely that the double dose of *H19* expressed in MatDi7 embryos (Ferguson-Smith, Sasaki, Cattanach and Surani, submitted) at least contributes to their lethality.

The human genetic disorders, Prader–Willi syndrome (PWS) and Angelman syndrome (AS) are linked to the region of syntenic homology in the mouse just distal to the breakpoint on chromosome 7 in the T(7;15)9H translocation. PWS is associated with maternal duplication/paternal deficiency and AS reciprocally with paternal duplication/maternal deficiency of human chromosome 15, as shown by pedigree studies of patients with loss of heterozygosity. PWS is associated with growth anomalies and both disorders are associated with different types of behavioural anomalies (Nicholls *et al.*, 1992). Mapping studies in man have identified two closely linked critical regions, one for each syndrome. Therefore one might assume that in the mouse there are two closely linked imprinted genes in this region of chromosome 7. Very recently, it has been shown that a gene encoding a protein component of a small ribonucleoprotein complex, the *SnrpnN* gene, is located in this region and is expressed solely from the paternally inherited allele. Therefore in mice inheriting maternal duplication/paternal deficiency of this region of chromosome 7, the gene is repressed (Barr *et al.*, 1992; Cattanach *et al.*, 1992). Interestingly, this *SnrpnN* gene-product is neural specific and is therefore a good candidate for the PWS gene.

The activities of the three imprinted genes so far identified on mouse chromosome 7 are illustrated schematically in Fig. 1.

6. Imprinting and the chromosome

Mechanism of parental imprinting

We assume that the signals which allow the transcriptional machinery of the cell to activate or repress an allele are set down at some stage in the germline during gametogenesis when the parental chromosomes are separated. Then it is supposed that subsequent events occur at and/or after fertilization, and progressively render the imprints complete and stable. Analysis of the properties of imprinted genes in AG, PG and normal embryonic stem cells will be useful in assessing the stability of parental imprints. These epigenetic events responsible for the initiation, establishment and maintenance of imprints are not understood. However, evidence is beginning to accumulate that suggests that each imprinted gene may be subject to different epigenetic modifications. DNA methylation has been implicated in parental imprinting (Chaillet, 1992). CpG-dinucleotide methylation is a dynamic and heritable process, long associated in some instances with gene inactivity (Cedar, 1988). During gametogenesis and early embryogenesis, genome wide changes in DNA methylation occur (Kafri *et al.*, 1992; Monk *et al.*, 1987). In addition, DNA methylation has been shown to play a vital role in embryogenesis (Li *et al.*, 1992).

Initial studies on the imprinting of randomly integrated transgenes identified two insertions which showed germline specific modifications. These two lines, RSV-*myc* and MPA434, showed DNA methylation specific to passage through the female germline with the paternally inherited loci remaining undermethylated. Both transgenes remained methylated at the maternal allele throughout embryogenesis. There are no definitive examples of transgene methylation after paternal transmission. The preintegration site of MPA434 has been cloned but does not show parental origin-specific methylation (Sasaki *et al.*, 1992a). There-

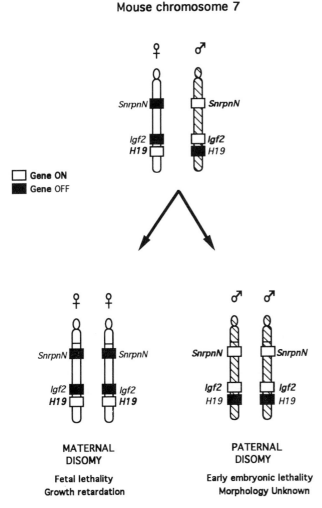

Mouse chromosome 7

Figure 1. Schematic representation of the imprinted genes identified on mouse chromosome 7. Altered gene dosage in monoparental duplication/deficiency (maternal/paternal disomy) is indicated. *Igf2* = insulin-like growth factor II; *H19* = H19; *SnrpnN* = small nuclear RNA-associated protein, N.

fore, the imprinting of transgenes may have limitations as a model system to study the regulation of endogenous genes by imprinting.

One can also draw parallels between X-inactivation, the mechanism of X chromosome dosage compensation in female mammals, and parental imprinting. Firstly, the extraembryonic tissues of female mammals and all somatic tissues of female marsupials undergo preferential inactivation of the paternal X chromosome (Sharman, 1971; Takagi and Sasaki, 1975). In addition, X-linked genes with CpG-rich promoters usually show methylation of these sequences on the inactive X (Grant and Chapman, 1988). It is important to note, however, that there are other forms of epigenetic modification associated with chromatin structure that would affect gene expression. The inactive X, for example, undergoes chromatin condensation and the chromatin in promoter portions of genes on the inactive X lose DNase I hypersensitivity and show reduced accessibility to nucleases (Riley *et al.*, 1984). The temporal relationship between DNA methylation and chromatin condensation on the inactive X is not clear. However, in an investigation of the imprinting of endogeneous genes, analysis of DNA methylation and chromatin condensation may be informative. We have compared the methylation status and chromatin structure of the *Igf2* and *H19* genes between MatDi7 and normal embryos. In this way, epigenetic differences identified between the two embryo types could be attributed to the paternal allele absent in MatDi7 material.

The Igf2 *gene*. The mouse *Igf2* gene encodes an embryonal mitogen widely expressed in embryonic and extra-embryonic tissues (Lee *et al.*, 1990). It is a complex gene of six exons with three promoters (P1, 2 and 3) and multiple transcripts arising through differential splicing (Ikejiri *et al.*, 1991; Rotwein and Hall, 1990). P2 is a typical CpG-island and P2 and P3 are known to be developmentally regulated and subject to imprinting (DeChiara *et al.*, 1991). A detailed analysis of the methylation and chromatin structural differences between MatDi7 and normal embryos was carried out. No methylation differences were detectable between the maternal and paternal alleles in the promoters and coding portions of the gene at sites susceptible to cleavage by methylation sensitive restriction enzymes. Therefore there was no widespread methylation of the maternally repressed *Igf2* allele (Sasaki *et al.*, 1992b). We cannot rule out the possibility that more subtle differences in methylation exist between the two alleles at sites not accessible in this analysis. Likewise, there was no difference in nuclease accessibility of the chromatin of the active or repressed alleles. DNase I hypersensitive sites were present in both MatDi7 and normal embryos. Indeed, it was shown that the repressed maternal *Igf2* allele showed a very low but significant level of activity as measured by a sensitive quantitative RT-PCR assay (Sasaki *et al.*, 1992b). This expression was not attributed to the absence of imprinting in a small subset of cells. The absence of methylation or chromatin compaction at the repressed locus may be due to this low level activity, although there is no evidence to link the two observations. A region 3 kb upstream of the promoters shows minor differences in DNA methylation with the paternally inherited sites being slightly more methylated than maternally inherited ones (Sasaki *et al.*, 1992b). Further studies are needed to elucidate the significance, if any, of this finding for the imprinted regulation of *Igf2*. Thus the mechanism regulating the parental origin specific activity of the *Igf2* gene remains an enigma. Possibly, the imprinting control regions lie at sites more distant to those examined in the aforementioned study. Indeed the secret of the mechanism of *Igf2* imprinting may lie in the neighbouring and reciprocally imprinted *H19* gene which shows quite different epigenetic characteristics.

The H19 *gene.* The closely linked and paternally repressed *H19* gene consists of a single defined promoter, five exons and two downstream enhancers and is located 90 kb downstream of *Igf2* (Yoo-Warren *et al.*, 1988; Zemel *et al.*, 1992). The gene encodes a single, abundant developmentally regulated transcript of unknown function (Pachnis *et al.*, 1988). The promoter is a CpG island (Ferguson-Smith *et al.*, submitted). Similar studies to those described above for *Igf2* were carried out for the *H19* gene. The results clearly demonstrate that all sites susceptible to cleavage by methylation sensitive enzymes in the promoter of *H19* are unmethylated when maternally inherited. The paternally inherited promoter is hypermethylated. These differences extend into the first exon; however, sites in the rest of the gene and at the 3′ end in the region of the enhancers do not show differential methylation. Analysis of sperm DNA shows absence of methylation in the promoter, implying that the methylation of the repressed paternal allele is a post-fertilization event. Further studies have shown that the allele is methylated by d9 of gestation (Ferguson-Smith *et al.*, submitted). It will be of interest to determine when this differential methylation occurs. Also, in contrast to the repressed *Igf2* allele, the chromatin of the repressed allele of *H19* shows reduced accessibility to nucleases compared to the active maternal alleles in MatDi7 embryos. These studies reveal that there are epigenetic differences between the repressed and active promoters of the *H19* gene. The epigenetic status of the *H19* gene is therefore more in line with observations associated with genes on the inactive X chromosome; however, the question still remains as to the exact nature of the initial imprint and whether these post-zygotic differences in methylation are a secondary response to allelic inactivity. The differences observed between the paternal allele of *H19* and the maternal allele of *Igf2* may reflect differences in the mechanisms of repression of maternally versus paternally inherited genes. This question will be resolved when comparisons can be made with the epigenetic status of the paternally repressed *Igf2r* gene and the maternally repressed *SnrpnN* gene. Alternatively, the absence of gross epigenetic differences between the alleles of the *Igf2* gene prompts speculations on whether the imprinting of *Igf2* is a secondary consequence of the reciprocal imprinting of *H19*. A number of theories have been proposed to address this question (and see later). A schematic representation of the epigenetic differences between the maternal and paternal alleles of *Igf2* and *H19* are illustrated in Fig. 2.

Relationships between imprinted genes. An interesting pattern is emerging regarding the organization of imprinted genes, that even at this speculative stage, merits note. Although this pattern may not be expected to hold true for all imprinted genes, it seems that many of the imprinted genes have a pairwise relationship. The *H19* and *Igf2* genes are 90 kb apart and reciprocally imprinted (Zemel *et al.*, 1992). One might consider a functional relationship between this pair of imprinted

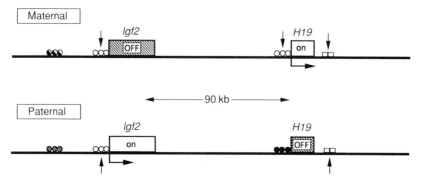

Figure 2. Schematic representation *Igf2*-H19 domain. The differences in epigenetic modification observed between the paternally and maternally inherited loci are shown. Both genes are in the same transcriptional orientation. Closed circles indicate hypermethylation and open circles, absence of methylation. Partially filled circles indicate partial methylation. DNase I hypersensitive sites are indicated by a single arrow. The small squares represent the two downstream enhancers of *H19*. (The schematic is not drawn to scale.)

genes in three ways. Firstly the genes may share common regulatory elements associated with the function of their gene products. For example, both *H19* and the *Igf2* gene are expressed in the same tissues at around the same developmental stages. The temporal and spatial specific elements controlling this activity may be shared. This may be unrelated to their reciprocal imprinting and has been proposed previously for the clustering of other non-imprinted genes with common tissue or temporal specificity (Ruddle *et al.*, 1987). Secondly, and perhaps more intriguingly, the pairing of these reciprocally imprinted genes may indicate a relationship in the mechanism of their imprinting. For example, in order for the *Igf2* gene to be active on the paternally inherited chromosome, the neighbouring *H19* gene must be inactive on the same chromosome and *vice versa* on the maternally inherited chromosome. In other words, they may be sharing common imprinting control elements (also discussed in Sasaki *et al.*, 1992). This has some bearing on the evolution of the *Igf2-H19* domain and how and why the imprinting signals arose, a third consideration. Maintaining a balance between the ratio of paternally to maternally imprinted genes might be considered an evolutionary advantage to a sexual population. Such *cis*-effects would not necessarily be confined to regions as small as 90 kb, as epigenetic events governing gene regulation may affect long range chromatin structure and act at quite a distance. Thus these same three considerations could apply to the reciprocally imprinted PWS and AS genes. Requirement for an evolutionary balance between maternally and paternally imprinted chromosomes has already been implicated in the reciprocal imprinting of the *Igf2* and *Igf2r* genes and invokes a functional relationship in terms of their biological roles (rather than a mechanistic one governing their imprinting) as the pressure maintaining the balance (Haig and Graham, 1991; Moore and Haig, 1991). Clearly, in order to address such relationships between pairs of imprinted genes, others must be identified and their functions elucidated and addressed in an evolutionary context.

7. Conclusion

It is clear that, to date, only the surface of the intriguing phenomenon of parental imprinting has been scratched. Further endogenous imprinted genes need to be identified together with the molecular basis of this unusual form of gene regulation. Understanding the developmental roles of imprinted genes will require detailed molecular and developmental scrutiny of mice carrying parental duplications/deficiencies. Such analyses will provide valuable insight into the human genetic disorders exhibiting parental origin effects in their patterns of inheritance and perhaps provide mouse models for some of these syndromes. Studies using mutations generated by targeted mutagenesis will provide valuable information on functions of imprinted genes both in terms of their roles during embryogenesis and in terms of potential relationships in their mechanisms of imprinting. Analysis of primary signals and the roles of subsequent nucleocytoplasmic interactions and other DNA/protein interactions in the control of parental origin-specific gene regulation continues to make the study of parental imprinting one of the best experimental systems in the analysis of epigenetic inheritance in a wider context.

References

Barlow DP, Stoger R, Herrmann BG, Saito K, Schweifer N. (1991) The mouse insulin-like growth factor type-2 receptor is imprinted and closely linked to the *Tme* locus. *Nature*, **349:** 84-87.

Barr J, Jones J, Beechey C, Cattanach B. (1992) Imprinting studies of the central region of mouse chromosome 7 (abstract). Mammalian Genetics and Development Workshop, Birkbeck College, London. November 1992. *Genet. Res. (Cambridge)* in press.

Bartolomei MS, Zemei S, Tilghman SM. (1991) Parental imprinting of the mouse *H19* gene. *Nature*, **351:** 153-155.

Barton SC, Surani MA, Norris ML. (1984) Role of paternal and maternal genomes in mouse development. *Nature*, **311:** 374-376.

Barton SC, Adams CA, Norris ML, Surani MA. (1985) Development of gynogenetic and parthenogenetic inner cell mass and trophectoderm tissues in reconstituted blastocysts in the mouse. *J. Embryol. Exp. Morphol.* **90:** 267-285.

Barton SC, Ferguson-Smith AC, Fundele R, Surani MA. (1991) Influence of paternally imprinted genes on development. *Development*, **113:** 679-688.

Brunkow ME, Tilghman SM. (1991) Ectopic expression of the *H19* gene in mice causes prenatal lethality. *Genes Dev.* **5:** 1092-1101.

Cattanach BM, Beechey CV. (1990) Autosomal and X-chromosome imprinting. *Development Suppl.:* 63-72.

Cattanach BM, Barr JA, Evans EP, Burtenshaw M, Beechey CV, Leff SE, Brannan CI, Copeland NG, Jenkins NA, Jones J. (1992) A candidate mouse model for Prader–Willi syndrome which shows an absence of SnrnpN expression. *Nature Genet.* **2:** 270–274.

Cedar H. (1988) DNA methylation and gene activity. *Cell*, **53:** 3-4

Chaillet JR. (1992) DNA methylation and genomic imprinting in the mouse. *Semin. Dev. Biol.* **3:** 99-105.

Davis RL, Weintraub H, Lassar AB. (1987) Expression of a single transfected cDNA converts fibroblasts to myoblasts. *Cell*, **51:** 987-1000.

DeChiara TM, Efstratiadis A, Robertson EJ. (1990) A growth-deficiency phenotype in heterozygous mice carrying an insulin-like growth factor II gene disrupted by targeting. *Nature*, **345:** 78-80.

DeChiara TM, Robertson EJ, Efstratiadis A. (1991) Parental imprinting of the mouse insulin-like growth factor II gene. *Cell*, **64:** 849-859.

Ferguson-Smith AC, Reik W, Surani MA. (1990) Genomic imprinting and cancer. *Cancer Surv.* **9:** 487-503.

Ferguson-Smith AC, Cattanach BM, Barton SC, Beechey CV, Surani MA. (1991) Embryological and molecular investigations of parental imprinting on mouse chromosome 7. *Nature*, **351:** 667-670.

Forejt J, Gregorova S. (1992) Genetic analysis of genomic imprinting: an *Imprintor-1* gene controls inactivation of the paternal copy of the mouse *Tme* locus. *Cell*, **70:** 443-450.

Fundele R, Norris ML, Barton SC, Reik W, Surani MA. (1989) Systematic elimination of parthenogenetic cells in mouse chimaeras. *Development*, **106:** 20-35.

Fundele RH, Norris ML, Barton SC, Fehlau M, Howlett SK, Mills WE, Surani MA. (1990) Temporal and spatial selection against parthenogenetic cells during development of fetal chimeras. *Development*, **108:** 203-211.

Gardner RL, Barton SC, Surani MAH. (1990) Use of triple tissue blastocyst reconstitution to study the development of diploid parthenogenetic primitive ectoderm in combination with fertilization-derived trophectoderm and primitive endoderm. *Genet. Res. (Cambridge)* **56:** 209-222.

Grant SG, Chapman VM. (1988) Mechanism of X-chromosome regulation. *Annu. Rev. Genet.* **22:** 199-233.

Haig D, Graham T. (1991) Genomic imprinting and the strange case of the insulin-like growth factor II receptor. *Cell*, **64:** 1045-1046.

Ikejiri K, Furuichi M, Ueno T, Matsuguchi T, Takahashi K, Endo H, Yamamoto M. (1991) The presence and active transcription of three independent leader exons in the mouse insulin-like growth factor II gene. *Biochim. Biophys. Acta* **1089:** 77-82.

Johnson DR. (1974) Hairpin-tail: a case of post-reductional gene action in the mouse egg? *Genetics*, **76:** 795-805.

Johnson DR. (1975) Further observations on the Hairpin-tail (Thp) mutation in the mouse. *Genet. Res. (Cambridge)* **24:** 207-213.

Kafri T, Ariel M, Brandeis M, Shemer R, Urven L, McCarrey J, Cedar H, Razin A. (1992) Developmental pattern of gene-specific DNA methylation in the mouse embryo and germ line. *Genes Dev.* **6:** 705-714.

Lee J, Pintar J, Efstratiadis A. (1990) Pattern of the insulin-like growth factor II gene expression during early mouse development. *Development*, **110:** 151-159.

Li E, Bestor TH, Jaenisch R. (1992) Targeted mutation of the DNA methyltransferase gene result in embryonic lethality. *Cell*, **69:** 915-926.

Monk M, Boubelik M, Lehnert S. (1987) Temporal and regional changes in DNA methylation in the embryonic, extraembryonic and germ cell lineages during mouse embryo development. *Development*, **99:** 371-382.

Moore T, Haig D. (1991) Genomic imprinting in mammalian development: a parental tug-of-war. *Trends Genet.* **7:** 45-49.

Morgan DO, Edman JC, Standring DN, Fried VA, Smith MC, Roth R, Rutter WJ. (1987) Insulin-like growth factor II receptor as a multifunctional binding protein. *Nature*, **329:** 301-307.

Nicholls RD, Rinchik EM, Driscoll DJ. (1992) Genomic imprinting in mammalian development: Prader–Willi and Angelman syndromes as disease models. *Semin. Dev. Biol.* **3:** 139-152.

Pachnis V, Brannan CI, Tilghman SM. (1988) The structure and expression of a novel gene activated in early mouse embryogenesis. *EMBO J.* **7:** 673-681.

Poirier F, Chan C-TJ, Timmons P, Robertson EJ, Evans MJ, Rigby PWJ. (1991) The murine *H19* gene is activated during embryonic stem cell differentiation in vitro and at the time of implantation in the developing embryo. *Development*, **113:** 1105-1114.

Riley DE, Canfield TK, Gartler SM. (1984) Chromatin structure of active and inactive human X chromosomes. *Nucl. Acids Res.* **12:** 1829-1845.

Rotwein P, Hall LJ. (1990) Evolution of insulin-like growth factor II: Characterisation of the mouse *IgfII* gene and identification of two pseudo-exons. *DNA Cell Biol.* **9:** 725-735.

Ruddle FH, Hart CP, Rabin M, Ferguson-Smith AC, Pravtcheva D. (1987) Comparative genetic analysis of homeobox genes in mouse and man. In: *New Frontiers in the Study of Gene Functions* (Poste G, Crooke ST, eds). New York: Plenum Publishing. pp 73-86.

Rudnicki MA, Braun T, Hinuma S, Jaenisch R. (1992) Inactivation of MyoD in mice leads to up-regulation of the myogenic HLH gene *myf-5* and results in apparently normal muscle development. *Cell,* **71:** 383-390.

Sasaki H, Allen ND, Surani MA. (1992a) DNA methylation and genomic imprinting in mammals. In: *DNA Methylation: Biological Significance* (Jost JP, Saluz H, eds). Basel: Birkhauser Verlag.

Sasaki H, Jones PA, Chaillet JR, Ferguson-Smith AC, Barton SC, Reik W, Surani MA. (1992b) Parental imprinting: potentially active chromatin of the repressed maternal allele of the mouse insulin-like growth factor II (*Igf2*) gene. *Genes Dev.* **6:** 1843-1856.

Searle AG, Beechey CV. (1990) Genome imprinting phenomena on mouse chromosome 7. *Genet. Res. (Cambridge)* **56:** 237-244.

Senior PV, Byrne S, Brammar WJ, Beck F. (1990) Expression of the IGF-II/mannose-6-phosphate receptor mRNA and protein in the developing rat. *Development,* **109:** 67-73.

Sharman GB. (1971) Late replication in the paternally derived X chromosome of female kangaroos. *Nature,* **230:** 231-232.

Solter D. (1988) Differential imprinting and expression of maternal and paternal genomes. *Annu. Rev. Genet.* **22:** 127-146.

Surani MA, Barton SC, Norris ML. (1986) Nuclear transplantation in the mouse: heritable differences between parental genomes after activation of the embryonic genome. *Cell,* **45:** 127-136.

Takagi N, Sasaki M. (1975) Preferential inactivation of the paternally derived X-chromosome in the extraembryonic membranes of the mouse. *Nature,* **256:** 640-642.

Winking H, Silver LM. (1984) Characterization of a recombinant mouse t haplotype that express a dominant maternal effect. *Genetics,* **108:** 1013-1020.

Yoo-Warren H, Pachnis V, Ingram RS, Tilghman SM. (1988) Two regulatory domains flank the mouse *H19* gene. *Mol. Cell. Biol.* **8:** 4707-4715.

Zemel S, Bartolomei MS, Tilghman SM. (1992) Physical linkage of two mammalian imprinted genes, *H19* and *Igf2*. *Nature Genet.* **2:** 61-65.

Chapter 16

The physical organization of interphase nuclei

JS (Pat) Heslop-Harrison, Andrew R Leitch and Trude Schwarzacher

1. Chromosomes at metaphase and interphase

The chromosome is a highly organized structure, showing a hierarchy of coiling, supercoiling and looping (Manuelidis and Chen, 1990). Studies of metaphase chromosome morphology have been valuable for understanding karyotypes and their evolution, the localization of genes, and for understanding the composition and structure of chromosomes (Gautier *et al.*, 1992; Harauz *et al.*, 1987). At interphase, when the chromosomes are active in RNA transcription and DNA replication, chromosomes 'decondense' to become longer. After making reconstructions of interphase nuclei from conventionally-stained electron micrographs of serially sectioned nuclei from root-tips, we were unable to distinguish discrete chromosomes, and most of the chromatin present formed a continuous network (Heslop-Harrison *et al.*, 1988, and unpublished data). Hence the organization of chromosomes within the nuclear envelope at interphase becomes extremely difficult to study.

Different species and cell types have different organizations of the DNA within interphase nuclei (Fig. 1; see also Nagl and Capesius, 1977). Models of interphase nucleus organization must account for both quantitative and qualitative differences in the DNA of different species and the appearance of their nuclei. There is a greater than 20 000-fold variation in DNA content between different eukaryotes: ranging from 14 Mb or 0.014 pg 1C DNA content in yeast (see Oliver *et al.*, Chapter 17, this volume), through 130 000 Mb in some angiosperm plants (see Bennett *et al.*, 1982), and perhaps up to 300 000 Mb in some Protozoa (see Cavalier-Smith, 1985). Within nuclei, the decondensation and distribution of DNA varies, with some domains of the nuclear volume being largely filled with chromatin, while other domains have very little chromatin.

Examples of interphase organization

Figure 1 shows electron micrographs of interphase nuclei of three contrasting species. *Arabidopsis thaliana*, Arabidopsis, is a small, rapidly growing plant of the

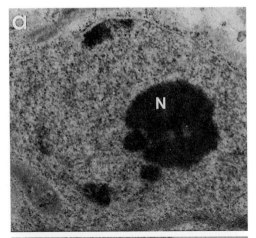

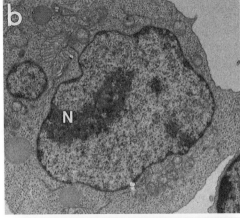

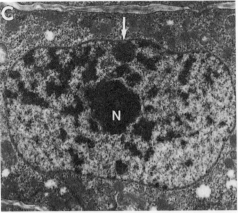

Figure 1. Electron micrographs of interphase nuclei from three species show contrasting packing of the DNA into nuclei. (a) *Arabidopsis thaliana* ovule wall nucleus, genome size 120 Mb (× 21 000); (b) human fibroblast nucleus, genome size 3000 Mb (× 5500); (c) barley (*Hordeum vulgare*) root-tip nucleus, genome size 5500 Mb (× 7900; provided by Dr Kesara Anamthawat-Jónsson); arrow, centromere. The nuclear envelope surrounds the nuclei, and darker, electron dense chromatin and nucleoli (N) are visible within. Light micrographs of the same species are shown in Figs 2d, 3d and 4d (see pp.178–180).

Brassica family that is widely used for molecular genetic studies (see Flavell *et al.*, Chapter 18, this volume). It has a genome size of about 120 Mb and interphase cells from the ovule wall show that much of the chromatin is decondensed (Fig. 1a).

Nevertheless, there are some chromatin segments lying against the nucleolus and nuclear envelope which show little decondensation from the metaphase structure. Human interphase nuclei (1C genome size 3000 Mb; Fig. 1b) from fibroblasts tend to have condensed chromatin against the nuclear envelope, and a few condensed sites within the nucleus. The peripheral condensed sites most probably represent DNA sequences which are not expressed and are constitutive or facultative heterochromatin (see also Henikoff *et al.*, Chapter 14, this volume). Within the rest of the nuclear volume, most chromatin is decondensed and, where visible, detected only as diffuse threads. The root-tip nuclei of cereal plants, such as barley (5500 Mb), have some 30% of the nuclear volume filled with condensed chromatin (Fig. 1c). The nuclei also include larger condensed segments corresponding to condensed rRNA genes. There is often a gradient across the nuclear volume in the proportion that is filled by condensed chromatin – from 50% near the centromeric pole of the nucleus to only 15% at the telomeric end (Anamthawat-Jónsson and Heslop-Harrison, 1990; Heslop-Harrison *et al.*, 1988). Interestingly and perhaps importantly (see Section 6), the less filled region of the nucleus resembles much of the nucleus of Arabidopsis or human.

The results above show that chromatin is not uniformly distributed within the volume of the interphase nucleus between the nuclear envelope and the nucleolus. The remainder of this chapter will discuss the location of genomes, chromosomes, repetitive DNA sequences and genes at interphase, and relate the results to interphase nuclear organization.

2. Locations of complete genomes in the nucleus

The haploid genome, as inherited from each parent in a diploid organism, is a basic unit of genome organization. Within species, imprinting effects may cause differential behaviour of the two genomes (see Ferguson-Smith and Surani, Chapter 15, this volume) and whole parental genomes may behave as units, either becoming heterochromatinized (see Nur, 1990, for review) or else being eliminated (see Finch and Bennett, 1983; Hunter *et al.*, 1993).

Hybrids made between different wild and cultivated cereal plant species are a valuable system in which to study differential genome behaviour. Such hybrids are relatively easy to make sexually by emasculation of the female parent, cross-pollination and subsequent embryo rescue. The resulting plant remains diploid and may live for many years. At metaphase, chromosomes can be identified by their sizes and morphology in spread preparations or reconstructions of sectioned nuclei. Many of the hybrid plants show parental genome separation at mitotic metaphase (Fig. 2a, see p. 178), which is maintained for many years (see Bennett, 1984, 1988; Finch *et al.*, 1981; Linde-Laursen and Jensen, 1991; Schwarzacher *et al.*, 1992b).

DNA:DNA *in situ* hybridization, using labelled total genomic DNA as a probe to chromosomes, enables identification of the parental origin of all chromosomes in some cereal hybrids throughout the cell cycle (Anamthawat-Jónsson *et al.*, 1993; Schwarzacher *et al.*, 1989) in both spread preparations (Fig. 2b, see p. 178) and three-dimensional reconstructions of sectioned nuclei (Fig. 2c, see p. 178). The nuclei show parental genome separation which is maintained throughout the cell cycle. Neither the mechanism for maintaining genome separation, nor the reason – perhaps functional – for its occurrence, are known (Leitch *et al.*, 1991). Genome separation is not a residual effect of the separation present after fertilization, since we noted no tendency to increased intermixing over several years. Neither is the separation an effect of chromosome size. One hybrid has different separation in two different tissues (Finch, 1983), indicating that genome localization is under genotypic control. Furthermore, there are correlations of gene expression with genome position (Bennett, 1984; Heslop-Harrison, 1990; Heslop-Harrison and Bennett, 1984, 1990). As yet, the frequency of parental genome separation in hybrids and species of plants and animals is unknown, although there is evidence for the phenomenon in species including mammals and insects (see Leitch *et al.*, 1991).

3. Locations of individual chromosomes in the nucleus

In situ hybridization enables the study of chromosome location during interphase. Pooled probes which are dispersed along the length of a chromosome or chromosome arm can be used to show its location (Cremer *et al.*, 1991; Lichter *et al.*, 1988). In humans, for example, chromosome 8 can be labelled specifically (Fig. 2d, see p. 178); most interphase nuclei show two discrete domains; the two chromosomes are not together, indicating that homologues are not associated, and the chromosomes are not decondensed into long fibres which run throughout the nucleus. Relative positions of several human chromosomes at interphase have been studied using probes for specific chromosomal regions. As indicated above for chromosome 8, the results show that each chromosome occupies a distinct domain within the interphase nucleus and that homologues are not associated (Emmerich *et al.*, 1989; Manuelidis and Borden, 1988; Popp *et al.*, 1990). Analysis of centromere positions in human brain, though, has shown that some centromeres are associated (Arnoldus *et al.*, 1989, 1991), and chromosome positions may alter in some diseases (Borden and Manuelidis, 1988), so there are mechanisms for the control of chromosome position within interphase nuclei.

The positions of individual chromosomes with respect to the nuclear envelope and each other have also been investigated in human cells. *In situ* hybridization data show that chromosome 1, the largest, is more peripheral and associated with the nuclear envelope (Emmerich *et al.*, 1989; van Dekken *et al.*, 1989). Our data from metaphase reconstructions indicate that larger chromosomes, including 1, are more peripheral than smaller chromosomes (Mosgöller *et al.*, 1991).

In wheat, we have investigated the locations of chromosomes or chromosome arms introduced from other, alien species (Heslop-Harrison *et al.*, 1990; Heslop-Harrison and Schwarzacher, 1993; Schwarzacher *et al.*, 1992a). Such lines result from backcrosses of intergeneric hybrids such as those discussed above and currently include the highest yielding and most widely grown wheats in the UK (Baum and Appels, 1991). The individual alien cereal chromosomes or chromosome arms tend to occupy discrete, elongated domains at interphase and prophase (Fig. 2e, f, see p. 178). Pairs of homologous chromosome arms may show similar patterns of coiling or condensation, although, as in human, there is no evidence for universal association of homologues. Homologous chromosomes show such concerted patterns regardless of their physical separation in the nucleus (Fig. 2e, f, see p. 178). Hence the condensation is probably a property of the chromatin structure related to the genes or DNA sequences carried, which include protein binding sites and nuclear scaffold attachment sites.

4. Sequence locations at interphase

Heterochromatin sequences

Figure 3(a–d, see p. 179) shows nuclei of Arabidopsis stained with the DNA fluorochrome DAPI and after *in situ* hybridization with a cloned sequence associated with the centromere, pAL1 (Martinez-Zapater *et al.*, 1986). The *in situ* results show ten bright hybridization sites; these are around the centromeres at metaphase and co-localize with bright sites in the DAPI-stained interphase nucleus (Fig. 3a–d, see p. 179). Most of the condensed chromatin seen near the periphery of the nucleus in the electron microscope (Fig. 1a) probably corresponds to the same sequence. Other major condensed sites correspond to unexpressed rRNA genes (see Section 5; Maluszynska and Heslop-Harrison, 1991; see also Shaw *et al.*, Chapter 12, this volume). There are no other major condensed sites, indicating that the rDNA and pAL1 (or closely related) sequences together account for much of the condensed DNA within the genome. Some 40% of the DNA in Arabidopsis is repetitive (Leutwiler *et al.*, 1984); together the sequences above account for only 10% of the total DNA. While sequences other than pAL1 may be included in the DNA in the peripheral condensed zones, the results indicate that there may be additional repetitive DNA that is not heavily condensed at interphase and is dispersed within the genome. Such dispersed repetitive sequences are likely to make molecular analysis of the genome difficult and reduce the efficiency of overlapping clone libraries and walking to genes from linked markers (see Flavell *et al.*, Chapter 18, this volume).

The cell shown in the electron micrograph in Fig. 1a is from a dividing population and probably diploid, although many cells of Arabidopsis become polyploid (often through chromosome endoreduplication) during differentiation

(Fig. 3c, d, see p. 179). Reduplication of chromosomes occurs in many cells with specialized, particularly secretory or nutritional, functions (e.g. *Drosophila* salivary glands and some gut cells, or in tapetal cells which are associated with developing pollen grains in plants). Increases in nuclear size may be an essential part of differentiation in species with low DNA contents (Galbraith *et al.*, 1991) or other species, and in certain cells with specialized functions.

Telomere sequences and sub-telomeric repeats

Synthetic oligomers to the consensus sequence for the telomeres can be used as probes *in situ*. We have used the sequence isolated from Arabidopsis by Richards and Ausubel (1988; see Richards *et al.*, Chapter 7, this volume) as a probe to chromosome preparations from barley and rye (Schwarzacher and Heslop-Harrison, 1991). The probe hybridizes to both ends of all seven chromosomes (Fig. 3e, f, see p. 179) but not regularly to any other sites within the genome. Although *in situ* hybridization is not a fully quantitative technique, the strength of hybridization is highly variable both between chromosome types within one cell and between chromosomes of the same type in different cells indicating that the copy number of the telomere repeat is variable. At interphase, the hybridization sites are found to be clustered towards one end of the nucleus (Fig. 3e, see p. 179). DAPI staining and fluorescence from non-specific hybridization show that the region containing the telomeres corresponds to the region of the nucleus with a lower proportion of DNA (cf. Fig. 1c).

Rye (*Secale cereale*) has a repetitive DNA sequence, pSc119 (Jones and Flavell, 1982), located near the telomeres of many chromosome arms. The sequence can be located at interphase by *in situ* hybridization, and is found predominantly in the region of the nucleus which has a lower proportion of DNA (Fig. 4e, see p. 180). pSc119 and other associated sequences are visible in DAPI-stained interphase nuclei as brightly staining heterochromatin sites which correlate with condensed chromatin in the low density region in electron micrographs (Anamthawat-Jónsson and Heslop-Harrison, 1990). Hence the gradient found across the interphase nucleus of cereals corresponds to the structure of the chromosomes at interphase. Centromeres can be identified in electron micrographs (Fig. 1c) and lie in the region with the higher chromatin proportion, while telomeres lie in the other region (Figs 3e and 4e, see pp. 179 and 180).

5. Location of rDNA in the nucleus

The rRNA genes (rDNA) are localized in tandem arrays of hundreds of repeats of the basic gene unit at one or a few distinct loci in most organisms. Nucleoli form at the sites of rRNA gene transcription and are easily visible within interphase nuclei by light or electron microscopy (Shaw *et al.*, Chapter 12, this volume; Wachtler *et al.*, in press; Figs 1, 2, see p. 178, Fig. 4, see p. 180, and Fig. 5). Nucleoli and residual nucleolar proteins on metaphase chromosomes can be silver stained (Goodpasture and Bloom, 1975) to emphasize their structural domains or show the number of active loci of gene transcription.

Gene expression occurs within the nucleolus, but the nucleolus or nucleoli are located in different parts of the nucleus in different species and cell types (Bennett, 1984; Bourgeois and Hubert, 1988). For example, nucleoli in human fibroblasts (Fig. 1b) or male meiocytes of cereals (Fig. 4a, see p. 180) tend to be adjacent to the nuclear envelope, while nucleoli from cereal root-tips tend to lie towards the centre of the nuclei (Figs 1c, 2c, see p. 178, and 4b, see p. 180; Bennett, 1984). Nucleolar morphology varies extensively with cell activity and between cell types (Schwarzacher and Wachtler, 1983); wheat nuclei contain variable numbers of nucleoli depending on stage of the cell cycle and their activity (Fig. 4c, f, see p. 180).

Within cereal nucleoli, active genes are decondensed in particular regions, while inactive genes can be present as condensed chromatin which may be within nucleoli, or as large sites in perinucleolar or other nuclear locations. The condensation of the rRNA genes can vary within and between species. In root-tip cells, we observed unexpressed perinucleolar rDNA in both wheat and rye, but smaller sites of condensed rDNA were found only within the nucleolus of wheat and barley (cf. Fig. 4c–e, see p. 180; see also Leitch *et al.*, 1992). Hence, the patterns of gene expression and locus decondensation is different in the related species: in rye, the gene units at the distal end of the locus are most often expressed, with unexpressed genes at the proximal end of the locus occurring outside the nucleolus. In wheat and barley, active genes were dispersed along each locus, with the expressed sites separated by condensed chromatin within the nucleolus (Figs 4c, d, see p. 180, and 5a). Condensed rRNA genes within the nucleolus were also observed by Shaw *et al.* in *Vicia faba* (Chapter 12, this volume), who reported that the sites of perinucleolar condensed genes were larger in differentiated and presumably less transcriptionally active cells than in undifferentiated cells. The precise location of active genes in the subnucleolar components remains controversial (Leitch *et al.*, 1992; Thiry and Goessens, 1992; Wachtler *et al.*, 1990, 1993; see Shaw *et al.*, Chapter 12, this volume). The electron micrographs in Fig. 5 show rDNA localization within nucleoli of wheat and human cells using two methods of high resolution *in situ* hybridization. In both cases, the hybridization signal from diffuse chromatin is located over the dense fibrillar component.

6. Nuclear architecture, chromosome condensation and gene rich domains

Data comparing the physical separations and genetic distances between markers along chromosomes indicate that there are substantial discrepancies (Gardiner *et*

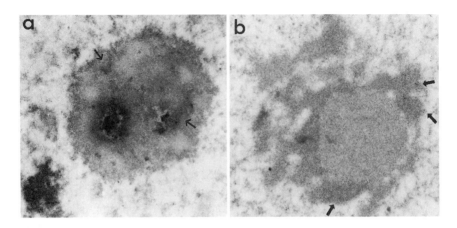

Figure 5. Electron micrographs showing the location of rRNA genes in wheat (a; × 17 000) and human (b; × 26 000) nucleoli. Decondensed, and hence active genes, shown by electron-dense diaminobenzidine reaction product (a; arrowed) or gold label (b; arrowed), are localized at the dense fibrillar component of the nucleolus (see A. R. Leitch and Heslop-Harrison, 1993; Wachtler *et al.*, 1993). Unexpressed, labelled, condensed rDNA sites are present in the nucleus and nucleolus of the wheat nucleus (a). (b) Kindly provided by Dr Franz Wachtler and Dr Christian Schöfer.

al., 1990; Heslop-Harrison, 1991; I. J. Leitch and Heslop-Harrison, 1993). In cereals, most recombination seems to occur in distal regions of the chromosomes. Thus, in the rye nucleolar organizing chromosome, 85% of the genetic length is accounted for by 15% of the physical length of the metaphase chromosome distal to the NOR. Of this 15% segment, half are repetitive, heterochromatic sequences. At interphase, it is evident that there is differential decondensation of the chromosome arm, and the recombination-rich, distal non-heterochromatic region decondenses to occupy between 20 and 40% of the length of the chromosome (Fig. 4g, see p. 180; see also Heslop-Harrison and Schwarzacher, 1993). RFLP mapping of cDNA and genomic DNA (Wang *et al.*, 1991) indicates that the distal region is not only the site of frequent recombination, but also rich in genetic markers (Heslop-Harrison, 1991). Hence there is a correlation between chromatin decondensation and gene expression, which is related to *both* the gradient across the nucleus in proportion filled with DNA *and* the location of centromeres and telomeres within the nucleus (Heslop-Harrison and Schwarzacher, 1993). The less filled region includes the telomeres (Figs 3e and 4e, see pp. 179 and 180) and expressed genes, while the filled region includes little genetic length of chromosomes but most of their physical length and the centromeres.

In mouse cells, de Graaf *et al.* (1990) have recognized domains of DNase I-sensitive chromatin which were preferentially, although not exclusively, localized at the nuclear periphery. Similar compartmentalization was found in both embryonal and differentiated cells which they investigated, and hence is probably a general characteristic of the nucleus. Since DNase I-sensitive sites are associated with actively transcribed chromatin (for discussion see Dillon *et al.*, Chapter 11, this volume), they concluded that their data showed that active transcribing chromatin was compartmentalized preferentially in the periphery of the nucleus.

In the cereal nuclei, we suggest that the telomeric hemisphere represents the most transcriptionally active domain. The simultaneous *in situ* hybridization of several probes to compare the absolute distances between marker DNA sequences along metaphase and interphase chromosomes show considerably more decondensation and extension of the fibres in the telomeric hemisphere than the centromeric hemisphere. Interestingly, comparisons of cereal RFLP maps with chromosome maps of sequences mapped by *in situ* hybridization (I. J. Leitch and Heslop-Harrison, 1993) show that most low copy sequences occur in the distal segment of cereal chromosomes (Heslop-Harrison, 1991). At interphase, these sequences must be predominantly occurring in the telomeric hemisphere of the nucleus where DNA decondensation is the greatest.

Bennett (1984) examined cereal meristematic metaphase cells and, based on the volume measurements of the chromosomes, predicted additional compartmentalization of the nucleus. By comparing the absolute volumes of homologous chromosomes (measured from cells reconstructed from electron micrographs), he was able to show in rye that chromosome volume increased with the peripherality of the chromosome within the cell (up to 23% larger than average when peripherally located, and 19% smaller when centrally located). He speculated that this radial gradient of increasing chromosome volume reflected greater chromosomal decondensation at the preceding interphase and hence greater activity.

Whether models of chromosome decondensation in gene-rich regions can be applied to species other than cereals is unknown. Certainly there are regions of frequent recombination (hot spots) in many species (Dooner, 1986; Grotewold *et al.*, 1991; Ponticelli and Smith, 1992) which probably occur in gene-rich regions, and it is likely that heterochromatin and chromosome regions influenced by heterochromatin are unexpressed (see Henikoff, Chapter 14, this volume). The architecture of the interphase nucleus and the condensation of DNA sequences within are clearly important for nuclear function. Modulation of nuclear structure could provide a sensitive method for control of gene expression, and perhaps in differentiation of different cell types. It is important to compare different organisms – mammals and plants, low and high DNA amounts, or species and hybrids – to isolate universal features of nuclear architecture and to help link the structure and behaviour of the nucleus.

Acknowledgements

We thank Terry E. Miller and Ralf Kynast for providing wheat lines and Prof. Jola Maluszynska, Dr Franz Wachtler, Dr Kesara Anamthawat-Jónsson, Dr Willi Mosgöller and Dr Christian Schöfer for assistance and providing micrographs. We thank Shi Min, Gill E. Harrison, Dr Ilia J. Leitch and Prof. Michael D. Bennett for helpful discussions, and Venture Research International, BP and AFRC for support.

References

Anamthawat-Jónsson K, Heslop-Harrison JS. (1990) Centromeres, telomeres and chromatin in the interphase nucleus of cereals. *Caryologia*, **43**: 205-213.

Anamthawat-Jónsson K, Schwarzacher T, Heslop-Harrison JS. (1993) Behavior of parental genomes in the hybrid *Hordeum vulgare* × *H. bulbosum*. *J. Hered.* **84**: 78–82.

Arnoldus EPJ, Noordermeer IA, Peters ACB, Raap AK, Ploeg M. (1991) Interphase cytogenetics reveals somatic pairing of chromosome 17 centromeres in normal human brain tissue, but not trisomy 7 or sex-chromosome loss. *Cytogenet. Cell Genet.* **56**: 214-216.

Arnoldus EPJ, Peters ACB, Bots GTAM, Raap AK, Ploeg M. (1989) Somatic pairing of chromosome 1 centromeres in interphase nuclei of human cerebellum. *Hum. Genet.* **83**: 231-234.

Baum M, Appels R. (1991) The cytogenetic and molecular architecture of chromosome 1R – one of the most widely utilized sources of alien chromatin in wheat varieties. *Chromosoma*, **101**: 1-10.

Bennett MD. (1988) Parental genome separation in F_1 hybrids between grass species. In: *Kew Chromosome Conference III* (Brandham, PE, ed.). London: Allen and Unwin, pp. 195-208.

Bennett MD, Smith JB, Heslop-Harrison JS. (1982) Nuclear DNA amounts in angiosperms. *Proc. R. Soc. Lond. B* **216**: 179-199.

Bennett MD. (1984) Nuclear architecture and its manipulation. *Gene Manipulation in Plant Improvement* I: *Stadler Genet. Symp.*, pp. 469-502.

Bourgeois CA, Hubert J. (1988) Spatial relationship between the nucleolus and the nuclear envelope: structural and functional significance. *Int. Rev. Cytol.* **111**: 1-52.

Borden J, Manuelidis L. (1988) Movement of the X chromosome in epilepsy. *Science*, **242**: 1687-1691.

Cavalier-Smith T. (1985) Eukaryote gene numbers, non-coding DNA and genome size. In: *The Evolution of Genome Size* (Cavalier-Smith T, ed.). London: John Wiley, pp. 69-103.

Cremer T, Remm B, Kharboush I, Jauch A, Wienberg J, Stelzer E, Cremer C. (1991) Non-isotopic *in situ* hybridization and digital image analysis of chromosomes in mitotic and interphase cells. *Rev. Eur. Technol. Biomed.* **13**: 50-54.

van Dekken H, Pinkel D, Mullikin J, Trask B, van den Engh G, Gray J. (1989) Three dimensional analysis of the organization of human chromosome domains in human and human–hamster hybrid interphase nuclei. *J. Cell Sci.* **94**: 299-306.

Dooner HK. (1986) Genetic fine structure of the *Bronze* locus in maize. *Genetics*, **113**: 1021-1036.

Emmerich P, Loos P, Jauch A, Hopman AHN, Wiegant J, Higgins MJ, White BN, van der Ploeg M, Cremer C, Cremer T. (1989) Double *in situ* hybridization in combination with digital image analysis – a new approach to study interphase chromosome topography. *Exp. Cell Res.* **181**: 126-140.

Finch RA. (1983) Tissue-specific elimination of alternative whole parental genomes in one barley hybrid. *Chromosoma*, **88**: 386-393.

Finch RA, Bennett MD. (1983) The mechanism of somatic chromosome elimination in *Hordeum*. In: *Kew Chromosome Conference III* (Brandham PE, ed.). London: Allen and Unwin, pp. 147-154.

Finch RA, Smith JB, Bennett MD. (1981) *Hordeum* and *Secale* mitotic genomes lie apart in a hybrid. *J. Cell Sci.* **52**: 391-403.

Galbraith DW, Harkins KR, Knapp S. (1991) Systemic endopolyploidy in *Arabidopsis thaliana. Plant Physiol.* **96**: 985-989.

Gardiner K, Horisberger M, Kraus J, Tantravahi U, Korenberg J, Rao V, Reddy S, Patterson D. (1990) Analysis of human chromosome 21: correlation of physical and cytogenetic maps; gene and CpG island distributions. *EMBO J.* **9**: 25-34.

Gautier T, Masson C, Quintana C, Arnoult J, Hernandez-Verdun D. (1992) The ultrastructure of the chromosome periphery in human cell lines. An *in situ* study using cryomethods in electron microscopy. *Chromosoma,* **101**: 502-510.

Goodpasture C, Bloom SE. (1975) Visualization of nucleolar organiser regions in mammalian chromosomes using silver staining. *Chromosoma,* **53**: 37-50.

de Graaf A, van Hemert F, Linnemans WAM, Brakenhoff GJ, de Jong L, van Renswoude J, van Driel R. (1990) Three-dimensional distribution of DNase I-sensitive chromatin regions in interphase nuclei of embryonal carcinoma cells. *Eur. J. Cell Biol.* **52**: 135-141.

Grotewold E, Athma P, Peterson T. (1991) A possible hot spot for Ac insertion in the maize *P* gene. *Mol. Gen. Genet.* **230**: 329-331.

Harauz G, Borland L, Bahr GF, Zeitler E, van Heel M. (1987) Three-dimensional reconstruction of a human metaphase chromosome from electron micrographs. *Chromosoma,* **95**: 366-374.

Heslop-Harrison JS. (1990) Gene expression and parental dominance in hybrid plants. *Devel. Suppl.*: 21-28.

Heslop-Harrison JS. (1991) The molecular cytogenetics of plants. *J. Cell Sci.* **100**: 15-21.

Heslop-Harrison JS, Bennett MD. (1984) Chromosome order – possible implications for development. *J. Embryol. Exp. Morph.* **83** (Suppl.): 51-73.

Heslop-Harrison JS, Bennett MD. (1990) Nuclear architecture in plants. *Trends Genet.* **6**: 401-405.

Heslop-Harrison JS, Huelskamp M, Wendroth S, Atkinson MD, Leitch AR, Bennett MD. (1988) Chromatin and centromeric structures in interphase nuclei. In: *Kew Chromosome Conference III* (Brandham PE, ed.). London: Allen and Unwin, pp. 209-217.

Heslop-Harrison JS, Leitch AR, Schwarzacher T, Anamthawat-Jónsson K. (1990) Detection and characterization of 1B/1R translocations in hexaploid wheat. *Heredity,* **65**: 385-392.

Heslop-Harrison JS, Schwarzacher T. (1993) Molecular cytogenetics – biology and applications in plant breeding. In: *Chromosomes Today,* Vol.11 (Chandley AC, Sumner A, eds). London: Allen and Unwin, pp. 191–198.

Hunter MS, Nur U, Werren JH. (1993) Origin of males by genome loss in an autoparasitoid wasp. *Heredity* **70**: 162-171.

Jones JDG, Flavell RB. (1982) The mapping of highly-repeated DNA families and their relationship to C-bands in chromosomes of *Secale cereale. Chromosoma,* **86**: 595-612.

Leitch AR, Heslop-Harrison JS. (1993) Ribosomal RNA gene expression and localization in cereals. In: *Chromosomes Today,* Vol.11 (Chandley AC, Sumner A, eds). London: Allen and Unwin, pp. 91–100.

Leitch AR, Mosgöller W, Shi M, Heslop-Harrison JS. (1992) Different patterns of rDNA organization at interphase in nuclei of wheat and rye. *J. Cell Sci.* **101**: 751-757.

Leitch AR, Schwarzacher T, Mosgöller W, Bennett MD, Heslop-Harrison JS. (1991) Parental genomes are separated throughout the cell cycle in a plant hybrid. *Chromosoma,* **101**: 206-213.

Leitch IJ, Heslop-Harrison JS. (1993) Physical mapping of four sites of 5S ribosomal DNA sequences and one site of the alpha-amylase 2 gene in barley (*Hordeum vulgare*). *Genome,* in press.

Leutwiler LS, Hough-Evans BR, Meyerowitz EM. (1984) The DNA of *Arabidopsis thaliana. Mol. Gen. Genet.* **194**: 15-23.

Lichter P, Cremer T, Borden J, Manuelidis L, Ward DC. (1988) Delineation of individual human chromosomes in metaphase and interphase cells by *in situ* suppression hybridization using recombinant DNA libraries. *Hum. Genet.* **80**: 224-234.

Linde-Laursen I, Jensen J. (1991) Genome and chromosome disposition at somatic metaphase in a *Hordeum* × *Psathyrostachys* hybrid. *Heredity*, **66**: 203-210.

Maluszynska J, Heslop-Harrison JS. (1991) Localization of tandemly repeated DNA sequences in *Arabidopsis thaliana*. *Plant J.* **1**: 159-166.

Maluszynska J, Heslop-Harrison JS. (1993) Molecular cytogenetics of the genus *Arabidopsis: In situ* localization of rDNA sites, chromosome numbers and diversity in centromeric heterochromatin. *Ann. Bot.* in press.

Manuelidis L, Borden J. (1988) Reproducible compartmentalization of individual chromosome domains in human CNS cells revealed by *in situ* hybridization and three-dimensional reconstruction. *Chromosoma*, **96**: 397-410.

Manuelidis L, Chen TL. (1990) A unified model of eukaryotic chromosomes. *Cytometry*, **11**: 8-25.

Martinez-Zapater JM, Estelle MA, Somerville CR. (1986) A highly repeated DNA sequence in *Arabidopsis thaliana*. *Mol. Gen. Genet.* **204**: 417-423.

Mosgöller W, Leitch AR, Brown JKM, Heslop-Harrison JS. (1991) Chromosome arrangements in human fibroblasts at mitosis. *Hum. Genet.* **88**: 27-33.

Nagl W, Capesius I. (1977) Repetitive DNA and heterochromatin as factors of karyotype evolution in phylogeny and ontogeny of orchids. *Chromosomes Today* **6**: 141–150.

Nur U. (1990) Heterochromatization and euchromatization of whole genomes in scale insects (Coccoidea: Homoptera). *Devel. Suppl.*: 29-34.

Ponticelli AS, Smith GR. (1992) Chromosomal context dependence of a eukaryotic recombinational hot spot. *Proc. Natl. Acad. Sci. USA* **89**: 227-231.

Popp S, Scholl HP, Loos P, Jauch A, Stelzer E, Cremer C, Cremer T. (1990) Distribution of chromosome 18 and X centric heterochromatin in the interphase nucleus of cultured human cells. *Exp. Cell Res.* **189**: 1-12.

Richards EJ, Ausubel FM. (1988) Isolation of a higher eukaryotic telomere from *Arabidopsis thaliana*. *Cell*, **53**: 127-136.

Schwarzacher HG, Wachtler F. (1983) Nucleolus organizer regions and nucleoli. *Hum. Genet.* **63**: 89-99.

Schwarzacher T, Anamthawat-Jónsson K, Harrison GE, Islam AKMR, Jia JZ, King IP, Leitch AR, Miller TE, Reader SM, Rogers WJ, Shi M, Heslop-Harrison JS. (1992a) Genomic *in situ* hybridization to identify alien chromosomes and chromosome segments in wheat. *Theor. Appl. Genet.* **84**: 778-786.

Schwarzacher T, Heslop-Harrison JS, Anamthawat-Jónsson K, Finch RA, Bennett MD. (1992b) Parental genome separation in reconstructions of somatic and premeiotic metaphases of *Hordeum vulgare* × *H. bulbosum*. *J. Cell Sci.* **101**: 13-24.

Schwarzacher T, Heslop-Harrison JS. (1991) *In situ* hybridization to plant telomeres using synthetic oligomers. *Genome*, **34**: 317-323.

Schwarzacher T, Leitch AR, Bennett MD, Heslop-Harrison JS. (1989) *In situ* localization of parental genomes in a wide hybrid. *Ann. Bot.* **64**: 315-324.

Thiry M, Goessens G. (1992) Where, within the nucleolus, are the rRNA genes located? *Expl. Cell Res.* **200**: 1-4.

Wachtler F, Mosgöller W, Schöfer C, Sylvester J, Hozak P, Derenzini M, Stahl A. (1993) Ribosomal genes and nucleolar morphology. In: *Chromosomes Today*, Vol.11 (Chandley AC, Sumner A, eds). London: Allen and Unwin.

Wachtler F, Mosgöller W, Schwarzacher, HG. (1990) Electron microscopic *in situ* hybridization and autoradiography: localization and transcription of rDNA in human lymphocyte nucleoli. *Expl. Cell Res.* **187**: 346-348.

Wang ML, Atkinson MD, Chinoy CN, Devos KM, Harcourt RL, Liu CJ, Rogers WJ, Gale MD. (1991) RFLP-based genetic map of rye (*Secale cereale* L.) chromosome 1R. *Theor. Appl. Genet.* **82**: 174-178.

Chapter 17

The Great Wall: sequencing the yeast genome

Stephen G Oliver, Carolyn M James, Manda E Gent and Keith J Indge

1. Introduction

Genetics, latterly combined with recombinant DNA methodology, has proved to be the most powerful analytical tool in biology. Detailed genetic maps, obtained by mutation and linkage analysis, have been compiled for many organisms (O'Brien, 1990). These maps are especially detailed for microorganisms, such as the bacteria *Escherichia coli* (Bachmann, 1990) and *Bacillus subtilis* (Piggot and Hoch, 1985; Zeigler and Dean, 1985), and for fungi, such as *Neurospora crassa* (Perkins, 1990; Perkins *et al.*, 1982) and *Saccharomyces cerevisiae* (Mortimer *et al.*, 1989, 1992). Impressive as the achievements of classical genetics are, it is evident that the number of genes which have been identified by conventional means falls far short of the number which might be predicted to exist based on our current knowledge of the size of genomes and the structure and function of their genes. There appears to be a certain circularity in the logic of genetics: our ability to identify genes depends on being able to recognise the phenotypes of mutant alleles. Mutant isolation and gene identification advances our understanding of biology. However, our ability to recognise mutant phenotypes is, itself, dependent on that understanding. In order to break out of this impasse (or, at the very least, to assure ourselves that the problem is more apparent than real) we require some systematic way of identifying all of the genes contained within a particular genome. The technique of DNA sequencing (Sanger *et al.*, 1977) offers just such a systematic approach and genome sequencing, especially if combined with facile techniques of functional analysis, promises to provide new insights in biology. It may even reveal areas for study which we did not realize existed and thus free us from the circularity of the paradigm of classical genetics.

2. The yeast *Saccharomyces cerevisiae* as an object for genome sequencing

The limitations of the classical genetic approach are very evident for eukaryotic organisms. Indeed, the total number of genes identified for *E. coli* (1403; Bach-

233

mann, 1990; see Blattner, Chapter 4, this volume) may not be far short of the total which a circular genome of about 4 Mb is likely to contain. The budding yeast *Saccharomyces cerevisiae* has a genome size of about 14 Mb, less than 4 × the size of that of *E. coli*. Such a genome, by analogy with the bacterium, might be expected to contain some 5000–6000 genes. By 1989, just 769 had been identified by conventional (function-oriented) genetic approaches (Mortimer *et al.*, 1989). The promoters and terminators of protein encoding genes in yeast are, admittedly, larger than those in *E. coli* (Oliver and Warmington, 1989). However, given the fact that introns are both rare and small in genes of yeast, it is evident that there is a considerable shortfall in our knowledge of even the small and intensively studied genome of this microbial eukaryote.

Saccharomyces cerevisiae is a very suitable subject for the systematic approach to genome analysis which DNA sequencing offers. The yeast genome contains less repetitive DNA than is common in the larger genomes of more complex eukaryotes (see Olson, 1991). Indeed, the only highly repetitive DNA sequences are ca. 120 copies of the genes encoding rRNA which are found in a large tandem array on chromosome XII (Petes and Botstein, 1977) and the Ty transposons and their long terminal repeats (more of these later). The *S. cerevisiae* genome is divided between 16 chromosomes (Table 1). No yeast chromosome is as big as that of *E. coli* and many are of a similar size to the DNA molecules of bacterial viruses. The technique of pulsed field gel electrophoresis (Carle and Olson, 1984)

Table 1. The size distribution of *S. cerevisiae* chromosomes

Chromosome	Approximate size (kb)
I	220
VI	260
III	320
IX	410
V	540
VIII	540
XI	630
X	700
XIV	750
II	770
XIII	870
XVI	900
VII	1030
XV	1040
IV	1510
XII	1010 + rDNA (2 Mb)
Total	11 500
+rDNA	2 000
	13 500

These size estimates are somewhat lower than previously published data and are based on data from chromosome III and other physical mapping studies.

allows yeast chromosomes to be separated, thus enabling the construction of chromosome specific gene banks.

The proficiency, and the accuracy, with which yeast carries out genetic recombination has allowed the construction of a good genetic map (Mortimer *et al.*, 1989, 1992). The 16 linkage groups it defines have been correlated with the 16 chromosomes which can be physically resolved. The organism also has excellent recombinant DNA technology (see Broach *et al.*, 1991) and the aforementioned characteristics of yeast's recombination system allow the targetting of cloned fragments to their homologous sites in the chromosome, thus allowing gene disruption or replacement (Rothstein, 1983) experiments to be carried out with ease. This facility is central to any strategy for revealing the function of unknown genes discovered by systematic genome sequencing. Transcript analyses suggest that the bulk of the yeast genome is transcribed and, therefore, might be assumed to be functional. However, a study of random disruption mutations (Goebl and Petes, 1986) suggested that about 70% of genes were not required for normal growth on glucose-containing media. There is an apparent paradox here. Either there is a large amount of redundant information in this very small genome or our knowledge of yeast physiology and cell biology is very incomplete.

It is just such a paradox which systematic genome sequencing, combined with a rational strategy to discover the function of any novel genes revealed by that sequence, is designed to resolve. However, the sequencing of genomic DNA, as opposed to cDNA clones should reveal much more. It should yield important information about genome organization and evolution, and about the rules governing the replication, recombination and segregation of chromosomes. The possession of the complete DNA sequence of a single yeast chromosome is thus a highly desirable and achievable objective. Moreover, the successive sequencing of individual chromosomes, until the entire yeast genome has been obtained, represents a sensible and scientifically profitable strategy.

3. The choice of chromosome III

The first yeast chromosome to be sequenced in its entirety was chromosome III (Oliver *et al.*, 1992). It appeared the obvious choice for a number of reasons: it is small, coincidentally the third smallest chromosome, and estimates of its size from electrophoretic analyses ranged from 300 to 360 kb. It has been well-mapped genetically; prior to the determination of the sequence, 37 genes had been assigned to specific positions on chromosome III and there were no apparent ambiguities in gene order (Mortimer *et al.*, 1989).

Chromosome III has been the object of especial interest by yeast geneticists, not least because it contains the three genetic loci involved in mating-type control: *MAT*, *HML* and *HMR* (Strathern *et al.*, 1989 and see Fig. 1). The possession of these three loci enabled the construction of the first chromosome-specific gene bank from yeast since intramolecular recombination events between homologous

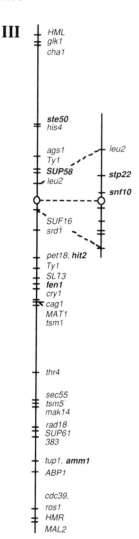

Figure 1. Genetic map of *Saccharomyces cerevisiae* chromosome III (reproduced from Mortimer *et al.*, 1989, with permission from John Wiley & Sons Ltd.).

sequences in these three sites can generate circular chromosomes which are physically isolable entities (Strathern *et al.*, 1979). It was the gene bank generated by Carol Newlon (Newlon *et al.*, 1986, 1991) from the small ring derivative of chromosome III, which arose from a cross-over between *HML* and *HMR*, which formed the basis for the initial work on sequence determination. The European Yeast Genome Sequencing Consortium (see later) which completed the chromosome III sequence also benefitted from gifts of clones from Linda Riles and Maynard Olson (Olson *et al.*, 1986) as well as from Akio Yoshikawa and Katsumi Isono (Yosihikawa and Isono, 1990).

4. The 'Great Wall' (or network) approach to genome sequencing

The title of this contribution is a reference to the short story 'The Great Wall of China' by Franz Kafka (1931) in which he describes the strategy employed by the ancient Chinese to construct that awesome edifice. The approach was to establish building operations at a large number of sites along the planned route. Workers were taken to remote sites to build the wall and, on the return journeys to their home villages, they saw activity all along the way and were encouraged that the enormous task of constructing the wall would actually be completed. The parallels with, and lessons for, genome sequencing are plain to see. The determination of the DNA sequence will require the collaboration of large numbers of researchers and a successful outcome is more likely if individual groups are given achievable objectives and are kept in regular contact with their collaborators. The application of this 'Great Wall' strategy (which he prefers to call the Network approach) to genome sequencing was the vision of André Goffeau and was adopted by DGXII of the Commission of the European Community under their Biotechnology Action Plan. The European Consortium which sequenced chromosome III comprised 35 laboratories from 10 of the 12 nations of the European Community. The paper which announced the sequence (Oliver et al., 1992) had 147 authors. The rest of this contribution is devoted to a discussion of the results which have arisen from the Yeast Genome Sequencing Project.

5. The relationship between the genetic and the physical map

It might be suggested that genome sequencing will make the construction of genetic maps by classical recombination analysis a redundant activity. This is not true, rather the genetic map acquires a different emphasis: it now becomes a tool with which to study recombination and the dynamics of chromosome evolution. The comparison of the genetic and physical maps of chromosome III is shown in Table 2. It can be seen that there is no difference in the gene order presented in the two maps except for the reversal of the positions of the *glk1* and *cha1* genes. (This is not a serious contradiction since these two genes had never been mapped to one another, only to the centromere; R K Mortimer, personal communication.) However, there is a large variation in intergenic distances between the two maps. In other words, the frequency of genetic recombination varies by at least 10-fold along the length of the chromosome. The ratio of genetic distance (in cM) to physical distance (in kb) ranges from 0.03 in the interval between *SUF2* and *pgk1* to 1.27 between *MAT* and *thr4*. The average ratio for chromosome III is 0.51 cM/kb which is significantly greater than the estimated average, of 0.34 cM/kb, for the entire yeast genome (Mortimer et al., 1989). A higher relative frequency of genetic recombination seems to be a characteristic of small yeast chromosomes. The estimated cM/kb ratios for the two smallest chromosomes, I and VI, are 0.62

Table 2. Relation between genetic and physical map distances

Gene interval	Genetic distance (cM)	Physical distance (kb)	cM/kb
HML–his4	27.0	50.9	0.53
his4–Ty2	13.0	16.1	0.81
Ty2–leu2	4.0	9.3	0.43
leu2–cen3	5.0	21.7	0.23
cen3–SUF2	1.8	8.4	0.21
SUF2–pgk1	0.2	6.0	0.03
pgk1–pet18	14.8	20.8	0.71
pet18–cry1	9.0	19.6	0.46
cry1–MAT	4.0	18.6	0.22
MAT–thr4	21.0	17.5	1.20
thr4–rad18	16.0	14.6	1.10
rad18–tup1	10.0	32.1	0.31
tup1–ABP1	3.0	0.6	5.00
ABP1–HMR	14.0	28.9	0.48

Total physical distance from *HML to HMR* = 272.6 kb. Total genetic distance from *HML to HMR* = 142.8 cM. Average cM/kb = 0.52.

and 0.55, respectively (Mortimer *et al.*, 1989). Kaback *et al.* (1989) have suggested that small chromosomes have higher frequencies of recombination relative to their physical length in order to ensure that at least one cross-over occurs on these chromosomes in each meiosis. It is believed that crossing over is essential to the correct disjunction of chromosomes at meiotic nuclear division. In some organisms, e.g. *Drosophila*, non-recombining chromosomes segregate from each other at the first meiotic nuclear division by a size-dependent process known as distributive disjunction; this mechanism is about 99.9% efficient in *Drosophila*. Although distributive disjunction does occur in yeast (Guacci and Kaback, 1991; Kaback, 1989), it is only about 90% efficient (Guacci and Kaback, 1991) and therefore the enhanced frequency of crossing over on the small yeast chromosomes is absolutely essential to the correct segregation of homologues.

In addition to helping us to understand the process of genetic recombination in yeast, chromosome sequence data also provide evidence of past recombination events which may have helped shape the organization of the *S. cerevisiae* genome. The telomeres of yeast contain a series of 100–200 repeats of the sequence $C_{1-3}A$ (Shampay *et al.*, 1984; Walmsley *et al.*, 1984; Walmsley and Petes, 1985). All yeast chromosomes have a sequence element called X adjacent to these repeats (Chan and Tye, 1980; Link and Olson, 1991). In some chromosomes, there is an additional element, called Y', which also has ARS activity and which is inserted into the array of $C_{1-3}A$ repeats (Link and Olson, 1991; Louis and Haber, 1990a). The telomeres of chromosome III both have X elements immediately adjacent to them, with no intervening Y' sequence.

The X elements show considerable sequence divergence (8–18%) both within and between strains, whilst the Y' sequences are as well-conserved as the single copy genes within the yeast genome (Louis and Haber, 1992). Y' elements contain two open reading frames, one of which is expressed and whose presumed protein product shows amino acid sequence similarity to RNA helicases (Louis and Haber, 1992). The junction between the X and Y' elements usually represents $C_{1-3}A$ repeats, but on some chromosomes (IX and X in strain YP1, and IV and X in strain A364A), these repeats together with parts of the flanking elements are replaced by a sequence which is identical to the fourth intron of the yeast mitochondrial gene encoding cytochrome b (Louis and Haber, 1991). This class I intron shares structural features with known transposable introns (Dujon, 1989). In this case, it appears to have inserted into the telomere of chromosome X and then spread to other chromosomes. The spread of telomere-associated sequence elements and genes may occur via ectopic recombination events at mitotic nuclear division involving either Y' (Louis and Haber, 1990b) or X elements. Such events are believed to have enabled the spread of multiple copies of the *SUC* (sucrose fermentation) genes through the yeast genome (Carlson *et al.*, 1985). However, recombination between telomeric elements on different chromosomes may have had even more profound effects on the evolution of the yeast genome.

An unexpected finding from the chromosome III sequence was that there are additional copies of X at internal sites of the chromosome (Oliver *et al.*, 1992). One occurs about 25 kb from the telomere on the right arm while the other is very close (about 4 kb) to the left telomere. In addition, the sequence analysis of chromosome XI currently being carried out by the European consortium has revealed the presence of both X and Y' sequences about half way between the left telomere and the centromere (Dujon *et al.*, unpublished). The true nature of this XY' motif in chromosome XI was difficult to recognize until physical mapping studies (Dujon, personal communication) made it obvious that there is a large segment of the left arm (about 240 kb) inverted between the genetic and physical maps. There is no easily apparent explanation for this discrepancy; however, the physical data are unambiguous. The presence of these pseudo-telomeric elements suggests that perhaps yeast chromosomes were once even shorter than they are now and they have 'grown' by recombination events involving their telomeres. Both the instability of linear plasmids and small artificial chromosomes (Newlon, 1988) in yeast, as well as the Kaback rule (Kaback *et al.*, 1989), suggest that there is selection against short chromosomes in *S. cerevisiae*.

The pseudo-X element close to the left telomere of chromosome III is of particular interest. Voytas and Boeke (1992), following a detailed analysis of the ORFs (open reading frames) between the pseudo-X and the true, telomeric, X element have suggested that the two X sequences represent the long terminal repeats (LTRs) of a new class of yeast transposon called Ty5. The possible existence of such a transposon had been suggested by Oliver *et al.* (1992) based on the similarity of the open reading frame YCL74w with the polyprotein of the tomato *copia*-like transposon Tnt1 (Grandbastien *et al.*, 1989). Voytas and Boeke (1992) pointed out similarities between YCL76w/YCL75w and YCL74w with retroviral acid protease and reverse transcriptase-RNaseH, respectively. The representative of Ty5 on chromosome III is evidently a non-functional element and neither it, nor the pseudo-X on the right arm (which Voytas and Boeke interpret as a solo LTR) are flanked by the chromosomal target site duplications which typically result from transposition events (Farabaugh and Fink, 1980; Gafner and Philippsen, 1980). Thus, until a truly mobile element is identified, the designation of Ty5 as a new yeast transposon is somewhat tentative. Nevertheless, it is interesting to note that transposons are often found on the healed ends of broken chromosomes in *Drosophila* (Biessmann *et al.*, 1992; Danilevskaya *et al.*, 1992). The synthesis of new telomeric repeats by telomerase (see Grieder *et al.*, Chapter 8, this volume) is essentially a reverse transcription process and we might speculate that current methods of telomere biogenesis in many eukaryotes had their origins in the activities of retroposons or retroviruses. Whatever the merits of these speculations, or the final interpretation of the origins of Ty5-1, it is evident that many of the polymorphisms observed between homologous chromosomes in different strains of *S. cerevisiae* are due to transposition events or recombination between transposons and/or their LTRs.

Fortunately for genome analysis, these spontaneous transposition events do not appear to occur randomly along the length of individual chromosomes. Instead, transposons insert preferentially in specific regions called transposition hot-spots. So far, three such hot-spots have been identified on chromosome III: the left-arm hot-spot (LAHS; Warmington *et al.*, 1986) at ca. 83 kb, the right-arm hot-spot (RAHS; Stucka *et al.*, 1989 ; Warmington *et al.*, 1987) at about 149 kb and the far right-arm hot-spot (FRAHS; Wicksteed *et al.*, submitted) at about 169 kb. A consideration of the two transposition hot-spots on the right arm of the chromosome will serve to illustrate the role which they may play in chromosome evolution.

The RAHS has been analysed in three different, but related, laboratory strains of *S. cerevisiae* (Warmington *et al.*, 1987) and also in a brewing strain (Stucka *et al.*, 1989). The data obtained provide ample evidence that this site is a major source of polymorphisms for chromosome III. The work of Stucka *et al.* (1989) on the RAHS provided the first example of an intact Ty4 element in the yeast genome; solo copies of its LTR, called tau, had been found previously (Genbauffe *et al.*, 1984). Our own data (Warmington *et al.*, 1987) can be used to reconstruct a presumptive evolutionary order for the transposition and recombination events occuring at the RAHS (see Fig. 2). Strain CN31C has no complete Ty elements at the site but does have three solo δ elements and a δ fragment in this region. The next two events are inferred, since no strains currently exist to fill this part of the molecular 'archaeological' record. The two presumptive events were the insertion of a Ty1 element, which we now call Ty1-3R, into the δ fragment and a second transposition which introduced a Ty1 (called Ty1-161; Kingsman *et al.*, 1981) into an ORF. The insertion of a Ty element directly into an ORF is a very unusual event. Tys are most commonly found in promoter regions

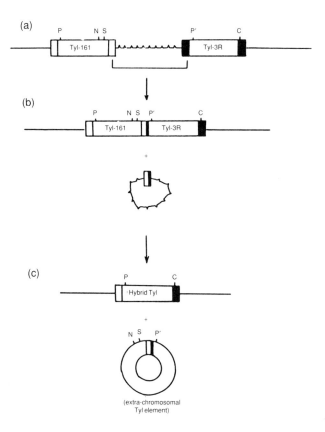

Figure 2. The generation of Ty-associated polymorphisms at the right-arm transposition hot-spot (RAHS) of chromosome III (from Warmington *et al.*, 1987). Two independent transposition events inserted Ty-161 and Ty1-3R into the RAHS of strain CN31C. (a) A δ × δ recombination produced the tandem Ty fusion found in XJ24-24a and the small ring chromosome. This event excised 3 kb of unique chromosomal DNA including the open reading frame YCR189c and a tRNA^Lys gene. (b) Finally an ε × ε recombination event reduced the tandem pair of Tys to the single, hybrid, element found in strain CF4-16B (c).

and are frequently associated with tRNA genes (Eigel and Feldmann, 1982; Genbauffe *et al.*, 1984). These two transpositions were followed by a δ × δ recombination event, to generate a tandem pair of Ty elements joined via a single LTR. It is this arrangement which is found in the small ring version of chromosome III (Newlon *et al.*, 1986, 1991).

It should be noted that this last recombination event excised a unique portion of the chromosome containing an incomplete ORF and a tRNA gene (see Fig. 2). These sequences are included in the complete sequence of the chromosome presented by Oliver *et al.* (1992). The circular fragment generated by the recombination (Fig. 2) itself represents a translocatable element. It contains a solo δ element and so has about 130 sites in the yeast genome at which it could integrate back into a chromosome by homologous recombination. Such an integration, followed by a mating, would duplicate that fragment of the genome. Thus the frequent association of Ty elements with tRNA genes may have come about since events of this type would permit the amplification of such genes through the yeast genome. (There are about 360 tRNA genes in yeast; Feldmann, 1976.) The chances of such an excised chromosomal fragment reinserting into the genome would be greatly enhanced if the circular product contained a potential origin of replication, an ARS element (Newlon, 1988). The fragment excised from chromosome III does not contain an ARS.

The final event in our archaeological progression was the reduction in the number of Ty elements at the RAHS from two to one, as is found in strain CF4-16B (Warmington et al., 1987). This reduction occurred by a recombination between the two tandem Ty elements. However, this time, an $\varepsilon \times \varepsilon$, rather than a $\delta \times \delta$, recombination was involved. Such events are probably very rare but require sequence analysis for their detection and also the good fortune of appropriately placed polymorphisms between the two Tys. In this particular case, those polymorphisms allowed us to fix the cross-over site to within a few base pairs (Warmington et al., 1987).

The ring form of chromosome III contains a solo δ element, distal to the RAHS and close to the gene CRY1 (Newlon et al., 1991; Oliver et al., 1992). In strain S288C, an intact Ty is found at this site, which we dub the FRAHS (far right-arm transposition hot-spot; Wicksteed et al., submitted). There are a number of yeast strains, such as SR112-1A and YNN214, which have a long version of chromosome III (Roeder et al., 1984; Sikorski and Hieter, 1989). Our analysis of strain SR112-1A (Wicksteed et al., submitted) has demonstrated that this long version of the chromosome has the region between the RAHS and the FRAHS duplicated (Wicksteed et al., submitted). This presumably occurred via an unequal crossing over event as shown in Fig. 3. Conversely, strains carrying the pet18 mutation; a pleiotropic lesion first recognized by a respiratory deficient phenotype (Leibowitz and Wickner,1978) have a deletion in this region (Toh-e and Sahashi, 1985). Thus the analysis of these two polymorphic sites (the RAHS and the FRAHS) demonstrates how redundancy can be both generated in, and removed from, the yeast genome. Such redundancy is the raw material used by evolution to produce novel functions (Ohno, 1970).

6. New ORFs and new challenges

The DNA sequence analysis of chromosome III (Oliver et al., 1992) has revealed a total of 182 ORFs which would encode proteins of >100 amino acids in length. Prior to the determination of the sequence, only 37 of these were known. Thus, two years of systematic sequencing has revealed about five times as many genes as the previous 40 years of classical genetic analysis (Mortimer et al., 1989). Why had all these genes been missed previously? One possibility is that only a few of

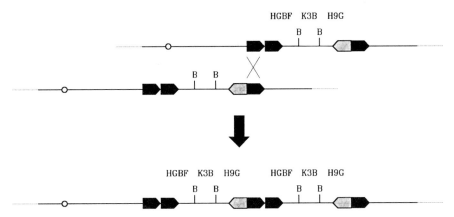

Figure 3. An unequal crossing over event between δ elements in the RAHS and FRAHS duplicates part of the chromosome, creating the long form of chromosome III found in strain SR112-1A. (From Wicksteed *et al.*, submitted.)

these novel ORFs are actually transcribed. This seems unlikely; a detailed transcript analysis of chromosome III carried out by Yoshikawa and Isono (1990) indicated the presence of 160 protein-encoding genes in addition to the Ty elements. This is likely to be an underestimate as these workers ran their Northern blots only on poly(A)$^+$ RNA extracted from cells growing in rich medium. Nevertheless, the agreement between the ORF map derived from the sequence and the experimental transcript map is very high. Therefore the bulk of the 145 novel ORFs discovered are likely to represent functional genes and it is a major challenge to determine just what those functions are.

A start has been made on the functional analysis by the 35 contributing laboratories in the EC Consortium. A total of 55 of these ORFs have so far been disrupted or deleted. This revealed three new essential genes. Of the remainder, 42 disruptants were tested for other effects on the activity of yeast and, in 214 cases, some phenotype was observed. This leaves 28 of 45 genes tested for which there is no obvious function.

One easy approach to functional determination is to scan the sequence databases to discover whether any of these unknown genes are similar to genes sequenced from other organisms. The results of such an analysis, using Pearson and Lipman's (1988) FASTA algorithm are shown in Fig. 4. The data imply that between 53% and 77% of the genes on chromosome III have not previously been isolated and sequenced from any other organism. The humbling conclusion is that not only yeast molecular geneticists, but all biologists, are working on just a subset of the total genes which their experimental organisms contain.

The discovery of the function of these unknown genes will not be an easy task and approaches every bit as systematic and efficient as DNA sequencing itself will have to be devised. It is likely that there is considerable functional redundancy in the yeast genome, at least when the organism is grown under laboratory conditions. Thus the emerging sequence database from the yeast genome will enable us

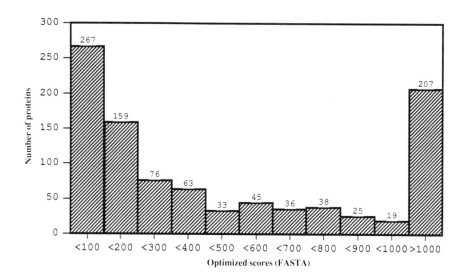

(a)   **FASTA scores of yeast proteins compared to MIPSX**

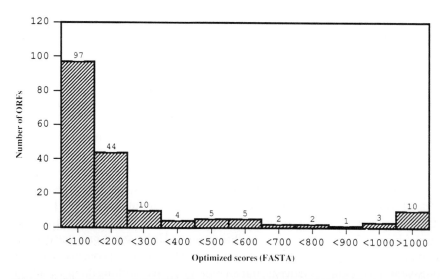

(b)   **FASTA scores of chromosome III ORFS compared to MIPSX**

Figure 4. Distribution of FASTA (Pearson and Lipmann, 1988) similarity scores for all yeast ORFs compared with non-*Saccharomyces* proteins (a) and chromosome III ORFs compared with the same data-set (b). (From Oliver *et al.*, 1992.)

to identify similar genes and make double, or multiple, disruptants to discover whether these have a 'synthetic' lethal, or other, phenotype. The proteins encoded by the unknown genes may be antigenically tagged by creating fusion proteins and immunomicroscopy used to discover whether a given gene product is localized to a particular cellular compartment or is expressed only at a specific stage of the yeast life cycle, e.g. during mating or meiosis (Snyder and Roeder, personal communication). An analogous, but more general, approach is to identify the protein products of the unknown genes on two-dimensional polyacrylamide gels. This can be done by either inactivating the gene or overexpressing it. Radioisotopic labelling may then be used to correlate the synthesis of particular proteins with specific physiological states or transitions (e.g. starvation or heat shock). This approach is being pioneered by Helion Boucherie and his colleagues in Bordeaux (Thoraval *et al.*, 1990). Finally, it may be possible to ablate the function of closely related proteins by using electroporation to introduce antisense RNA or antibody molecules into the yeast cell. It is likely that all of these approaches, and more, will be needed to elucidate the function of all of the genes in the *S. cerevisiae* genome. However, when the task is complete, we will have completely escaped the circularity of classical genetic analysis and laid a firm foundation for understanding the full genetic potential of higher organisms – including humans.

Acknowledgements

Genome sequencing work at UMIST has been supported by the Biotechnology Directorate of the CEC within the framework of their Biotechnology Action Programme (BAP). Further support came from a NATO Exchange Award to SGO and Carol S. Newlon. The following are thanked for help, advice and useful discussions: André Goffeau, John Sgouros, Carol Newlon, Alistair Brown, Bernard Dujon, Linda Riles, Maynard Olson, Katsumi Isono and Richard Walmsley.

References

Bachmann BJ. (1990) Linkage map of *Escherichia coli* K-12, Edition 8. *Microbiol. Rev.* **54:** 130-197.

Biessmann H, Valgeirdottir K, Lofsky A, Chin C, Ginther B, Levis RW, Pardue M-L. (1992) Het-A, a transposable element specifically involved in healing broken chromosome ends in *Drosophila melanogaster*. *Mol. Cell. Biol.* **12:** 3910-3918.

Broach JR, Pringle JR. Jones EW. (eds) (1991) *The Molecular and Cellular Biology of the Yeast* Saccharomyces cerevisiae. Cold Spring Harbor, New York: Cold Spring Harbor Laboratory Press.

Carle GF, Olson MV. (1984) Separation of chromosomal DNA molecules by orthogonal field alternation gel electrophoresis. *Nucl. Acids Res.* **12:** 5647-5664.

Carlson M, Celenza JL, Eng FJ. (1985) Evolution of dispersed SUC gene family in *Saccharomyces cerevisiae* by rearrangements of chromosome telomeres. *Mol. Cell. Biol.* **5:** 2894-2902.

Chan C, Tye B-K. (1980) Autonomously replicating sequences in *Saccharomyces cerevisiae*. *Proc. Natl Acad. Sci. USA* **77:** 6329–6333.

Danilevskaya O, Lofsky A, Pardue M-L. (1992) Het-A, a transposable element with an apparent role in the telomeres of *Drosophila. Mol. Cell. Biol.* **35S:** 83.

Dujon B. (1989) Group I introns as mobile genetic elements, facts and mechanistic speculations – a review. *Gene,* **82:** 91-114.

Eigel A, Feldmann H. (1982) Ty1 and δ elements occur next to several transfer RNA genes in yeast. *EMBO J.* **1:** 1245-1250.

Farabaugh PJ, Fink GR. (1980) Insertion of the eukaryotic transposable element Ty1 creates a 5 base pair duplication. *Nature,* **286:** 352-356.

Feldmann H. (1976) Arrangement of transfer RNA genes in yeast. *Nucl. Acids Res.* **3:** 2379-2386.

Gafner J, Philippsen P. (1980) The yeast transposon Ty1 generates duplications of target DNA on insertion. *Nature,* **286:** 414-418.

Genbauffe FS, Chisholm GE, Cooper TG. (1984) Tau, sigma and delta. A family of repeated elements in yeast. *J. Biol. Chem.* **259:** 10 518-10 525.

Goebl MG, Petes TD. (1986) Most of the yeast genomic sequences are not essential for cell growth and division. *Cell,* **46:** 983-992.

Grandbastien M-A, Spielman A, Caboche M. (1989) Tnt1, a mobile retroviral-like transposable element of tobacco isolated by plant cell genetics. *Nature,* **337:** 376-380.

Guacci V, Kaback DB. (1991) Distributive disjunction of authentic chromosomes in *Saccharomyces cerevisiae. Genetics,* **127:** 475-488.

Kaback DB. (1989) Meiotic segregation of circular plasmid-minichromosomes from intact chromosomes in *Saccharomyces cerevisiae. Curr. Genet.* **15:** 385-392.

Kaback DB, Steensma HY, De Jonge P. (1989) Enhanced meiotic recombination of the smallest chromosome of *Saccharomyces cerevisiae. Proc. Natl Acad. Sci. USA* **86:** 3694-3698.

Kafka F. (1931) *Biem Bau der Chinesischen Mauer.* London: Martin Secker and Warburg.

Kingsman AJ, Gimlich RL, Clarke L, Chinault RC. Carbon J. (1981) Sequence variation in dispersed repetitive sequences in *Saccharomyces cerevisiae. J. Mol. Biol.* **145:** 619-632.

Leibowitz MJ, Wickner RB. (1978) *pet18:* A chromosomal gene required for cell growth and for the maintenance of mitochondrial DNA and the killer plasmid of yeast. *Mol. Gen. Genet.* **165:** 115-121.

Link AJ, Olson MV. (1991) Physical map of *Saccharomyces cerevisiae* genome at 110-kilobase resolution. *Genetics,* **127:** 681-698.

Louis EJ, Haber JE. (1990a) Mitotic recombination among the subtelomeric Y′ repeat family in *Saccharomyces cerevisiae:* an experimental system for repeated sequence evolution. *Genetics,* **124:** 533-546.

Louis EJ, Haber JE. (1990b) Mitotic recombination among subtelomeric Y′ repeats in *Saccharomyces cerevisiae. Genetics,* **124:** 547-559.

Louis EJ, Haber JE. (1991) Evolutionarily recent transfer of a group I mitochondrial intron to telomere regions in *Saccharomyces cerevisiae. Curr. Genet.* **20:** 411-415.

Louis EJ, Haber JE. (1992) The structure and evolution of subtelomeric Y′ repeats in *Saccharomyces cerevisiae. Genetics,* **131:** 559-574.

Mortimer RK, Schild D, Contopoulou CR, Kans J. (1989) Genetic map of *Saccharomyces cerevisiae,* Edition 10. *Yeast,* **5:** 321-404.

Mortimer RK, Contopoulou CR, King JS. (1992) Genetic and physical maps of *Saccharomyces cerevisiae,* Edition 11. *Yeast,* **8:** 817-902.

Newlon CS. (1988) Yeast chromosome replication and segregation. *Microbiol. Rev.* **52:** 568-601.

Newlon CS, Green R, Hardeman K, Kim K, Lipchitz L, Palzkill T, Synn S, Woody ST. (1986) Structure and organization of yeast chromosome III. *UCLA Symp. Mol. Cell. Biol.* **33:** 211-223.

Newlon CS, Lipchitz LR, Collins I, Deshpande A, Devenish RJ, Green RP, Klein HL, Palzkill TG, Ren R, Synn S, Woody ST. (1991) Analysis of a circular derivative of *Saccharomyces cerevisiae* chromosome III: a physical map and identification of ARS elements. *Genetics,* **129:** 343-357.

O'Brien SJ. (ed.) (1990) *Genetic Maps 5: Locus Maps Of Complex Genomes*. Cold Spring Harbor, New York: Cold Spring Harbor Laboratory Press.

Ohno S. (1970) *Evolution Through Gene Duplication*. London: Allen and Unwin.

Oliver SG, Aart QJM van der, Agostoni-Carbone ML, *et al*. (1992) The complete DNA sequence of yeast chromosome III. *Nature*, **357**: 38-46.

Oliver SG, Warmington JR. (1989) Transcription. In: *The Yeasts*, vol. 3 (Rose AH, Harrison JS, eds). London: Academic Press, pp. 117-160.

Olson MV. (1991) Genome structure and organisation in *Saccharomyces cerevisiae*. In: *The Molecular and Cellular Biology of the Yeast* Saccharomyces cerevisiae (Broach JR, Pringle JR, Jones EW, eds). Cold Spring Harbor, New York: Cold Spring Harbor Laboratory Press, pp 1-39.

Olson MV, Dutchik JE, Graham MY, Brodeur GM, Helms C, Frank M, MacCollin M, Scheinman R, Frank T. (1986) Random clone strategy for genome restriction mapping in yeast. *Proc. Natl Acad. Sci. USA* **83**: 7826-7830.

Pearson WR, Lipman DJ. (1988) Improved tools for biological sequence comparison. *Proc. Natl Acad. Sci. USA* **85**: 2444-2448.

Perkins DD. (1990) *Neurospora crassa* (chromosomal genes). In: *Genetic Maps 5* (O'Brien SJ, ed.). Cold Spring Harbor, New York: Cold Spring Harbor Laboratory Press, pp. 35-46.

Perkins DD, Radford A, Newmeyer D, Bjoerkman M. (1982) Chromosomal loci of *Neurospora crassa*. *Microbiol. Rev.* **46**: 426-570.

Petes TD, Botstein D. (1977) Simple Mendelian inheritance of the reiterated ribosomal DNA of yeast. *Proc. Natl Acad. Sci. USA* **74**: 5091-5095.

Piggot PJ, Hoch JA. (1985) Revised genetic linkage map of *Bacillus subtilis*. *Microbiol. Rev.* **32**: 101-134.

Roeder GS, Smith M, Lambie EJ. (1984) Intrachromosomal movement of genetically marked *Saccharomyces cerevisiae* transposons by gene conversion. *Mol. Cell. Biol.* **4**: 703-711.

Rothstein RJ. (1983) One-step gene disruption in yeast. *Methods Enzymol.* **101**: 202-211.

Sanger F, Nicklen S, Coulson AR. (1977) DNA sequencing with chain terminating inhibitors. *Proc. Natl Acad. Sci. USA* **86**: 3694-3698.

Shampay J, Szostak JW, Blackburn EH. (1984) DNA sequences of telomeres maintained in yeast. *Nature*, **310**: 154-157.

Sikorski RS, Hieter P. (1989) A system of shuttle vectors and host strains designed for efficient manipulation of DNA in *Saccharomyces cerevisiae*. *Genetics* **122**: 19-27.

Strathern JN, Newlon CS, Herskowitz I, Hicks JB. (1979) Isolation of a circular derivative of yeast chromosome III: implications for the mechanism of mating-type conversion. *Cell*, **18**: 309-319.

Stucka R, Lochmuller H, Feldmann H. (1989) Ty4, a novel low copy number element in *Saccharomyces cerevisiae* – one copy is located in a cluster of Ty elements and transfer RNA genes. *Nucl. Acids Res.* **17**: 4993-5001.

Thoraval D, Regnacq M, Neuville P, Boucherie H. (1990) Functional analysis of the yeast genome: use of 2-dimensional gels to detect genes in randomly cloned DNA sequences. *Curr. Genet.* **18**: 281-286.

Toh-e, A, Sahashi Y. (1985) The *pet18* locus of *Saccharomyces cerevisiae*: a complex locus containing multiple genes. *Yeast*, **1**: 159-172.

Voytas DF, Boeke JD. (1992) Yeast retrotransposon revealed. *Nature*, **358**: 717.

Walmsley RM, Chan CSM, Tye B-K, Petes TD. (1984) Unusual DNA sequences associated with the ends of chromosomes. *Nature*, **310**: 157-160.

Walmsley RM, Petes TD. (1985) Genetic control of chromosome length in yeast. *Proc. Natl Acad. Sci. USA* **82**: 506-510.

Warmington JR, Anwar R, Newlon CS, Waring RB, Davies RW, Indge KJ, Oliver SG. (1986) A 'hot-spot' for Ty transposition on the left arm of chromosome III. *Nucl. Acids Res.* **14**: 3475-3485.

Warmington JR, Green RP, Newlon CS, Oliver SG. (1987) Polymorphisms on the right arm of yeast chromosome III associated with Ty transposition and recombination events. *Nucl. Acids Res.* **15**: 8963-8982.

Wicksteed BL, Collins I, Dershowitz A, Stateva LI, Green RP, Oliver SG, Brown AJP, Newlon CS. (1993) A physical comparison of chromosome III in six strains of *Saccharomyces cerevisiae*. (Submitted).

Yoshikawa A, Isono K. (1990) Chromosome III of *Saccharomyces cerevisiae* – an ordered clone bank, a detailed restriction map and analysis of transcripts suggest the presence of 160 genes. *Yeast*, **6:** 383-401.

Zeigler DR, Dean DH. (1985) Revised genetic map of *Bacillus subtilis 168*. *FEMS Microbiol. Rev.* **32:** 101-134.

Chapter 18

Exploring plant genomes, small and large

Richard B Flavell, Caroline Dean and Graham Moore

1. Molecular maps – progress and problems

Formal genetics started with Mendel and his experiments on peas in the middle of
the last century. This introduced the concept of units determining phenotypic
characters being linked or unlinked. In the following 100 years, genetic maps of a
few plant species were developed using allelic variation affecting morphological
characteristics. Subsequently, genes encoding proteins, identified after electrophoretic separation, were added to the maps but since the mid 1970s a new era of
genetic analysis, gene mapping and chromosome manipulation has emerged. This
has greatly accelerated the pace of genome mapping and its application to address
fundamental questions and commercial problems.

The recent progress has been built upon identifying and mapping variation
within the DNA of the chromosome rather than within the phenotype. Variation
in DNA is readily recognized by restriction endonucleases or by short DNA
fragments whose ability to form a stable duplex with chromosomal fragments is
very sensitive to single base changes. In the first approach, variation in a restriction endonuclease cleavage site generates variation in the length of the DNA
fragment (restriction fragment length polymorphism, RFLP) that can be recognized by size separation of the DNA fragments, Southern blotting and hybridization to a labelled fragment homologous to sequences on the fragment (Chang *et
al.*, 1988). In the second approach, short DNA primers are hybridized to the
DNA, the intervening DNA amplified by the polymerase chain reaction (PCR)
and the products (random amplified polymorphic DNAs, RAPDs) observed after
separation by electrophoresis (Reiter *et al.*, 1992). Failure to bind a primer due to
a mutation results in loss of an amplified fragment or to one of a variant size. Very
recently a third approach (amplified fragment length polymorphism, AFLP)
which combines attributes of both RFLP and RAPD methods has been derived
(M Zabeau and P Vos, pers. comm.). This has the merit of identifying restriction
fragment polymorphism, but via PCR amplified fragments, and can survey hundreds of loci simultaneously. In all approaches the chromosomal locations of the
genomic variation can be determined by analysing segregating F_2 or subsequent

recombinant inbred populations derived from parents showing enzyme or primer recognition site differences. RFLP or RAPD based genetic maps have now been produced for many species. The genetic maps of plants such as tomato, wheat, barley, soybean, *Arabidopsis thaliana*, maize, pea, potato and others have recently been collated (O'Brien, 1990, 1993).

DNA sequence markers useful for making RFLP maps need not be a gene or a transcribed sequence. However, it is desirable for them to be present in only one or a few copies per genome. Hence cDNA copies of genes have been widely used as probes to identify RFLPs. In addition, in some species (tomato, maize, barley, wheat) short fragments of chromosomes cleaved with a restriction endonuclease (e.g. *Pst*I) that recognizes regions enriched with unmethylated cytosines and preferentially cleaves close to genes, have also been used extensively as probes. In pea, a dispersed repeated sequence has been used (Ellis *et al.*, 1992; Lee *et al.*, 1990) and, interestingly, has revealed the presence of translocations between lines.

The frequencies of molecular polymorphism differ widely between species. Maize is very polymorphic while tomato, for example, has perhaps only 5% as many variants. This is likely to be due to the amount of DNA (repetitive?) not under tight selection, whether the species is outbreeding or inbreeding and the concentration of active mutagenic systems within the species. Maize, for example, has been selected by man to harbour transposable elements that are very mutagenic and create polymorphisms when they move. The frequency of polymorphisms has significant consequences in genome analysis. In particular, a low frequency requires many more sites to be explored to establish a map with sufficient markers and therefore limits the value of the mapping approach for both genetic analysis and plant breeding. In establishing maps, very divergent parents can be selected in order to boost the frequency of polymorphic sites in the segregating mapping population, but the elite material of plant breeders may be less polymorphic.

Recently a high frequency of polymorphisms due to changes in cytosine methylation has been recognized in wheat – a species with relatively few RFLP polymorphisms at other sites. While the frequency of polymorphisms in cytosine methylation is too high and unstable for many purposes, it is useful for studying segregants from precisely known and available parental genomes (Flavell and O'Dell, 1990).

In the human genome short tandem arrays of repeats exist at many locations – products of slippage errors in replication or other localized amplification events. These so-called 'minisatellites' are highly polymorphic due to frequent unequal crossing over or other deletion-duplication events. They have been used extensively as probes in RFLP analyses of DNAs from individuals and genome mapping (Jefferys *et al.*, 1985; Royle *et al.*, 1988). Similar minisatellites exist in plant genomes and are beginning to be used in genome mapping in a similar way (Dallas, 1988; Winberg *et al.*, 1993).

The provision of genetic maps and the tools to examine them are stimulating new research projects. For example, chromosome and genome evolution can be studied by comparing maps of related species. Bonierbale *et al.* (1988), Gebhardt *et al.* (1991) and Tanksley *et al.* (1992) have shown by RFLP mapping that large segments of the potato genome can be aligned with segments of the tomato genome, indicating conservation of gene order during divergence of these species from a common ancestor in the Solanaceae family. Tanksley and colleagues have also shown that single copy DNA sequences lie in the same linear order along the chromosomes of rice and maize, two very distantly related grasses. The cDNA sequences on maize chromosome 9 are co-linear with those on rice chromosomes 6 and 3, while those on the rice chromosome 1 are co-linear with those on maize chromosomes 3 and 8.

The synteny – conservation of gene order – between rice and maize chromosomes indicates that genes which are mapped in rice are likely to have a counterpart in maize (and other grasses) in a similar relative map location and vice versa. Thus it should be possible to predict the genetic architecture and genes involved in a complex phenotype of one species from another or, even more exciting, synthesize a composite map for all the species in a family by re-organizing homologous segments of the genome into a single map. There are likely to be pieces that do not fit well because they are the products of multiple small re-arrangements but overall it looks likely that an ancestral genetic map can be created for each taxonomic family (Flavell *et al.*, 1993; Moore *et al.*, 1993a).

This conclusion is likely to be especially valuable to plant geneticists who work, world-wide, on a large number of species, with too few studying each except for the most popular species. If composite maps can be created, then each map can benefit from the maps of related species, using new methods of cross-referencing databases and maps. Here is a challenge for future research.

While the genetic maps between related species are likely to be co-linear over evolutionarily related chromosomes segments, all the DNA sequences in homologous segments of the genome are not. This is because most plant genomes contain substantial amounts of non-coding DNA that evolves rapidly in evolution (Flavell, 1982, 1985). This kind of DNA is predominantly in the form of repeated sequences whose amount is related to the total genome size (Flavell, 1980). Species such as Arabidopsis with an especially small genome have only a few per cent of repetitive non-coding DNA (Pruitt and Meyerowitz, 1986) while in pea, for example, with a relatively large genome, it constitutes over 90% of the total DNA (Flavell, 1980). During evolution, DNA amplification and deletion occur at high frequency to result in the continual turnover and hence change of non-coding DNA (Flavell, 1985). Although it is unknown why different amounts of repetitive DNA become stabilized in different species, it is likely that there is selection for an optimum genome size (Flavell, 1986). Whatever the reason, repetitive DNA is likely to contribute to major distortions between the genetical and physical maps within and between species, and large differences in the distances between genes within and between species (Flavell *et al.*, 1993; Snape *et al.*, 1985). Such studies in tomato suggest that the physical relationship to the genetic map can vary between 80 and 1000 kb/cM in different chromosomal regions.

Such findings have substantial implications for interpreting genetic maps and for the isolation of genes by 'genome walking' procedures. As described later in Arabidopsis genome research (section 2) it is feasible to 'walk' efficiently from one gene to another or from an RFLP/RAPD marker to a gene, for reasonable cost, if the distance is 500 kb or so. If in a cereal genome that has 50 times the amount of DNA as in Arabidopsis, the distance between genes is proportionally longer, walking over equivalent genetic distances will be very expensive. Further-more, where the intervening DNA is repetitive, procedures using intervening repetitive DNA as a hybridization probe will not be useful because repetitive DNA from one location will be represented at a large number of other locations too. For these sorts of reasons the isolation of genes by genome walking at reasonable cost seems much more attractive in species with small genomes. This poses a substantial constraint for research into the crop species that have relatively large genomes (Bennett and Smith, 1991). However, the preliminary discoveries of synteny in gene order between segments of a small genome and a related large genome may offer a new hope and a new way ahead. If genetic maps can be aligned between a small genome model and large genome species from which a derived allele is required, then it might be possible to either 'walk' to the homo-logous gene in the model species or to use unique sequences in the model genome to jump along the much larger chromosome segment carrying homologous unique sequences but much repetitive DNA too. The situation envisaged is illustrated schematically in Fig. 1 where rice and wheat chromosome segments are shown with a conserved gene order in spite of the total DNA content being very different within the homologous segments. Clearly, being able to jump from A to B to C would be much more feasible in rice than in wheat.

In the remainder of this chapter we describe some of the recent progress in analysing the small genome of a dicotyledonous (dicot) plant (Arabidopsis) and the large genome of a monocot plant (wheat) whose detailed analysis could be aided greatly by studies on the relatively small genome of the monocotyledonous (monocot), rice (*Oryza sativa*).

2. The International Arabidopsis Genome Project

A few years ago *Arabidopsis thaliana* (thale cress or Arabidopsis) was selected as the principal model plant species for the goal of complete characterization of a plant genome (Meyerowitz *et al.*, 1991). Arabidopsis is a very small plant with a rapid life cycle producing lots of seeds and so is very conveniently handled in laboratories. It is also especially convenient because it has one of the smallest higher plant genomes (10^8 bp) and can be transformed with exogenous genes either directly (Damm *et al.*, 1989) or via the soil bacterium *Agrobacterium tumefaciens* (Lloyd *et al.*, 1986; Valvekens *et al.*, 1988). In the first approach the genes are transfected into protoplasts and the transformed protoplasts are cultured and regenerated into plants. In the second method cells in cotyledon or root explants are transformed with T-DNA from a plasmid in *Agrobacterium* and the transformed cells cultivated under selection are regenerated into whole plants. In order to isolate genes by chromosome walking, a dense map with many RFLP markers is being created. Two maps have been published (Chang *et al.*, 1988; Nam *et al.*, 1989), with over 350 markers (Meyerowitz *et al.*, 1991) and these have been statistically integrated (Hauge *et al.*, 1993).

Genomic libraries of cloned Arabidopsis DNA have been created in cosmids and yeast artificial chromosomes (YACs) and work is progressing well to create the complete genome in overlapping DNA fragments. This project is being undertaken by a few laboratories, principally in the US and UK, in a co-ordinated way to minimize duplication and to enable others who map loci or isolate chromosome fragments to add information to the growing database. For example, Dean and co-workers in Norwich, UK, are focusing on the top halves of chromosomes 4 and 5. Stock centres for seeds of the wild-type accessions and mutant strains have been set up in Nottingham, UK, and Ohio, USA, and resource centres for DNA chromosome fragments, RFLP markers, genes, etc., have been established in Köln, Germany, and Ohio, USA. Computer databases (e.g. AAtDB which is based on ACeDB devised for *Caenorhabditis elegans*) are being developed to provide comprehensive coverage of, and access to, all the genome information known.

Several different approaches have been initiated to reproduce the Arabidopsis genome in the form of overlapping DNA fragments. That initiated first, by Goodman and colleagues (Hauge *et al.*, 1991) was based on cosmids. A very large number (20 000) of cosmids have been analysed to search for those which contain homologous segments and consequently are contiguous and overlapping in the genome. This resulted in the identification of 750 segments (contigs) of overlapping pieces of DNA with an average length of 120 kb. Joining of the cosmid

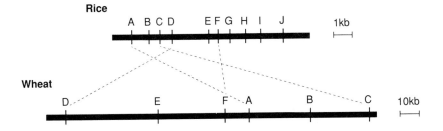

Figure 1. Hypothetical chromosome segments of rice and wheat showing synteny of genes (denoted by letters along segments) and re-arrangement of segments during chromosome evolution. The 50 times as much DNA projected to be in the wheat chromosome segments is likely to consist entirely of repeated sequences.

contigs will now rely on alternative methods (see later). Another approach has been to use larger pieces of Arabidopsis DNA cloned in YACs. Several such libraries with average insert sizes of 150 and 250 kb have been constructed and exploited (Ecker, 1990; Grill and Somerville, 1991; Ward and Jen, 1990). To focus on a specific chromosomal region, YACs are selected from the library by hybridization to mapped RFLP probes and can be positioned on the map directly (see Fig. 2, p. 173). Two problems have been encountered frequently; the presence of recombinant (chimeric) YACs containing plant DNA from more than one chromosomal region and the presence of similar sequences in multiple sites in the genome. When multiple YACs hybridize to an RFLP marker, an extended contig can usually be created with confidence. Using 45 RFLP markers from the top halves of chromosomes 4 and 5 more than 10 Mb distributed in numerous contigs have been assigned to these chromosomes (Schmidt and Dean, 1993; see also Fig. 2, p. 173). This represents approximately 10% of the whole genome.

Contigs can be extended and joined to close-by contigs by selecting sequences at or near the ends of each contig and using these to identify YACs containing the adjacent plant DNA. Several rounds of this procedure (chromosome walking) are facilitating the construction of large contigs. For example Schmidt and Dean (1993) have generated two contigs containing 1000 and 1300 kb involving more than 80 YACs. To build these, five smaller contigs previously identified by seven RFLP markers had to be joined using 60 end probes derived from 47 YACs. Pieces of DNA from such large contigs can be used to add more markers to the genetic map and to confirm the co-linearity of the reconstructed map with the *in vivo* map.

More than 30 walking experiments have been described for Arabidopsis (Meyerowitz *et al.*, 1991). Many more (>50) fine mapping experiments have been initiated to localize mutations on to the physical map. Accurate integration of the RFLP, RAPD, mutation maps and the overlapping fragment library is now required to complete the resource base for the efficient isolation of mapped genes. The experience to date has indicated that chromosome walking via RFLP markers and YACs is a sound approach but is inefficient at present due to the YACs being too short with respect to the distances between mapped markers and also to some YACs being chimeric. There is therefore a need to obtain say ten times as many DNA markers mapped on the genome to aid in YAC selection and verification of contig structure and to obtain a library of YACs with much larger non-chimeric plant DNA inserts (say >500 kb). These steps are in hand and should lead to completion of the resource base for the whole genome in 2 to 3 years.

A major aim of the International Arabidopsis Genome Project is to discover all the genes in the species and their function (see also Blattner *et al.*, Chapter 4, this volume, for the equivalent *Escherichia coli* project and Oliver *et al.*, Chapter 17, this volume, for the yeast project). Some have been recognized by isolating mutants with visible or screenable phenotypic alterations, following saturation mutagenesis. Others are being recognized by transposon or T-DNA tagging

(Feldmann, 1992; see later). Others will only be found by cDNA analysis and by sequencing of genomic DNA. Mutations noted to date include those affecting morphological and physiological characters, essential embryo and seedling development factors, and metabolism. Recently, co-ordinated approaches to sequencing, either genomic fragments or cDNAs, have been initiated, in the EC, the USA (e.g. Hauge et al., 1991) and Australia with the aim of gaining the complete sequence of the genome soon after the year 2000. This co-ordinated and vigorous attack on the Arabidopsis genome is being undertaken with the conviction that knowledge of the complete genome of a plant species will lead to a massive increase in the understanding of genes, gene expression and plant cellular and developmental biology when the defined genes are in the hands of cellular and developmental biologists.

3. The rice genome

The genome of rice (*Oryza sativa*) is currently receiving more attention than that of any other monocot species. This is because rice is the world's most important food crop. The genome of rice is relatively small (4×10^8 bp) and consists of 12 chromosomes and linkage groups (Khush and Kinoshita, 1991). It can be transformed, albeit inefficiently as yet (Cao *et al.*, 1991) with exogenous genes. These attributes together with the volume of research being done on its genome make it a major model for plant research. Over 161 loci defining traits such as pigmentation, chlorophyll production, height, ear morphology, grain size and structure, leaf shape, male sterility, disease and insect resistance have been mapped on to the linkage groups together with 33 loci specifying isoenzymes. Several RFLP maps are being constructed. One published by McCouch and Tanksley (1991) shows 230 markers comprising 130 markers from a *Pst*I genomic library and about 100 cDNA markers. The map was constructed using segregating populations from an interspecific cross between an African land race of *O. sativa* and an accession of *O. longistaminata* from Botswana. The level of polymorphisms between these parents far exceeds that in the intraspecific crosses evaluated to date. On a per restriction enzyme basis 60% of the marker sequences surveyed produced polymorphic patterns. The linkage data obtained so far suggest that gene order is identical to that gained from the studies on intraspecific crosses. However, recombination appears repressed in the interspecific cross because the recombination frequency is only 75% of that scored in intraspecific crosses when averaged throughout the genome (1800 cM versus 2300 cM).

A map based on an interspecific cross having some 235 markers showed areas without markers and unlinked markers, not expected if recombination was distributed well between single copy sequences (McCouch and Tanksley, 1991). These difficulties were probably due to monomorphic genomic segments in both parents of the crosses and emphasize the value of utilizing wide crosses for creating linkage maps.

Japanese laboratories have made a major commitment to studying the rice genome. They have sequenced some 1500 cDNAs and are generating a dense RFLP map which currently includes over 500 sequenced cDNAs (Minobe, 1992; Saito et al., 1991). As various RFLP maps emerge, it will be important to merge them together, as is being done for Arabidopsis maps (Hauge et al., 1993), and also to the map of morphologically scored loci, to create a comprehensive map with several thousand loci. This should provide a basis for isolating YACs of rice DNA from existing libraries that correspond to known sites in the genome and for reconstructing chromosome segments from overlapping YACs as described earlier for Arabidopsis (see Fig. 2, p. 173). If the RFLP and other markers are sufficiently dense and large YACs are isolated there will be little need for long walking strategies to isolate mapped genes of interest. However, if physical distances between mapped markers are large, relative to the sizes of YACs, then 'linking' libraries or other techniques for walking to genes will need to be adopted to walk to desired genes. Chromosome walking should be relatively straight-forward in the rice genome, as described earlier for Arabidopsis (section 2), because there is no high percentage of dispersed repetitive DNA to create ambiguities in the walking process.

The premier importance of rice as a food crop provides strong motivation to map a very large number of genes of known phenotypic effect, in order to understand the complexity of genes affecting a single trait, to isolate known genes from genomic libraries and to assemble panels of markers that are closely linked to genes of agronomic interest. Such 'gene tagging' can be of considerable utility in marker-assisted plant breeding programmes where breeders have the resources to utilize molecular markers. Many experiments to find and exploit such markers are in progress. For example, two markers linked to the valuable gene conferring resistance to the white-backed plant hopper have been identified (McCouch and Tanksley, 1991). Comprehensive maps and panels of marker probes are also useful for identifying different seed lots and for helping assess which seed lots should be conserved in seed banks when there are logistical or financial difficulties in maintaining duplicates or lines with duplicated chromosome segments.

4. Gene isolation by transposon and T-DNA tagging

The isolation of genes by genome walking from linked markers is a relatively time-consuming way to isolate genes. Direct isolation without walking would be preferable. This is facilitated by tagging a gene with a known piece of DNA that can be isolated with the associated gene of interest. The isolation of genes following the generation of mutations by 'transposon tagging', first developed in bacteria, is now well developed in maize and *Antirrhinum majus*, thanks to the pioneering studies by Barbara McClintock in maize. The first genes to be cloned in both systems encoded components of the anthocyanin biosynthetic pathway (Fedoroff *et al.*, 1984; Martin *et al.*, 1985). Subsequently, genes controlling plant morphology have been isolated by tagging (e.g. the *knotted* locus – Hake *et al.*, 1989; *Floricaula* – Coen *et al.*, 1990). Many mutants had been generated by insertion of transposable elements into genes in both species long before a molecular definition of the transposable elements was known. Now many families of such elements have been described in each species and representatives sequenced. Using the appropriate mutant stocks and the appropriate element as a DNA probe, genes of known phenotypic effect have been cloned and sequenced and their biology is being explored. This is clearly a powerful route for gene isolation which depends only on being able to recognize the mutant phenotype. For those species not containing well-characterized transposons, insertional mutagenesis systems are being developed either by introducing the maize or *Antirrhinum majus* elements into a new species or by utilizing the piece of DNA (T-DNA) introduced into the plant chromosome by *Agrobacterium*. Most success in the development of heterologous transposon systems has been achieved using the maize elements *Ac* and *Ds*. These elements transpose in all the species into which they have been introduced (reviewed by Bhatt and Dean, 1992) and in Arabidopsis a *Ds*-induced mutant has recently been isolated (I Bancroft, JDG Jones and C Dean, manuscript in preparation). T-DNA tagging has also been very successful in Arabidopsis: many mutations have been characterized and a large number of independent transformants carrying T-DNAs inserted all over the genome are now available for mutant screening (Feldmann, 1992). Successful transposon and T-DNA tagging in Arabidopsis is a very significant step because it provides a direct system for gene identification and isolation in the model species being adopted worldwide for extensive genome characterization.

5. Towards a molecular model of the wheat chromosome

The reconstruction and subsequent sequence analysis of rice and Arabidopsis chromosomes will teach us many of the molecular details of monocot, and dicot, architecture. Accumulating equivalent information for chromosomes from a large genome will take much longer, because the amount of DNA is so much greater. To gain a molecular description of a cereal chromosome a range of approaches have been adopted. Long arrays of repeated sequences have been mapped with respect to the centromere, telomeres and chromosome arms (Flavell *et al.*, 1987); RFLP maps have been created (Chao *et al.*, 1989; Devos and Gale, 1992; Devos *et al.*, 1992; Heun *et al.*, 1991; Wang *et al.*, 1991); individual chromosomes have been separated and their components cloned (Wang *et al.*, 1992); physical maps are being compared with genetical maps (Lawrence and Appels, 1986;

Leitch and Heslop-Harrison, 1992, 1993; Linde-Laursen, 1982; Snape *et al.*, 1985); regions enriched with unmethylated cytosine residues have been mapped with respect to genes (Moore *et al.*, 1993b); the long range organization of multicopy DNA units defined by unmethylated regions has been investigated and the distribution of recombination sites determined (Moore *et al.*, 1991a, 1993b). A schematic model of a cereal chromosome is given in Fig. 3 (see also Moore *et al.*, 1993 a, b, and Flavell *et al.*, 1993, for further details).

The 1C DNA content of diploid wheats and barley is about 5×10^9 base pairs and each species has seven chromosomes per haploid genome. Approximately 80% of these large cereal genomes consists of highly repeated sequences. Around the centromeres, near the telomere and at certain interstitial sites there are regions that stain as C- or N-bands (heterochromatin) which are the sites of very long tandem arrays of a few sequences that can account for more DNA than the entire Arabidopsis genome. The number of repeats and hence the size of the heterochromatin regions vary between chromosomes within and between species. These repeats are useful markers of individual chromosomes because they can be used for *in situ* hybridization to recognize chromosomes (Flavell *et al.*, 1987). Most of the repetitive DNA is dispersed over the genome in complex permutations that must have been generated during evolution by the amplification and transposition of small segments of DNA (Flavell, 1982). Many of these dispersed repeat units belong to families of retrotransposons – structures of DNA that are amplified by transcription to mRNA followed by reverse transcription by a self-encoded reverse transcriptase, production of double-stranded DNA copies and integration of the copies into new sites in the chromosomes. Two families have been characterized in wheat in detail – Wire I and Wis 2. Wire I elements are around 12.5 kb and Wis 2 elements are 8.6 kb long (Murphy *et al.*, 1992). Each has characteristic duplications at each end which arise from their mode of replication. The duplications at the ends of Wis 2 are particularly long (1755 bp) and contain many inverted repeats. It has been postulated that these and other variants have arisen as errors in the replication process by reverse transcription where template switching is required (Lucas *et al.*, 1992). Examination of a few DNA fragments containing genes has shown that members of these retrotransposon families lie close to genes and single/few copy sequences. Members of each family probably lie all over the genome.

When cereal DNAs are treated with restriction endonucleases that cleave only when cytosines in CpG or CpXpG residues in the recognition sequence are unmethylated (e.g. *Nru*I, *Mlu*I), many multicopy fragments of defined lengths are observed (Moore *et al.*, 1993b). Hence there must be reiterated spacing of such unmethylated sites in the genome (see Fig. 3) although the origin of this spacing is unclear. However, it is likely to be 'imposed' by some functional system, such as might be expected in a replication origin (see DePamphilis, Chapter 6, this volume) or scaffold attachment site. If such sites were 'imposed' at regular intervals then they could represent some product of natural selection on sequence

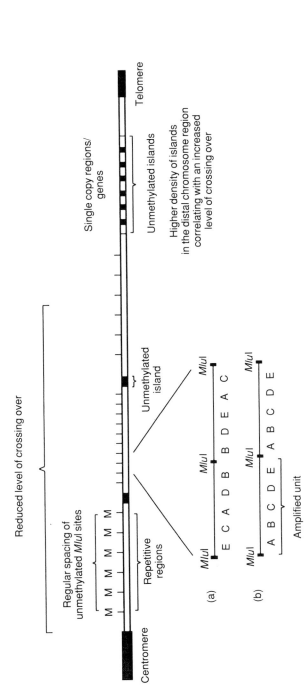

Figure 3. A model of sequence organization in a cereal chromosome. The proximal region close to the centromere is enriched for long stretches of repetitive DNA defined by regular spacing of unmethylated *Mlu*I sites. The repetitive sequences (A–E) are either randomly (a) or non-randomly (b) distributed in the multicopy *Mlu*I restriction fragments derived from the cleavage of these regions. Non-randomly distributed repetitive sequences would occur in tandem (b) or inverted (not shown) in these genomic regions. Other repeats are dispersed throughout the chromosome. The gene density increases from proximal to distal regions of the chromosomes as does the frequency of recombination sites. Genes are preferentially associated with unmethylated cytosines. (For further details see Moore *et al.*, 1993a, b.)

organization along the linear molecule of the chromosomal DNA. Alternatively, the reiterated spacing intervals could result from the amplification of long pieces of DNA (60 kb to several megabases) containing two or more unmethylated sites which are subsequently conserved. Surprisingly, some of the larger sizes of units generated by cleavage with *Mlu*I and *Nru*I are similar for wheat and barley. This implies that the unequal crossing-over, transposition and other genomic re-organizational events that have led to DNA sequence turnover during the divergence of wheat and barley from a common ancestor (Flavell, 1982) have not undermined the conservation of the unmethylated restriction sites (Moore *et al.*, 1993b). The units are localized preferentially in the half of the chromosome arm closest to the centromere in the two rye chromosomes assayed, as shown in Fig. 3 (see Moore *et al.*, 1993b).

If very long regions of repeated sequence exist then the single- or low-copy sequences will not be scattered at random but concentrated closer together. We expect this to be the case, with the clusters being localized preferentially towards the telomere where recombination is more frequent (see Fig. 3). Proof of this will require the physical mapping of more genes or single-copy sequences. Despite this progress, the presumed large physical distances between mapped markers, the relatively low recombination frequency in specific regions, the high content of repetitive DNA and the few scientists studying these species all make a complete description of a chromosome from wheat, barley or close relative beyond reach in the foreseeable future. This is not necessarily true of specific chromosomal regions though, as evidenced by the description of the large multi-genic locus encoding α-amylase genes (Cheung *et al.*, 1991) using pulse field gel electrophoresis and suitable hybridization probes.

The development of an effective transposon tagging system in cereals such as wheat and barley and more comprehensive cDNA analysis (Liang and Pardee, 1992) would also be helpful. More significant, however, are the implications of synteny between grass genomes as mentioned previously (Moore *et al.*, 1993a). The establishment of an ordered set of genomic DNA clones for the rice genome in combination with the ancestral map should provide a genetic framework for each and every grass genome used in the construction of the map. This could enable a gene controlling a particular trait to be cloned from these grasses irrespective of their genome size. The analysis will in turn lead to a greater understanding of the organization of large cereal chromosomes, such as those of wheat, as well as having major consequences for cereal genetics.

6. Concluding remarks

Molecular studies of plant chromosomes are now advancing rapidly due to the application of the techniques developed for the genomes of man, *C. elegans*, *E. coli* (Blattner *et al.*, Chapter 4, this volume) and yeast (Oliver *et al.*, Chapter 17, this volume). In the next year or two the number of genes recognized in plants from cDNA sequencing, genomic sequencing, transposon tagging and genome walking will increase rapidly, opening up research avenues in developmental and cellular biology. As whole chromosomes are sequenced, new kinds of information about genome structure and organization will emerge. Exploitation of this knowledge in crop improvement programmes will accelerate as molecular maps and methods for detecting variation become more efficient in the crop species, and efficient transformation procedures for the insertion of new genes become easier to use. There are few ethical or moral issues connected with the exploitation of transgenic plants but there is debate about the release of transgenic crops into the environment, on the scale required in routine agricultural use, and whether the products of transgenic crops should enter food chains. Until adequate confidence has accumulated about the use of transgenic plants in agriculture the full value of the new abilities to dissect and interfere with the plant genome is unlikely to be realized in helping to solve the major problem of feeding the world in the coming decades.

References

Bennett MD, Smith JB. (1991) Nuclear DNA amount in angiosperms. *Phil. Trans. R. Soc. Lond. B.* **334**: 309-345.

Bhatt AM, Dean C. (1992) Development of tagging systems in plants using heterologous transposons. *Curr. Opin. Biotech.* **3**: 152-158.

Bonierbale MW, Plaisted RL, Tanksley SD. (1988) RFLP maps based on a common set of clones reveal modes of chromosomal evolution in potato and tomato. *Genetics*, **120**: 1095-1103.

Cao J, Zhang W, McElroy D, Wu R. (1991) Assessment of rice genetic transformation techniques. In: *Rice Biotechnology* (Khush GS, Toenniessin G, eds). Wallingford: CAB International, pp.175-198.

Chang C, Bowman JL, DeJohn AW, Lander ES, Meyerowitz EM. (1988) Restriction fragment length polymorphism linkage map for *Arabidopsis thaliana*. *Proc. Natl Acad. Sci. USA* **85**: 6856-6860.

Chao S, Sharp PJ, Worland AJ, Warham EJ, Koebner RMD, Gale MD. (1989) RFLP-based genetic maps of wheat homoeologous group 7 chromosomes. *Theor. Appl. Genet.* **78**: 495-504.

Cheung WY, Chao S, Gale MD. (1991) Long-range mapping of the α-amylase-1 (α-amy-1) loci on homoeologous group 6 chromosomes of wheat. *Mol. Gen. Genet.* **229**: 373-379.

Coen E, Romero J, Doyle S, Elliot R, Murphy G, Carpenter R. (1990) *floricaula*: a homeotic gene required for flower development in *Antirrhinum majus*. *Cell*, **63**: 1311-1322.

Dallas JF. (1988) Detection of DNA fingerprints of cultivated rice by hybridisation with a human minisatellite probe. *Proc. Natl Acad. Sci. USA* **85**: 6831-6835.

Damm B, Schmidt R, Willmitzer L. (1989) Efficient transformation of *Arabidopsis thaliana* using direct gene transfer to protoplasts. *Mol. Gen. Genet.* **217**: 6-12.

Devos K, Atkinson M, Chinoy C, Liu C, Gale MD. (1992) RFLP based genetic map of homoeologous group 3 chromosomes of wheat and rye. *Theor. Appl. Genet.* **83**: 931-939.

Devos KM, Gale MD. (1992) The use of random amplified polymorphic markers in wheat. *Theor. Appl. Genet.* **84:** 567-572.

Ecker JR. (1990) PFGE and YAC analysis of the *Arabidopsis* genome. *Methods,* **1:** 186-194.

Ellis THN, Turner L, Hellens RP, Lee D, Harker CL, Enard C, Downey C, Davies DR. (1992) Linkage maps in pea. *Genetics,* **130:** 649-663.

Fedoroff NV, Furtek DB, Nelson OE. (1984) Cloning of the *bronze* locus in maize by a simple and generalizable procedure using the transposable controlling element activator (Ac). *Proc. Natl. Acad. Sci. USA* **81:** 3825-3829.

Feldmann KA. (1992) T-DNA insertion mutagenesis in *Arabidopsis*: mutational spectrum. *Plant J.* **1:** 71-82.

Flavell RB. (1980) The molecular characterisation and organisation of plant chromosomal DNA sequences. *Ann. Rev. Plant Physiol.* **31:** 569-596.

Flavell RB. (1982) Sequence amplification deletion and rearrangement: major sources of variation during species divergence. In: *Genome Evolution* (Dover GA, Flavell RB, eds). London: Academic Press, pp. 301-323.

Flavell RB. (1985) Repeated sequences and genome change. In: *Genetic Flux in Plants* (Hohn B and Dennis ES, eds). Vienna: Springer, pp. 129-156.

Flavell RB. (1986) Repetitive DNA and chromosome evolution in plants. *Phil. Trans. R. Soc. Lond. B.* **312:** 227-242.

Flavell RB, Bennett MD, Hutchinson-Brace J. (1987) Chromosome structure and organisation. In: *Wheat Breeding, its Scientific Basis* (Lupton F, ed). London: Chapman and Hall, pp. 211-268.

Flavell RB, Gale, MD, O'Dell M, Lucas H, Murphy G, Moore G. (1993) Molecular organisation of gene and repeats in the large cereal genomes and implications for the isolation of genes by chromosome walking. In: *Chromosomes Today*, Vol II (Chandley A, Sumner A, eds). London: Allen and Unwin.

Flavell RB, O'Dell M. (1990) Variation and inheritance of cytosine methylation patterns in wheat at the high molecular weight glutenin and ribosomal RNA loci. *Devel. Suppl.*: 15-20.

Gebhardt C, Ritter E, Barone A *et al.* (1991) RFLP maps of potato and their alignment with the homoeologous tomato genome. *Theor. Appl. Genet.* **83:** 49-57.

Grill E, Sommerville C. (1991) Construction and characterization of a yeast artificial chromosome library of *Arabidopsis* which is suitable for chromosome walking. *Mol. Gen. Genet.* **226:** 484-490.

Hake S, Volbrecht E, Freeling M. (1989) Cloning *knotted* the dominant morphological mutant in maize using Ds2 as a transposon tag. *EMBO J.* **8:** 15-22.

Hauge B, Hanley S, Cartinhour S, Cherry JM, Goodman H, Koormeef M, Stam P, Chang C, Kempin S, Medrano L, Meyerowitze EM. (1993) An integrated genetic/RFLP map of the *Arabidopsis thaliana* genome. *Plant J.* **3:** 745–754.

Hauge BM, Hanley S, Giraudat J, Goodman HM. (1991) Mapping the *Arabidopsis* genome. In: *Molecular Biology of Plant Development* (Jenkins GI, Schuch W, eds). Cambridge: Company of Biologists, pp. 45-56.

Heun M, Kennedy AE, Anderson J, Lapitan NLV, Sorrells ME, Tanksley SD. (1991) Construction of a restriction fragment length polymorphism map for barley (*Hordeum vulgare*). *Genome,* **34:** 437-447.

Jefferys AJ, Wilson V, Thein SL. (1985) Hypervariable 'minisatellite' regions in human DNA. *Nature,* **314:** 67-73.

Khush GS, Kinoshita T. (1991) Rice karyotype, marker genes and linkage groups. In: *Rice Biotechnology* (Khush GS, Toemmiersen G, eds). Wallingford: CAB International, pp. 83-108.

Lawrence GJ, Appels R. (1986). Mapping the nucleolus organiser seed protein loci and isozyme loci on chromosome 1R in rye. *Theor. Appl. Genet.* **71:** 742-749.

Lee D, Ellis THN, Turner L, Hellens RP, Cleary WG. (1990) A copia like element in *Pisum* demonstrates the use of dispersed repeated sequences in genetic analysis. *Plant Mol. Biol.* **15:** 707-772.

Leitch IJ, Heslop-Harrison JS. (1992) Physical mapping of the 18S-5.8S-26S rRNA genes in barley by *in situ* hybridization. *Genome*, **35:** 1013-1018

Leitch IJ, Heslop-Harrison JS. (1993) Physical mapping of four sites of 5S ribosomal DNA sequences and one site of the α-amylase 2 gene in barley (*Hordeum vulgare*). *Genome*, in press.

Liang P, Pardee AB. (1992) Differential display of eukaryotic messenger RNA by means of the polymerase chain reaction. *Science*, **257:** 967-970.

Linde-Laursen I. (1982) Linkage map of the long arm of barley chromosome 3 using c-bands and marker genes. *Heredity*, **49:** 27-35.

Lloyd AM, Barnason AR, Rogers SG, Byrne MC, Fraley RT, Horsch RB. (1986) Transformation of *Arabidopsis thaliana* with *Agrobacterium tumefaciens*. *Science*, **234:** 464-466.

Lucas H, Moore G, Murphy G, Flavell RB. (1992) Inverted repeats in the long-terminal repeats of the wheat retrotransposon WIS-2-1A. *Mol. Biol. Evol.* **9:** 716-728.

Martin C, Carpenter R, Sommer H, Saedler H, Coen ES. (1985) Molecular analysis of instability in flower pigmentation of *Antirrhinum majus* following isolation of the *pallida* locus by transposon tagging. *EMBO J.* **4:** 1625-1630.

McCouch SR, Tanksley SD. (1991) Development and use of restriction fragment length polymorphic in rice breeding and genetics. In: *Rice Biotechnology* (Khush GS, Toenniessen GH, eds). Wallingford: CAB International, pp. 109-134.

Meyerowitz E, Dean C, Flavell R, Goodman H, Koorneef M, Peacock J, Shimura Y, Somerville C, Van Montagu M. (1991) The multinational coordinated *Arabidopsis thaliana* genome research project. *Progress Report: Year One*, Publ. 91-60. Washington, DC: National Science Foundation.

Minobe Y. (ed.) (1992) *Rice Genome Research Programme*, Vol. 1. Ibaraki, Japan: NIAR.

Moore G, Cheung W, Schwarzacher T, Flavell RB. (1991a) Bis 1 a major component of the cereal genome and a tool for studying genomic organisation. *Genomics*, **10:** 469-476.

Moore G, Lucas H, Batty N, Flavell RB. (1991b) A family of retrotransposons and associated genomic variation in wheat. *Genomics*, **10:** 461-468.

Moore G, Gale MD, Karata N, Flavell RB. (1993a) Molecular analysis of small grain cereal genomes, current status and prospects. *Biotechnology*, in press.

Moore G, Abbo S, Cheung W, Foote T, Gale M, Koebner R, Leitch A, Leitch I, Money T, Stancombe P, Yano M, Flavell R. (1993b) Key features of cereal genome organisation as revealed by the use of cytosine methylation-sensitive restriction endonucleases. *Genomics*, **15:** 472–482.

Murphy GJP, Lucas H, Moore G, Flavell RB. (1992) Sequence analysis of WIS-2-1A, a retrotransposon-like element from wheat. *Plant. Mol. Biol.* **20:** 991-995.

Nam HG, Giraudat J, den Boer B, Moonan F, Loos WDB, Hauge BM, Goodman H. (1989) Restriction fragment length polymorphism linkage map of *Arabidopsis thaliana*. *Plant Cell* **1:** 699-705.

O'Brien SJ. (1990, 1993) (ed.) *Genetic Locus Maps of Complex Genomes of Plants*. New York: Cold Spring Harbor Laboratory Press.

Pruitt RE, Meyerowitz EM. (1986) Characterisation of the genome of *Arabidopsis thaliana*. *J. Mol. Biol.* **187:** 169-184.

Reiter RS, Williams JGK, Feldmann KA, Rafalski JA, Tingey SV, Scolnik PA. (1992) Global and local genome mapping in *Arabidopsis thaliana* by using recombinant inbred lines and random amplified polymorphic DNAs. *Proc. Natl Acad. Sci. USA* **89:** 1477-1481.

Royle NJ, Clarkson RE, Wong Z, Jefferys AJ. (1988) Clustering of hypervariable minisatellites in the proterminal regions of human autosomes. *Genomics*, **3:** 352-360.

Saito A, Yano M, Kishimoto N *et al.* (1991) Linkage map of restriction fragment polymorphism loci in rice. *Jpn. J. Breed.* **41:** 665-670.

Schmidt R, Dean C. (1993) Towards construction of an overlapping YAC library of the *Arabidopsis thaliana* genome. *BioEssays* **15:** 63-69.

Snape JW, Flavell RB, O'Dell M, Hughes WG, Payne PI. (1985) Interchromosomal mapping of the nucleolar organiser region relative to three marker loci on chromosome 1B of wheat (*Triticum aestivum*). *Theor. Appl. Genet.* **69:** 263-270.

Tanksley SD, Ganal MW, Prince JP, de Vicente MC, Bonierbale MW, Broun P, Fulton TM, Giovannoni JJ, Grandillo S, Martin GB, Messeguer R, Miller JC, Miller L, Paterson AH, Pineda O, Röder MS, Wing RA, Wu W, Young ND. (1992) High density molecular linkage maps of the tomato and potato genomes. *Genetics*, **132:** 1141-1160.

Valvekens D, Van Montagu M, Van Lijsebettens M. (1988) *Agrobacterium tumefaciens*-mediated transformation of *Arabidopsis thaliana* root explants by using kanamycin selection. *Proc. Natl Acad. Sci. USA* **85:** 5536-5540.

Wang ML, Atkinson MD, Chinoy CN, Devos KM, Liu C, Rogers WJ, Gale MD. (1991) RFLP-based map of rye *S. cereale* chromosome 1R. *Theor. Appl. Genet.* **82:** 174-178.

Wang ML, Leitch AR, Schwarzacher T, Heslop-Harrison JS, Moore G. (1992) Construction of a chromosome enriched *Hpa*II library from flow-sorted wheat chromosomes. *Nucl. Acids Res.* **20:** 1897-1901.

Ward ER, Jen GC. (1990) Isolation of single-copy-sequence clones from a yeast artificial chromosomes library of randomly-sheared *Arabidopsis thaliana* DNA. *Plant Mol. Biol.*, **14:** 561-568.

Winberg BC, Zhov Z, Dallas JF, McIntyre CL, Gustafson JP. (1993) Characterisation of minisatellite sequences from *Oryza sativa* L. Submitted.

Index